LONDON MATHEMATICAL SOCIETY LECTURE N͏ͭ͏͏ ͏͏

Managing Editor: Professor Endre Süli, Mathematical Institute,
Woodstock Road, Oxford OX2 6GG, United Kingdom

The titles below are available from booksellers, or from Cambri
www.cambridge.org/mathematics

London Mathematical Society Lecture Note Series: 474

Equivariant Topology and Derived Algebra

Edited by

SCOTT BALCHIN
Max-Planck-Institut für Mathematik, Bonn

DAVID BARNES
Queen's University Belfast

MAGDALENA KĘDZIOREK
Radboud Universiteit Nijmegen

MARKUS SZYMIK
Norwegian University of Science and Technology, Trondheim

CAMBRIDGE
UNIVERSITY PRESS

University Printing House, Cambridge CB2 8BS, United Kingdom

One Liberty Plaza, 20th Floor, New York, NY 10006, USA

477 Williamstown Road, Port Melbourne, VIC 3207, Australia

314–321, 3rd Floor, Plot 3, Splendor Forum, Jasola District Centre, New Delhi – 110025, India

103 Penang Road, #05–06/07, Visioncrest Commercial, Singapore 238467

Cambridge University Press is part of the University of Cambridge.

It furthers the University's mission by disseminating knowledge in the pursuit
of education, learning, and research at the highest international levels of excellence.

www.cambridge.org
Information on this title: www.cambridge.org/9781108931946
DOI: 10.1017/9781108942874

© Cambridge University Press 2022

First published 2022

Printed in the United Kingdom by TJ Books Limited, Padstow Cornwall

A catalogue record for this publication is available from the British Library.

ISBN 978-1-108-93194-6 Paperback

Table of Contents

Contributors

Omar Antolín-Camarena   *National Autonomous Uni. of Mexico*

David Barnes   *Queen's University Belfast*

Tobias Barthel   *Max Planck Institute for Mathematics, Bonn*

Dave Benson   *University of Aberdeen*

Julia E. Bergner   *University of Virginia*

Andrew J. Blumberg   *Columbia University*

Ivo Dell'Ambrogio   *University of Lille*

Paul G. Goerss   *Northwestern University, Illinois*

Michael A. Hill   *University of California, Los Angeles*

Michael J. Hopkins   *Harvard University*

Srikanth B. Iyengar   *University of Utah*

Magdalena Kędziorek   *Radboud University Nijmegen*

Henning Krause   *Universität Bielefeld*

Julia Pevtsova   *University of Washington*

David White   *Denison University, Ohio*

Preface

The chapters of the present volume cover a range of topics in equivariant topology and derived algebra, chosen to connect with major themes from John Greenlees' vast mathematical career.

Conference

The catalyst for these proceedings was a week-long conference held at NTNU (Trondheim) between the 29th of July and the 2nd of August 2019. This conference, entitled *Equivariant Topology and Derived Algebra*, was held in honor of John Greenlees' 60th birthday. The conference consisted of 15 invited talks, 11 contributed talks, and 13 shorter *gong show* style talks, and was attended by a diverse group of over 90 international participants. The mathematical content was enhanced by a customary hiking excursion and a hearty conference dinner with beautiful scenic views.

Summary of the chapters

We briefly outline the chapters appearing in these proceedings, while also taking the opportunity to connect them to the work of John Greenlees, which at the date of writing spans more than 90 papers and four research monographs [32, 45, 57, 73].

Comparing Dualities in the $K(n)$-local Category
P. G. Goerss and M. J. Hopkins

Duality is a recurring theme through the work of John Greenlees; the starting place is perhaps Spanier–Whitehead duality. In modern language, the Spanier–Whitehead dual of a spectrum X is the function spectrum $DX = F(X, \mathbb{S})$, which arises from the commutative monoidal structure of the stable homotopy category. A common calculation is to show that the dual of the Moore spectrum for $\mathbb{Z}/2$ is simply a shift of that spectrum. A detailed examination of functional duals and Moore spectra is the subject of Greenlees' first published work, [93]. It is natural to look for generalisations of (Spanier–Whitehead) duality, for example [86] and [88] consider duality in the equivariant stable homotopy category, while [31] and [41] look more generally at questions of duality.

The first chapter of this volume takes up this theme and investigates duality in the $K(n)$-local stable homotopy category, giving a full and detailed proof of a result relating $K(n)$-local Spanier–Whitehead duality to the more computable notion of Brown–Comenetz duality.

Axiomatic Representation Theory of Finite Groups by way of Groupoids
I. Dell'Ambrogio

A second major theme in the work of John Greenlees is representation theory, and in particular, the use and study of Mackey functors. The most immediate way Mackey functors appear in the work of Greenlees is via equivariant cohomology theories. These are a generalisation of cohomology theories that have G-spaces as input, and take G-Mackey functors as coefficients. The category of Mackey functors is also a rich and interesting category in its own right, as demonstrated in [40, 60, 73, 82, 85]. Indeed, three chapters of this volume consider Mackey functors at length.

This chapter considers very general notions of Mackey functors and gives a common conceptual framework for several different versions. It provides relations between these different versions and connects the theory to 2-categories and bisets.

Chromatic Fracture Cubes
O. Antolín-Camarena and T. Barthel

A sizable portion of John Greenlees' work is dedicated to the development of algebraic models for rational G-spectra, where G is a compact Lie group. Algebraic models for several groups have been established, including all finite groups, the circle, tori of arbitrary dimension, $O(2)$ and $SO(3)$, see [73, 57, 16, 11, 59, 52]. Greenlees has conjectured that an algebraic model (satisfying a list of key properties) exists for every compact Lie group G. A key tool for this project is an isotropy separation of the sphere spectrum in rational G-spectra. This separation is a pullback square similar to the arithmetic pullback square or the chromatic fracture square. As the sphere spectrum is the monoidal unit, the isotropy separation extends to a decomposition of the (homotopy) category of rational G-spectra into simpler building blocks. Recent work of Greenlees abstracts this machinery to the setting of axiomatic stable homotopy theory [1].

This chapter generalises the familiar chromatic fracture square in the $E(n)$-local stable homotopy category to a chromatic fracture cube. This cube categorifies to provide a combinatorial decomposition of the category into monochromatic pieces. The $E(n)$-local stable homotopy category can be reconstructed by taking a homotopy limit of these monochromatic pieces over a certain diagram of diagrams.

An Introduction to Algebraic Models for Rational G-Spectra
D. Barnes and M. Kędziorek

As mentioned above, a major project of Greenlees is the development of algebraic models for rational G-spectra, where G is a compact Lie group, see [73, 57, 16, 11, 59, 52]. The initial case is where G is a finite group, here the algebraic model for rational G-spectra is a finite product over conjugacy classes of subgroups $H \leq G$ of graded $\mathbb{Q}[W_G H]$-modules ($W_G H$ is the Weyl group of H in G). One of the ways to prove this result uses the idempotent splitting of rational G-Mackey functors, see Appendix A of [73]. There are many papers building upon that work, such as constructing an algebraic model for naive-commutative ring G-spectra [9].

This chapter gives an introduction to rational Mackey functors and summarises the main techniques used to obtain algebraic models for rational G-spectra, concentrating on the case of a finite group G. It discusses the topological and algebraic parallels of using idempotents to split the category of rational G-spectra and rational G-Mackey functors. It also briefly mentions the techniques to obtain algebraic models when G is not finite.

Monoidal Bousfield Localizations and Algebras over Operads
D. White

Commutative monoidal structures appear throughout John Greenlees' work, occurring with algebraic origins [63, 72, 74], topological origins [4, 9] and bridging the divide between algebra and topology: [13, 36, 37, 75]. Moreover, the construction of algebraic models for rational G-spectra often depends upon making use of (commutative) monoidal structures in both topology and algebra. For example, the isotropy separation arguments require that certain localizations of the sphere spectrum are still commutative monoids. This property is not automatic, even under suitable cofibrancy conditions.

This chapter characterizes those Bousfield localizations that respect (commutative) monoidal structures, and moreover proves that these localizations preserve algebras over cofibrant operads. This general machinery can be used to retrieve many classical results which have repeatedly been used in the work of Greenlees.

Stratification and Duality for Unipotent Finite Supergroup Schemes
D. Benson, S. B. Iyengar, H. Krause and J. Pevtsova

A recent direction in the work of Greenlees is the study of tensor-triangulated categories, triangulated categories with compatible symmetric monoidal product and function object. A central example is the stable homotopy category, arising from homotopy (co)fibre sequences and the smash product and function spectrum. The equivariant stable homotopy category for a compact Lie group G is an even richer example, see [1], [3] and [8]. An important problem in such contexts is to classify the tensor ideal localising subcategories, and also the tensor ideal thick subcategories of compact objects, and to study questions related to duality as in [33].

The purpose of this chapter is to survey some methods developed to address these problems, and to illustrate them by establishing such classifications and duality statements for the stable module category of a unipotent finite supergroup scheme.

Bi-incomplete Tambara Functors
A. J. Blumberg and M. A. Hill

One of the recent themes in equivariant homotopy theory is to understand commutative ring objects in G-spaces and G-spectra. This is reflected in John Greenlees' work through the research on commutativity described above and more directly, as in [16, 9]. The subtlety and complications of equivariant commutativity can be described using a certain class of G-operads, called N_∞ operads. Algebras over an N_∞ operad $\mathcal{O}$ in G-topological spaces correspond, roughly speaking, to a G-spectrum with transfers determined by $\mathcal{O}$. Thus, one might think of $\mathcal{O}$ as governing the additive structure of a G-spectrum. Algebras over an N_∞ operad $\mathcal{O}$ in G-spectra (as opposed to G-spaces) correspond, roughly speaking, to $\mathcal{O}$-commutative ring G-spectra, that is, ring G-spectra with norm maps on homotopy groups determined by $\mathcal{O}$. Thus, in this case one might think of $\mathcal{O}$ as governing the multiplicative ring structure of a G-spectrum. The natural question is: how one can mix the various additive and multiplicative structures?

This chapter investigates the compatibility conditions between incomplete additive transfers and incomplete multiplicative norms in the algebraic setting of G-Tambara functors and provides a full description of the possible interactions of these two classes of maps.

Homotopy Limits of Model Categories, Revisited
J. E. Bergner

A key observation of the paper [1] is that the algebraic models for rational G-equivariant spectra can be described as homotopy limits of diagrams of model categories. This observation developed from homotopy pullback constructions in [11] based on isotropy separation, building on machinery of Greenlees–Shipley [20, 21, 25]. Homotopy limits also occur in the chapter of Antolín-Camarena–Barthel (in the setting of $(\infty, 1)$-categories) further demonstrating their ubiquity.

The final chapter of this volume provides a comprehensive outline of the machinery required for constructing homotopy limits of diagrams of Quillen model categories and left Quillen functors between them, collecting previous work of the author. Moreover the chapter provides a wealth of important examples of this homotopy limit construction. The chapter also provides some warning on working with diagrams which come with a mix of left and right Quillen functors.

Acknowledgements

This volume of proceedings would not have been possible without the continued efforts of the contributors, and the eagle-eyed referees. Moreover, the conference was only possible due to generous financial support from the NSF, NWO (Netherlands Organization for Scientific Research), NTNU, the BFS/TFS Foundation, the Compositio Foundation and EPFL.

Bonn, Germany   *Scott Balchin*
Belfast, United Kingdom   *David Barnes*
Nijmegen, Netherlands   *Magdalena Kędziorek*
Trondheim, Norway   *Markus Szymik*

John Greenlees' published work to date

[1] S. Balchin and J. P. C. Greenlees. Adelic models of tensor-triangulated categories. *Adv. Math.*, 375:107339, 45, 2020.

[2] J. P. C. Greenlees and G. Stevenson. Morita theory and singularity categories. *Adv. Math.*, 365:107055, 51, 2020.

[3] T. Barthel, J. P. C. Greenlees, and M. Hausmann. On the Balmer spectrum for compact Lie groups. *Compos. Math.*, 156(1):39–76, 2020.

[4] J. P. C. Greenlees. Couniversal spaces which are equivariantly commutative ring spectra. *Homology Homotopy Appl.*, 22(1):69–75, 2020.

[5] J. P. C. Greenlees. Borel cohomology and the relative Gorenstein condition for classifying spaces of compact Lie groups. *J. Pure Appl. Algebra*, 224(2):806–818, 2020.

[6] D. Barnes, J. P. C. Greenlees, and M. Kędziorek. An algebraic model for rational toral G-spectra. *Algebr. Geom. Topol.*, 19(7):3541–3599, 2019.

[7] J. P. C. Greenlees and D.-W. Lee. The representation-ring-graded local cohomology spectral sequence for $BP\mathbb{R}\langle 3\rangle$. *Comm. Algebra*, 47(6):2396–2411, 2019.

[8] J. P. C. Greenlees. The Balmer spectrum of rational equivariant cohomology theories. *J. Pure Appl. Algebra*, 223(7):2845–2871, 2019.

[9] D. Barnes, J. P. C. Greenlees, and M. Kędziorek. An algebraic model for rational naïve-commutative G-equivariant ring spectra for finite G. *Homology Homotopy Appl.*, 21(1):73–93, 2019.

[10] J. P. C. Greenlees and V. Stojanoska. Anderson and Gorenstein duality. In *Geometric and topological aspects of the representation theory of finite groups*, volume 242 of *Springer Proc. Math. Stat.*, pages 105–130. Springer, 2018.

[11] J. P. C. Greenlees and B. Shipley. An algebraic model for rational torus-equivariant spectra. *J. Topol.*, 11(3):666–719, 2018.

[12] J. P. C. Greenlees. Four approaches to cohomology theories with reality. In *An alpine bouquet of algebraic topology*, volume 708 of *Contemp. Math.*, pages 139–156. Amer. Math. Soc., Providence, RI, 2018.

[13] J. P. C. Greenlees. Homotopy invariant commutative algebra over fields. In *Building bridges between algebra and topology*, Adv. Courses Math. CRM Barcelona, pages 103–169. Birkhäuser/Springer, 2018.

[14] W. Chachólski, T. Dyckerhoff, J. P. C. Greenlees, and G. Stevenson. *Building bridges between algebra and topology*. Advanced Courses in Mathematics. CRM Barcelona. Birkhäuser/Springer, 2018. Lecture notes from courses held at CRM, Bellaterra, February 9–13, 2015 and April 13–17, 2015, Edited by Dolors Herbera, Wolfgang Pitsch and Santiago Zarzuela.

[15] J. P. C. Greenlees and L. Meier. Gorenstein duality for real spectra. *Algebr. Geom. Topol.*, 17(6):3547–3619, 2017.

[16] D. Barnes, J. P. C. Greenlees, M. Kędziorek, and B. Shipley. Rational SO(2)-equivariant spectra. *Algebr. Geom. Topol.*, 17(2):983–1020, 2017.

[17] J. P. C. Greenlees. Rational equivariant cohomology theories with toral support. *Algebr. Geom. Topol.*, 16(4):1953–2019, 2016.

[18] J. P. C. Greenlees. Rational torus-equivariant stable homotopy III: Comparison of models. *J. Pure Appl. Algebra*, 220(11):3573–3609, 2016.

[19] J. P. C. Greenlees. Ausoni-Bökstedt duality for topological Hochschild homology. *J. Pure Appl. Algebra*, 220(4):1382–1402, 2016.

[20] J. P. C. Greenlees and B. Shipley. Homotopy theory of modules over diagrams of rings. *Proc. Amer. Math. Soc. Ser. B*, 1:89–104, 2014.

[21] J. P. C. Greenlees and B. Shipley. Fixed point adjunctions for equivariant module spectra. *Algebr. Geom. Topol.*, 14(3):1779–1799, 2014.

[22] J. P. C. Greenlees and B. Shipley. An algebraic model for free rational G-spectra. *Bull. Lond. Math. Soc.*, 46(1):133–142, 2014.

[23] D. J. Benson and J. P. C. Greenlees. Stratifying the derived category of cochains on BG for G a compact Lie group. *J. Pure Appl. Algebra*, 218(4):642–650, 2014.

[24] W. G. Dwyer, J. P. C. Greenlees, and S. B. Iyengar. DG algebras with exterior homology. *Bull. Lond. Math. Soc.*, 45(6):1235–1245, 2013.

[25] J. P. C. Greenlees and B. Shipley. The cellularization principle for Quillen adjunctions. *Homology Homotopy Appl.*, 15(2):173–184, 2013.

[26] D. J. Benson, J. P. C. Greenlees, and S. Shamir. Complete intersections and mod p cochains. *Algebr. Geom. Topol.*, 13(1):61–114, 2013.

[27] J. P. C. Greenlees, K. Hess, and S. Shamir. Complete intersections in rational homotopy theory. *J. Pure Appl. Algebra*, 217(4):636–663, 2013.

[28] J. P. C. Greenlees. Rational torus-equivariant stable homotopy II: Algebra of the standard model. *J. Pure Appl. Algebra*, 216(10):2141–2158, 2012.

[29] M. Ando and J. P. C. Greenlees. Circle-equivariant classifying spaces and the rational equivariant sigma genus. *Math. Z.*, 269(3-4):1021–1104, 2011.

[30] J. P. C. Greenlees and B. Shipley. An algebraic model for free rational G-spectra for connected compact Lie groups G. *Math. Z.*, 269(1-2):373–400, 2011.

[31] W. G. Dwyer, J. P. C. Greenlees, and S. B. Iyengar. Gross-Hopkins duality and the Gorenstein condition. *J. K-Theory*, 8(1):107–133, 2011.

[32] R. R. Bruner and J. P. C. Greenlees. *Connective real K-theory of finite groups*, volume 169 of *Mathematical Surveys and Monographs*. American Mathematical Society, Providence, RI, 2010.

[33] D. J. Benson and J. P. C. Greenlees. Localization and duality in topology and modular representation theory. *J. Pure Appl. Algebra*, 212(7):1716–1743, 2008.

[34] J. P. C. Greenlees and G. R. Williams. Poincaré duality for K-theory of equivariant complex projective spaces. *Glasg. Math. J.*, 50(1):111–127, 2008.

[35] J. P. C. Greenlees. Rational torus-equivariant stable homotopy. I. Calculating groups of stable maps. *J. Pure Appl. Algebra*, 212(1):72–98, 2008.

[36] J. P. C. Greenlees. First steps in brave new commutative algebra. In *Interactions between homotopy theory and algebra*, volume 436 of *Contemp. Math.*, pages 239–275. Amer. Math. Soc., Providence, RI, 2007.

[37] J. P. C. Greenlees. Spectra for commutative algebraists. In *Interactions between homotopy theory and algebra*, volume 436 of *Contemp. Math.*, pages 149–173. Amer. Math. Soc., Providence, RI, 2007.

[38] J. P. C. Greenlees. Algebraic groups and equivariant cohomology theories. In *Elliptic cohomology*, volume 342 of *London Math. Soc. Lecture Note Ser.*, pages 89–110. Cambridge Univ. Press, Cambridge, 2007.

[39] W. Dwyer, J. P. C. Greenlees, and S. Iyengar. Finiteness in derived categories of local rings. *Comment. Math. Helv.*, 81(2):383–432, 2006.

[40] J. P. C. Greenlees and J.-Ph. Hoffmann. Rational extended Mackey functors for the circle group. In *An alpine anthology of homotopy theory*, volume 399 of *Contemp. Math.*, pages 123–131. Amer. Math. Soc., Providence, RI, 2006.

[41] W. G. Dwyer, J. P. C. Greenlees, and S. Iyengar. Duality in algebra and topology. *Adv. Math.*, 200(2):357–402, 2006.

[42] J. P. C. Greenlees. Equivariant versions of real and complex connective K-theory. *Homology Homotopy Appl.*, 7(3):63–82, 2005.

[43] J. P. C. Greenlees. Rational S^1-equivariant elliptic cohomology. *Topology*, 44(6):1213–1279, 2005.

[44] J. P. C. Greenlees. Equivariant connective K-theory for compact Lie groups. *J. Pure Appl. Algebra*, 187(1-3):129–152, 2004.

[45] R. R. Bruner and J. P. C. Greenlees. The connective K-theory of finite groups. *Mem. Amer. Math. Soc.*, 165(785):viii+127, 2003.

[46] M. Cole, J. P. C. Greenlees, and I. Kriz. The universality of equivariant complex bordism. *Math. Z.*, 239(3):455–475, 2002.

[47] J. P. C. Greenlees. Local cohomology in equivariant topology. In *Local cohomology and its applications (Guanajuato, 1999)*, volume 226 of *Lecture Notes in Pure and Appl. Math.*, pages 1–38. Dekker, New York, 2002.

[48] W. G. Dwyer and J. P. C. Greenlees. Complete modules and torsion modules. *Amer. J. Math.*, 124(1):199–220, 2002.

[49] J. P. C. Greenlees. Multiplicative equivariant formal group laws. *J. Pure Appl. Algebra*, 165(2):183–200, 2001.

[50] J. P. C. Greenlees. Equivariant formal group laws and complex oriented cohomology theories. *Homology Homotopy Appl.*, 3(2):225–263, 2001. Equivariant stable homotopy theory and related areas (Stanford, CA, 2000).

[51] J. P. C. Greenlees. Tate cohomology in axiomatic stable homotopy theory. In *Cohomological methods in homotopy theory (Bellaterra, 1998)*, volume 196 of *Progr. Math.*, pages 149–176. Birkhäuser, Basel, 2001.

[52] J. P. C. Greenlees. Rational SO(3)-equivariant cohomology theories. In *Homotopy methods in algebraic topology (Boulder, CO, 1999)*, volume 271 of *Contemp. Math.*, pages 99–125. Amer. Math. Soc., Providence, RI, 2001.

[53] M. Cole, J. P. C. Greenlees, and I. Kriz. Equivariant formal group laws. *Proc. London Math. Soc. (3)*, 81(2):355–386, 2000.

[54] J. P. C. Greenlees and G. Lyubeznik. Rings with a local cohomology theorem and applications to cohomology rings of groups. *J. Pure Appl. Algebra*, 149(3):267–285, 2000.

[55] J. P. C. Greenlees. Equivariant forms of connective K-theory. *Topology*, 38(5):1075–1092, 1999.

[56] J. P. C. Greenlees and N. P. Strickland. Varieties and local cohomology for chromatic group cohomology rings. *Topology*, 38(5):1093–1139, 1999.

[57] J. P. C. Greenlees. Rational S^1-equivariant stable homotopy theory. *Mem. Amer. Math. Soc.*, 138(661):xii+289, 1999.

[58] J. P. C. Greenlees. Augmentation ideals of equivariant cohomology rings. *Topology*, 37(6):1313–1323, 1998.

[59] J. P. C. Greenlees. Rational O(2)-equivariant cohomology theories. In *Stable and unstable homotopy (Toronto, ON, 1996)*, volume 19 of *Fields Inst. Commun.*, pages 103–110. Amer. Math. Soc., Providence, RI, 1998.

[60] J. P. C. Greenlees. Rational Mackey functors for compact Lie groups. I. *Proc. London Math. Soc. (3)*, 76(3):549–578, 1998.

[61] J. P. C. Greenlees and H. Sadofsky. Tate cohomology of theories with one-dimensional coefficient ring. *Topology*, 37(2):279–292, 1998.

[62] J. P. C. Greenlees and J. P. May. Localization and completion theorems for MU-module spectra. *Ann. of Math. (2)*, 146(3):509–544, 1997.

[63] D. J. Benson and J. P. C. Greenlees. Commutative algebra for cohomology rings of classifying spaces of compact Lie groups. *J. Pure Appl. Algebra*, 122(1-2):41–53, 1997.

[64] D. J. Benson and J. P. C. Greenlees. Commutative algebra for cohomology rings of virtual duality groups. *J. Algebra*, 192(2):678–700, 1997.

[65] J. P. C. Greenlees and J. A. Pérez. Connected Lie groups that act freely on a product of linear spheres. *Bull. London Math. Soc.*, 28(6):634–642, 1996.

[66] J. P. C. Greenlees and H. Sadofsky. The Tate spectrum of v_n-periodic complex oriented theories. *Math. Z.*, 222(3):391–405, 1996.

[67] J. P. C. Greenlees. A rational splitting theorem for the universal space for almost free actions. *Bull. London Math. Soc.*, 28(2):183–189, 1996.

[68] J. P. C. Greenlees and J. A. Pérez. Connected Lie groups acting freely on a product of linear spheres. In *XXVII National Congress of the Mexican Mathematical Society (Spanish) (Queretaro, 1994)*, volume 16 of *Aportaciones Mat. Comun.*, pages 197–203. Soc. Mat. Mexicana, México, 1995.

[69] R. R. Bruner and J. P. C. Greenlees. The Bredon-Löffler conjecture. *Experiment. Math.*, 4(4):289–297, 1995.

[70] J. P. C. Greenlees and J. P. May. Equivariant stable homotopy theory. In *Handbook of algebraic topology*, pages 277–323. North-Holland, Amsterdam, 1995.

[71] J. P. C. Greenlees and J. P. May. Completions in algebra and topology. In *Handbook of algebraic topology*, pages 255–276. North-Holland, Amsterdam, 1995.

[72] J. P. C. Greenlees. Commutative algebra in group cohomology. *J. Pure Appl. Algebra*, 98(2):151–162, 1995.

[73] J. P. C. Greenlees and J. P. May. Generalized Tate cohomology. *Mem. Amer. Math. Soc.*, 113(543):viii+178, 1995.

[74] J. P. C. Greenlees. Tate cohomology in commutative algebra. *J. Pure Appl. Algebra*, 94(1):59–83, 1994.

[75] A. D. Elmendorf, J. P. C. Greenlees, I. Kříž, and J. P. May. Commutative algebra in stable homotopy theory and a completion theorem. *Math. Res. Lett.*, 1(2):225–239, 1994.

[76] J. P. C. Greenlees. The geometric equivariant Segal conjecture for toral groups. *J. London Math. Soc. (2)*, 48(2):348–364, 1993.

[77] J. P. C. Greenlees. *K*-homology of universal spaces and local cohomology of the representation ring. *Topology*, 32(2):295–308, 1993.

[78] T. Bier and J. P. C. Greenlees. The lattice spanned by the cosets of subgroups in the integral group ring of a finite group. *J. London Math. Soc. (2)*, 47(3):433–449, 1993.

[79] D. J. Benson and J. P. C. Greenlees. The action of the Steenrod algebra on Tate cohomology. *J. Pure Appl. Algebra*, 85(1):21–26, 1993.

[80] J. P. C. Greenlees and J. P. May. Completions of *G*-spectra at ideals of the Burnside ring. In *Adams Memorial Symposium on Algebraic Topology, 2 (Manchester, 1990)*, volume 176 of *London Math. Soc. Lecture Note Ser.*, pages 145–178. Cambridge Univ. Press, Cambridge, 1992.

[81] J. P. C. Greenlees. Homotopy equivariance, strict equivariance and induction theory. *Proc. Edinburgh Math. Soc. (2)*, 35(3):473–492, 1992.

[82] J. P. C. Greenlees. Some remarks on projective Mackey functors. *J. Pure Appl. Algebra*, 81(1):17–38, 1992.

[83] J. P. C. Greenlees and J. P. May. Derived functors of *I*-adic completion and local homology. *J. Algebra*, 149(2):438–453, 1992.

[84] J. P. C. Greenlees. Generalized Eilenberg-Moore spectral sequences for elementary abelian groups and tori. *Math. Proc. Cambridge Philos. Soc.*, 112(1):77–89, 1992.

[85] J. P. C. Greenlees and J. P. May. Some remarks on the structure of Mackey functors. *Proc. Amer. Math. Soc.*, 115(1):237–243, 1992.

[86] J. P. C. Greenlees. Equivariant functional duals and completions at ideals of the Burnside ring. *Bull. London Math. Soc.*, 23(2):163–168, 1991.

[87] J. P. C. Greenlees. The power of mod *p* Borel homology. In *Homotopy theory and related topics (Kinosaki, 1988)*, volume 1418 of *Lecture Notes in Math.*, pages 140–151. Springer, Berlin, 1990.

[88] J. P. C. Greenlees. Equivariant functional duals and universal spaces. *J. London Math. Soc. (2)*, 40(2):347–354, 1989.

[89] J. P. C. Greenlees. Topological methods in equivariant cohomology. In *Group theory (Singapore, 1987)*, pages 373–389. de Gruyter, Berlin, 1989.

[90] J. P. C. Greenlees. How blind is your favourite cohomology theory? *Exposition. Math.*, 6(3):193–208, 1988.

[91] J. P. C. Greenlees. Stable maps into free *G*-spaces. *Trans. Amer. Math. Soc.*, 310(1):199–215, 1988.

[92] J. P. C. Greenlees. Representing Tate cohomology of *G*-spaces. *Proc. Edinburgh Math. Soc. (2)*, 30(3):435–443, 1987.

[93] J. P. C. Greenlees. Functional duals and Moore spectra. *Bull. London Math. Soc.*, 17(1):43–48, 1985.

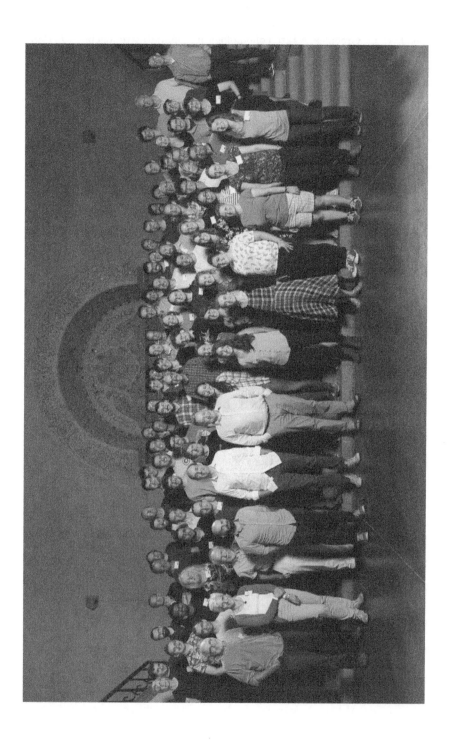

1
Comparing Dualities in the $K(n)$-local Category

Paul G. Goerss[a]

Michael J. Hopkins[b]

Abstract

In their work on the period map and the dualizing sheaf for Lubin-Tate space, Gross and the second author wrote down an equivalence between the Spanier-Whitehead and Brown-Comenetz duals of certain type n-complexes in the $K(n)$-local category at large primes. In the culture of the time, these results were accessible to educated readers, but this seems no longer to be the case; therefore, in this note we give the details. Because we are at large primes, the key result is algebraic: in the Picard group of Lubin-Tate space, two important invertible sheaves become isomorphic modulo p.

For John Greenlees, the master of duality.

Introduction

Fix a prime p and and an integer $n \geq 0$, and let $K(n)$ denote the nth Morava K-theory at the prime p. If $n \geq 1$, the $K(n)$-local stable homotopy category has two dualities. First, there is $K(n)$-local Spanier-Whitehead duality $D_n(-)$. This behaves very much like Spanier-Whitehead duality in the ordinary stable category: it has good formal properties, but it can be very hard to compute. Second, there is Brown-Comenetz duality $I_n(-)$, which behaves much like a Serre-Grothendieck duality and, in many ways, is much more computable. One of the key features of the

[a] Department of Mathematics, Northwestern University
[b] Department of Mathematics, Harvard University

$K(n)$-local category is that under some circumstances the two dualities are closely related.

Recall that a finite spectrum X is of type n if $K(m)_*X = 0$ for $m < n$. By [22], any type n spectrum has a $v_n^{p^k}$-self map; that is, there is an integer k and map

$$\Sigma^{2p^k(p^n-1)}X \to X$$

which induces multiplication by $v_n^{p^k}$ in $K(n)_*$. In their papers on the period map and the dualizing sheaf for Lubin-Tate space, Gross and the second author [20] wrote down the following result. Suppose X is a type n-spectrum with a $v_n^{p^k}$-self map and suppose further that p times the identity map of X is zero. Then if $2p > \max\{n^2 + 1, 2n + 2\}$ there is an equivalence in the $K(n)$-local category[1]

$$I_n X \simeq \Sigma^{2p^{nk}r(n)+n^2-n} D_n X \qquad (1.1)$$

where $r(n) = (p^n - 1)/(p - 1) = p^{n-1} + \cdots + p + 1$. This equivalence gives a conceptual explanation for many of the self-dual patterns apparent in the amazing computations of Shimomura and his coauthors. See, for example, [32], [31], [5], and [26].

The point of this note is to write down a linear narrative with this result at the center. In some sense, there is nothing new here, as the key ideas can be found scattered through the literature, and other authors have obliquely touched on this topic. A rich early example is in §5 of the paper [8] by Devinatz and the second author, and the important paper of Dwyer, Greenlees, and Iyengar [10] embeds many of the ideas here into a far-reaching and beautiful theory. In another sense, however, there is quite a bit to say, as there are any number of key technical ideas we need to access, some of which have not quite made it into print and others buried in ways that make them hard to uncover. In any case, the result is of enough importance that it deserves specific memorialization.

Here is a little more detail. We fix p and n and let $E = E_n$ be Morava E-theory for n and p. This represents a complex oriented cohomology theory with formal group law a universal deformation of the Honda formal group law H_n of height n. See §1 for more details. As always we write

$$E_*X = \pi_* L_{K(n)}(E \wedge X).$$

The E_*-module E_*X is a graded Morava module: it has a continuous

---

[1] The bound on p is very slightly different than in [20]; see Proposition 1.9.

and twisted action of the Morava stabilizer group $\mathbb{G}_n = \text{Aut}(H_n, \mathbb{F}_{p^n})$. See Remark 1.5.

There are two key steps to the equivalence (1.1). We have a $K(n)$-local equivalence $I_n X \simeq I_n \wedge D_n X$ where $I_n = I_n(S^0)$; thus, the first step is the identification of the homotopy type of I_n, at least for p large with respect to n. This is also due to Gross and the second author, with details laid out in [33]. The key fact is that I_n is dualizable in the $K(n)$-local category; by [21] this is equivalent to the statement that $E_* I_n$ is an invertible graded Morava module and, indeed, the main result of [33] (interpreting [20]) is that there is an isomorphism of Morava modules

$$E_* I_n \cong E_* S^{n^2-n}[\det]$$

where $S^0[\det] = S[\det]$ is a determinant twisted sphere in the $K(n)$-local category; see Remark 1.26. The number $r(n)$ in (1.1) is an artifact of the determinant; see (1.3.1).

The second key step is an analysis of the $K(n)$-local Picard group $\text{Pic}_{K(n)}$ of equivalence classes of invertible objects in the $K(n)$-local category. As mentioned, we know that a $K(n)$-local spectrum X is invertible if and only if $E_* X$ is an invertible graded Morava module. We also know that the group of invertible graded Morava modules concentrated in even degrees is isomorphic to the continuous cohomology group $H^1(\mathbb{G}_n, E_0^\times)$, where $E_0^\times$ is the group of units in the ring E_0. Hence, if we write $\text{Pic}_{K(n)}^0 \subseteq \text{Pic}_{K(n)}$ for the subgroup of objects X with $E_* X$ in even degrees, we get a map

$$e : \text{Pic}_{K(n)}^0 \longrightarrow H^1(\mathbb{G}_n, E_0^\times).$$

The map is an injection under the hypothesis $2p > \max\{n^2 + 1, 2n + 2\}$. See Proposition 1.9. This is the origin for the hypothesis on p and n in the equivalence of (1.1): it reduces that equivalence to an algebraic calculation.

It is an observation of [21] that the map $\mathbb{Z} \to \text{Pic}_{K(n)}^0$ sending k to S^{2k} extends to an inclusion of the completion of the integers

$$\mathfrak{z}_n \overset{\text{def}}{=} \lim_k \mathbb{Z}/(p^k(p^n - 1)) \to \text{Pic}_{K(n)}^0;$$

that is, for any $a \in \mathfrak{z}_n$ we have a sphere S^{2a}. (The phrase "p-adic sphere" is common here, but misleading: $\mathfrak{z}_n$ is not the p-adic integers. See Remark 1.23.) Now let $\lambda = \lim_k p^{nk} r(n) \in \mathfrak{z}_n$. The key algebraic result can now be deduced from Proposition 1.30 below: under the composition

$$\text{Pic}_{K(n)}^0 \overset{e}{\longrightarrow} H^1(\mathbb{G}_n, E_0^\times) \to H^1(\mathbb{G}_n, (E_0/p)^\times)$$

the spectra $S[\det]$ and $S^{2\lambda}$ map to the same element. The equivalence (1.1) follows once we observe that if X is type n and has a $v_n^{p^k}$-self map, then there is $K(n)$-local equivalence

$$S^{2\lambda} \wedge X \simeq \Sigma^{2p^{nk}r(n)} X.$$

See Theorems 1.42 and 1.43.

It is worth emphasizing that the algebraic result Proposition 1.30 only requires $p > 2$; it is the topological applications which require the more stringent restrictions on the prime. In fact, the equivalence of dualities in (1.1) can be false if the prime is small. See Remark 1.45.

The plan of this note is as follows: in the first section we give some homotopy theoretic and algebraic background, in the second section we give a discussion of the Picard group, lingering long enough to give details of the structure of $\mathrm{Pic}^0_{K(n)}$ as a profinite 3_n-module. See Proposition 1.18. In Section 3 we discuss the determinant and prove the key Proposition 1.30. In Section 4 we give some discussion of how Spanier-Whitehead and Brown-Comenetz duality behave in the Adams-Novikov Spectral Sequence. In the final section, we give the homotopy theoretic applications.

Acknowledgements

This project began as an attempt to find a conceptual computation of the self-dual patterns apparent in the Shimomura-Yabe calculation [32] of $\pi_* L_{K(2)} V(0)$ at $p > 3$. Then in a conversation with Tobias Barthel it emerged that there was no straightforward argument in print to prove the equivalence of dualities in (1.1). Later, as Guchuan Li was working on his thesis [27] it became apparent that he needed these results and, more, there were constructions once present in the general culture that were no longer easily accessible. Thus a sequence of notes, begun at MSRI in Spring of 2014, have evolved into this chapter. Many thanks for Agnès Beaudry and Vesna Stojanoska for reading through an early draft. Others have surely written down proofs for themselves as well; for example, Hans-Werner Henn once remarked that "the determinant essentially disappears mod p," which is a very succinct summary of Proposition 1.30.

Finally, the authors would like to the thank the referee for a very careful reading and helpful comments.

Support provided by the National Science Foundation under grant No. DMS–1810917 and by the Isaac Newton Institute for Mathematical

Sciences. The first author was in residence at the Newton Institute for the program "Homotopy Harnessing Higher Structures" supported by the EPSRC grant no EP/K032208/1.

1.1 Some background

In this section we gather together the basic material used in later sections. All of this is thoroughly covered in the literature and collected here only for narrative continuity.

1.1.1 The $K(n)$-local category

For an in-depth study of the technicalities in the $K(n)$-local category, see Hovey and Strickland [24]. Other introductions can be found in almost any paper on chromatic homotopy theory. We were especially thorough in [3] §2.

Fix a prime p and an integer $n > 0$. In order to be definite we define the nth Morava K-theory $K(n)$ to be the 2-periodic complex oriented cohomology theory with coefficients $K(n)_* = \mathbb{F}_{p^n}[u^{\pm 1}]$ with u in degree -2. The associated formal group law over $K(n)_0 = \mathbb{F}_{p^n}$ is the unique p-typical formal group law H_n with p-series $[p]_{H_n}(x) = x^{p^n}$. This is, of course, the nth Honda formal group law. For H_n we have

$$v_n = u^{1-p^n} \in K(n)_{2(p^n-1)}.$$

The $K(n)$-local category is the category of $K(n)$-local spectra.

We also have $K(0) = H\mathbb{Q}$, the rational Eilenberg-MacLane spectrum, and $K(0)$-local spectra are the subject of rational stable homotopy theory.

We define $\mathbb{G}_n = \mathrm{Aut}(H_n, \mathbb{F}_{p^n})$ to be the group of automorphisms of the pair $(H_n, \mathbb{F}_{p^n})$. Since H_n is defined over $\mathbb{F}_p$, there is a splitting

$$\mathrm{Aut}(H_n, \mathbb{F}_{p^n}) \cong \mathrm{Aut}(H_n/\mathbb{F}_{p^n}) \rtimes \mathrm{Gal}(\mathbb{F}_{p^n}/\mathbb{F}_p)$$

where the normal subgroup is the isomorphisms of H_n as a formal group law over $\mathbb{F}_{p^n}$. We write $\mathbb{S}_n = \mathrm{Aut}(H_n/\mathbb{F}_{p^n})$ for this subgroup.

To get a Landweber exact homology theory which captures more than Morava K-theory, we use the Morava (or Lubin-Tate) theory $E = E_n$. This theory has coefficients

$$E_* = \mathbb{W}[[u_1, \ldots, u_{n-1}]][u^{\pm 1}]$$

where again u is in degree -2 but the power series ring is in degree 0. The ring $\mathbb{W} = W(\mathbb{F}_{p^n})$ is the Witt vectors of $\mathbb{F}_{p^n}$.

Note that E_0 is a complete local ring with maximal ideal $\mathfrak{m}$ generated by the regular sequence $\{p, u_1, \ldots, u_{n-1}\}$. We choose the formal group law G_n over E_0 to be the unique p-typical formal group law with p-series

$$[p]_{G_n}(x) = px +_{G_n} u_1 x^p +_{G_n} \cdots +_{G_n} u_{n-1} x^{p^{n-1}} +_{G_n} x^{p^n}. \quad (1.1.1)$$

Thus $v_i = u_i u^{1-p^i}$, $1 \le i \le n-1$, $v_n = u^{1-p^n}$ and $v_i = 0$ if $i > n$. Note that G_n reduces to H_n modulo $\mathfrak{m}$.

We define $E_* X = (E_n)_* X$ by

$$E_* X = \pi_* L_{K(n)}(E \wedge X).$$

While not quite a homology theory, as it does not take wedges to sums, it is by far our most sensitive algebraic invariant in $K(n)$-local homotopy theory. The group $\mathbb{G}_n$ acts continuously on $E_* X$ making $E_* X$ into a *Morava module*. We will be more precise on this notion below in Remark 1.5.

A basic computation gives

$$E_0 E = \pi_0 L_{K(n)}(E \wedge E) \cong \mathrm{map}^c(\mathbb{G}_n, E_0)$$

where map^c denotes the continuous maps. See Lemma 10 of [33] for a proof. The $K(n)$-local E_n-based Adams-Novikov Spectral Sequence now reads

$$H^s(\mathbb{G}_n, E_t X) \Longrightarrow \pi_{t-s} L_{K(n)} X. \quad (1.1.2)$$

Cohomology here is continuous cohomology.

Remark 1.1 (Lubin-Tate theory) The pair (G_n, E_0) has an important universal property which is useful for understanding the action of $\mathbb{G}_n$.

Consider a complete local ring $(S, \mathfrak{m}_S)$ with $S/\mathfrak{m}_S$ of characteristic p. Define the groupoid of deformations $\mathrm{Def}_{H_n}(S)$ to be the category with objects (i, G) where $i : \mathbb{F}_{p^n} \to S/\mathfrak{m}_S$ is a morphism of fields and G is a formal group law over S with $q_* G = i_* H_n$. Here $q : S \to S/\mathfrak{m}_S$ is the quotient map. There are no morphisms $\psi : (i, G) \to (j, H)$ if $i \ne j$ and a morphism $(i, G) \to (i, H)$ is an isomorphism of formal groups laws $\psi : G \to H$ so that $q_* \psi$ is the identity. These are the $\star$-isomorphisms. By a theorem of Lubin and Tate [28] we know that if two deformations are $\star$-isomorphic, then there is a unique $\star$-isomorphism between them.

Put another way, the groupoid $\text{Def}_{H_n}(S)$ is discrete. Furthermore, E_0 represents the functor of $\star$-isomorphism classes of deformations:

$$\text{Hom}^c_{\mathbb{W}}(E_0, S) \cong \pi_0 \text{Def}_{H_n}(S).$$

Here $\text{Hom}^c_{\mathbb{W}}$ is the set of continuous $\mathbb{W}$-algebra maps. As a universal deformation we can and do choose the formal group law G_n over E_0 to be the p-typical formal group law defined above in (1.1.1).

Remark 1.2 (The action of the Morava stabilizer group)   We use Lubin-Tate theory to get an action of $\mathbb{G}_n$ on E_0. This exposition follows [18] §3.

Let $g = g(x) \in \mathbb{F}_{p^n}[[x]]$ be an element in $\mathbb{S}_n$. Choose any lift of $g(x)$ to $h(x) \in E_0[[x]]$ and let G_h be the unique formal group law over E_0 so that

$$h : G_h \to G_n$$

is an isomorphism. Since $g : H_n \to H_n$ is an isomorphism over $\mathbb{F}_{p^n}$, the pair (id, G_h) is a deformation of H_n. Hence there is a unique $\mathbb{W}$-algebra map $\phi = \phi_g : E_0 \to E_0$ and a unique $\star$-isomorphism $f : \phi_* G_n \to G_h$. Let ψ_g be the composition

$$\phi_* G_n \xrightarrow{\;f\;} G_h \xrightarrow{\;h\;} G_n \ . \tag{1.1.3}$$
$$\underbrace{\phantom{\phi_* G_n \xrightarrow{\;f\;} G_h \xrightarrow{\;h\;} G_n}}_{\psi_g}$$

Note that while G_h depends on choices, the map ϕ_g and the isomorphism ψ_g do not. The map $\mathbb{S}_n \to \text{Aut}(E_0)$ sending g to ϕ_g defines the action of $\mathbb{S}_n$ on E_0. The Galois action on $\mathbb{W} \subseteq E_0$ extends this to an action of all of $\mathbb{G}_n$ on E_0. The action can be extended to all of E_* be noting that $E_2 \cong \tilde{E}^0 S^2 \cong \tilde{E}^0 \mathbb{CP}^1$ is isomorphic to the module of invariant differentials on the universal deformation G_n. See (1.1.4) below for an explicit formula.

Remark 1.3 (Formulas for the action)   We make the action of $\mathbb{S}_n$ a bit more precise. By (1.1.3) we have an isomorphism $\psi_g : \phi_* G_n \to G_n$ of p-typical formal group laws over E_0. This can be written

$$\psi_g(x) = t_0(g) +_{G_n} t_1(g)x^p +_{G_n} t_2(g)x^{p^2} +_{G_n} \cdots .$$

This formula defines continuous functions $t_i : \mathbb{S}_n \to E_0$. As in Section 4.1 of [18] we have

$$g_* u = t_0(g)u. \tag{1.1.4}$$

The function t_0 is a crossed homomorphism $t_0 : \mathbb{S}_n \to E_0^\times$; that is,

$$t_0(gh) = [gt_0(h)]t_0(g).$$

Since the Honda formal group is defined over $\mathbb{F}_p$ we can choose the class u to be invariant under the action of the Galois group; hence t_0 extends to crossed homomorphism $t_0 : \mathbb{G}_n \to E_0^\times$ sending $(g, \phi) \in \mathbb{S}_n \rtimes \mathrm{Gal}(\mathbb{F}_{p^n}/\mathbb{F}_p) \cong \mathbb{G}_n$ to $t_0(g)$.

Remark 1.4 We record here some basic useful facts about the $K(n)$-local Adams-Novikov Spectral Sequence (1.1.2) which we will use later.

The first two statements are standard and are proved using the action of the center of $Z(\mathbb{G}_n) \subseteq \mathbb{G}_n$ on $E_* = E_* S^0$. There is an isomorphism $\mathbb{Z}_p^\times \cong Z(\mathbb{G}_n)$ sendings $a \in \mathbb{Z}_p^\times$ to the a-series $[a]_{H_n}(x)$ of the Honda formal group. The action of $Z(\mathbb{G}_n)$ on E_0 is trivial and the action on E_* is then determined by the fact that $t_0(a) = a$; that is, a acts on $u \in E_{-2}$ by multiplication by a.

1.) **Sparseness:** If $t \not\equiv 0$ modulo $2(p-1)$, then $H^*(\mathbb{G}_n, E_t) = 0$. If $p = 2$ this is not new information. If $p > 2$ let $C \subseteq Z(\mathbb{G}_n)$ be the cyclic subgroup of Teichmüller lifts of $\mathbb{F}_p^\times$. Then $E_t^C = 0$ and hence

$$H^*(\mathbb{G}_n, E_t) \cong H^*(\mathbb{G}_n/C, E_t^C) = 0.$$

2.) **Bounded torsion:** Suppose $p > 2$ and suppose

$$2t = 2p^k m(p-1) \neq 0$$

with m not divisible by p. Then we have

$$p^{k+1} H^*(\mathbb{G}_n, E_{2t}) = 0.$$

If $p = 2$ write $2t = 2^k(2m+1)$. Then we have

$$2H^*(\mathbb{G}_n, E_{2t}) = 0 \qquad \text{if } k = 1,$$

and

$$2^{k+1} H^*(\mathbb{G}_n, E_{2t}) = 0 \qquad \text{if } k > 1.$$

To get these bounds, first suppose $p > 2$. Let $K = 1 + p\mathbb{Z}_p \subseteq Z(\mathbb{G}_n)$ be the torsion-free subgroup and let $x \in K$ be a topological generator; for example, $x = 1 + p$. The choice of x defines an isomorphism $\mathbb{Z}_p \cong K$. Thus, there is an exact sequence

$$0 \to H^0(K, E_{2t}) \longrightarrow E_{2t} \xrightarrow{x^k - 1} E_{2t} \longrightarrow H^1(K, E_{2t}) \to 0.$$

Thus we see that $p^{k+1}H^1(K, E_{2t}) = 0$ and $H^q(K, E_{2t}) = 0$ if $q \neq 1$. Now use the Lyndon-Hochschild-Serre Spectral Sequence

$$H^p(\mathbb{G}_n/K, H^q(K, E_{2t})) \Longrightarrow H^{p+q}(\mathbb{G}_n, E_{2t})$$

to deduce the claim. At the prime 2 let $x \in \mathbb{Z}_2^\times$ be an element of infinite order which reduces to -1 modulo 4 – for example, $x = 3$ – and let K be the subgroup generated by x. The proof then proceeds in the same fashion.

Note that the arguments for parts (1) and (2) apply not only to $\mathbb{G}_n$, but also for any closed subgroup $G \subseteq \mathbb{G}_n$ which contains the center. In fact, for part (1) we need only have $C = \mathbb{F}_p^\times \subseteq G$.

3.) **There is a uniform and horizontal vanishing line at E_∞:** there is an integer N, depending only on n and p, so that in the Adams-Novikov Spectral Sequence (1.1.2) for any spectrum X

$$E_\infty^{s,*} = 0, \qquad s > N.$$

This can be found in the literature in several guises; for example, it can be put together from the material in Section 5 of [9], especially Lemma 5.11. See [3] §2.3 for references and explanation. See also [2] for even further explanation. If $p - 1 > n$, there is often a horizontal vanishing line at E_2. See Proposition 1.6 below.

1.1.2 Some local homological algebra.

Because E_0 is a complete local ring with maximal ideal $\mathfrak{m}$ generated by a regular sequence, we have a variety of tools from homological algebra. The classic paper here is Greenlees and May [15], but see also [24], Appendix A for direct connections to $E_*(-)$. Tensor product below will mean the $\mathfrak{m}$-completed tensor product. This is one place where the notation E_0 gets out of hand; thus we write $R = E_0$ in this subsection.

Let $u_0 = p$ and define a cochain complex $\Gamma_\mathfrak{m}$ by

$$\Gamma_\mathfrak{m} = \left(R \to R[\frac{1}{u_0}] \right) \otimes_R \left(R \to R[\frac{1}{u_1}] \right) \otimes_R \cdots \otimes_R \left(R \to R[\frac{1}{u_{n-1}}] \right)$$

and more generally we set

$$\Gamma_m(M) = M \otimes_R \Gamma_\mathfrak{m}.$$

Then $H^0_\mathfrak{m}(M) \overset{\text{def}}{=} H^0\Gamma_m(M)$ is the sub-module of $\mathfrak{m}$-torsion and we see that

$$H^s\Gamma_\mathfrak{m}(M) \overset{\text{def}}{=} H^s_\mathfrak{m}(M)$$

is the sth right derived functor of the $\mathfrak{m}$-torsion functor and thus independent of the choices. These are the local cohomology groups. If M is $\mathfrak{m}$-torsion, there is a composite functor spectral sequence

$$\operatorname{Ext}_R^p(M, H_\mathfrak{m}^q(N)) \Longrightarrow \operatorname{Ext}_R^{p+q}(M, N). \tag{1.1.5}$$

In the case $N = R$, this spectral sequence simplifies considerably. Note that $H_\mathfrak{m}^s(R) = 0$ unless $s = n$ and

$$H_\mathfrak{m}^n(R) \overset{\text{def}}{=} R/\mathfrak{m}^\infty \overset{\text{def}}{=} R/(p^\infty, u_1^\infty, \ldots, u_{n-1}^\infty) . \tag{1.1.6}$$

The R-module $R/\mathfrak{m}^\infty$ is an injective R-module and, in fact the injective hull of $R/\mathfrak{m}$. This is a consequence of Matlis duality for $(R, \mathfrak{m})$; see §12.1 of [7], especially Definition 12.1.2 and Remark 12.1.3.

Combining this observation with the spectral sequence (1.1.5) we have

$$\operatorname{Ext}_R^{p+n}(M, R) \cong \operatorname{Ext}_R^p(M, R/\mathfrak{m}^\infty) \cong \begin{cases} \operatorname{Hom}_R(M, R/\mathfrak{m}^\infty), & p = 0; \\ 0, & p \neq 0. \end{cases} \tag{1.1.7}$$

The module $R/\mathfrak{m}^\infty$ also arises in the theory of derived functors of completion. The completion functor

$$M \longmapsto \lim_k \left[M \otimes_R R/\mathfrak{m}^k \right]$$

is neither left nor right exact; however, it still has left derived functors $L_s^\mathfrak{m}(M)$. These vanish if $s > n$ and there is an isomorphism

$$L_n^\mathfrak{m}(M) \cong \lim \operatorname{Tor}_n^R(M, R/\mathfrak{m}^k)$$
$$\cong \lim \operatorname{Hom}_R(R/\mathfrak{m}^k, M)$$
$$\cong \operatorname{Hom}_R(R/\mathfrak{m}^\infty, M).$$

From this it follows that

$$L_s^\mathfrak{m}(M) \cong \operatorname{Ext}_R^{n-s}(R/\mathfrak{m}^\infty, M).$$

Remark 1.5 (Morava modules) If X is a spectrum we defined

$$E_*X = \pi_* L_{K(n)}(E \wedge X).$$

By [24], Proposition 8.4, the E_*-module E_*X is $L^\mathfrak{m}$-complete; that is, the map

$$E_*X \longrightarrow L_0^\mathfrak{m}(E_*X)$$

is an isomorphism. In particular, E_*X is equipped with the $\mathfrak{m}$-adic topology.

The action of $\mathbb{G}_n$ on E determines a continuous action of $\mathbb{G}_n$ on $E_t X$. This action is twisted in the sense that if $g \in \mathbb{G}_n$, $a \in E_0$ and $x \in E_t X$, then $g_*(ax) = g_*(a)g_*(x)$. We will call an $L^{\mathfrak{m}}$-complete E_0-module with a continuous and twisted $\mathbb{G}_n$ action a *Morava module*. Many (if not all) of our Morava modules will actually be $\mathfrak{m}$-complete; that is, the natural maps

$$M \longrightarrow L_0^{\mathfrak{m}} M \longrightarrow \lim M/\mathfrak{m}^k M$$

are all isomorphisms. For example, if M is $\mathfrak{m}$-complete, so is the induced module of continuous map $\mathrm{map}^c(\mathbb{G}_n, M)$. Hence the continuous cohomology of $\mathfrak{m}$-complete Morava modules can be constructed entirely in the category of $\mathfrak{m}$-complete Morava modules.

The graded Morava module module $E_* X$ is determined by $E_0 X$, $E_1 X$, and the isomorphism of $\mathbb{G}_n$-modules, for $n \in \mathbb{Z}$,

$$E_{t+2n} X \cong E_2^{\otimes n} \otimes_{E_0} E_t X.$$

The $\mathbb{G}_n$-action is the diagonal action. If $n \geq 0$, $E_2^{\otimes n} = E_2 \otimes_{E_0} \cdots \otimes_{E_0} E_2$ is a free of rank 1 over E_0. If $n < 0$, then $E_2^{\otimes n}$ is the dual $\mathbb{G}_n$-module to $E_2^{\otimes -n}$. This discussion gives an evident category of graded Morava modules.

We say a graded Morava module M_* is finitely generated if it is finitely generated as a graded E_*-module or, equivalently, if M_0 and M_1 are finitely generated as E_0-modules. We also say a graded Morava module M_* is finite if M_0 and M_1 are finite. If X is a finite CW spectrum then $E_* X$ is finitely generated. More generally, X is dualizable in the $K(n)$-local category if and only if $E_* X$ is finitely generated. See Theorem 8.6 of [24]. If X is also of type n, then $E_* X$ is finite.

Here is a key fact about Morava modules which we use often. The argument owes quite a good deal to the proof of Lemma 5 of [33].

Proposition 1.6 *Let $p - 1 > n$ and let M be an $\mathfrak{m}$-complete Morava module. Then for all $s > n^2$*

$$H^s(\mathbb{G}_n, M) = 0.$$

Proof Let $S_n \subseteq \mathbb{S}_n$ be the subgroup of automorphisms $g = g(x)$ of the Honda formal group H_n so that $g'(0) = 1$. Then S_n is a compact p-adic analytic group of dimension n^2 by §3.1.2 of [17]. Under the assumption $p - 1 > n$ the group S_n is torsion-free; see Theorem 3.2.1 of [17]. By a theorem of Lazard (combine Theorems 4.4.1 and 5.1.9. of [34]) we may conclude S_n is a Poincaré duality group of dimension n^2 and of cohomological dimension n^2. Since the index of S_n in $\mathbb{S}_n$ is finite and prime to p, the cohomological dimension of $\mathbb{S}_n$ is also n^2. So far we have not used the hypothesis on M.

Let $\mathrm{Gal} = \mathrm{Gal}(\mathbb{F}_{p^n}/\mathbb{F}_p)$. If M is an $\mathfrak{m}$-complete Morava module then $M^{\mathbb{S}_n}$ is a p-complete twisted $\mathbb{W}$-Gal-module; that is, if $g \in \mathrm{Gal}$ and $x \in M^{\mathbb{S}_n}$, then $g(ax) = g(a)g(x)$. Now we use a version of Galois descent – see Lemma 1.7 below – to conclude

$$H^*(\mathbb{G}_n, M) \cong H^*(\mathbb{S}_n, M)^{\mathrm{Gal}}$$

and we have the vanishing we need. $\qquad\square$

Lemma 1.7 *Let* $\mathrm{Gal} = \mathrm{Gal}(\mathbb{F}_{p^n}/\mathbb{F}_p)$. *Let* M *be a p-complete twisted* $\mathbb{W}$-Gal-*module. Then the inclusion* $M^{\mathrm{Gal}} \to M$ *of the invariants extends to an isomorphism of twisted* $\mathbb{W}$-Gal-*modules*

$$\mathbb{W} \otimes_{\mathbb{Z}_p} M^{\mathrm{Gal}} \cong M.$$

The functor $M \mapsto M^{\mathrm{Gal}}$ *from p-complete twisted* $\mathbb{W}$-Gal-*modules to p-complete modules is exact.*

Proof This can be proved using standard descent theory, but here is a completely explicit argument.

We are using the completed tensor product

$$\mathbb{W} \otimes_{\mathbb{Z}_p} N = \lim(\mathbb{W} \otimes_{\mathbb{Z}_p} N)/p^k \cong \lim(\mathbb{W}/p^k \otimes_{\mathbb{Z}/p^k} N/p^k N).$$

First, since inverse limits commute with invariants, we have

$$M^{\mathrm{Gal}} \cong (\lim M/p^k M)^{\mathrm{Gal}} \cong \lim(M/p^k M)^{\mathrm{Gal}}.$$

Next, the map $M^{\mathrm{Gal}} \to (M/p^k M)^{\mathrm{Gal}}$ factors as

$$M^{\mathrm{Gal}} \longrightarrow M^{\mathrm{Gal}}/p^k M^{\mathrm{Gal}} \longrightarrow (M/p^k M)^{\mathrm{Gal}}$$

with the second map an inclusion. This yields isomorphisms

$$M^{\mathrm{Gal}} \cong \lim(M^{\mathrm{Gal}}/p^k M^{\mathrm{Gal}}) \cong \lim(M/p^k M)^{\mathrm{Gal}}.$$

Since $\mathbb{W}$ is a finitely generated free $\mathbb{Z}_p$ module

$$\mathbb{W} \otimes_{\mathbb{Z}_p} M^{\mathrm{Gal}} \cong \lim(\mathbb{W}/p^k \otimes_{\mathbb{Z}_{p^k}} M^{\mathrm{Gal}}/p^k M^{\mathrm{Gal}})$$
$$\cong \lim(\mathbb{W}/p^k \otimes_{\mathbb{Z}_{p^k}} (M/p^k M)^{\mathrm{Gal}}).$$

Finally, $\mathbb{Z}/p^k \to \mathbb{W}/p^k$ is Galois with Galois group Gal we have

$$\mathbb{W}/p^k \otimes_{\mathbb{Z}/p^k} (M/p^k M)^{\mathrm{Gal}} \cong M/p^k M.$$

The exactness statement follows from the fact that $\mathbb{W}$ is a free and finitely generated $\mathbb{Z}_p$-module, so $N \mapsto \mathbb{W} \otimes_{\mathbb{Z}_p} N$ is exact.   □

1.2 Picard groups

The point of this section is to develop enough technology to pave the way for the key Proposition 1.30.

1.2.1 Some basics

Let $\mathrm{Pic}_{K(n)}$ denote the $K(n)$-local Picard group of weak equivalence classes of invertible elements. Here is an observation from [21]. If $X \in \mathrm{Pic}_{K(n)}$, then $K(n)_*X$ is an invertible $K(n)_*$-module and, since $K(n)_*$ is a graded field, it follows that $K(n)_*X$ is of rank 1 over $K(n)_*$. From this it follows from Proposition 8.4 of [24] that E_*X is also free of rank 1 over E_*.

Remark 1.8   Let $\mathrm{Pic}^0_{K(n)} \subseteq \mathrm{Pic}_{K(n)}$ be the subgroup of index 2 generated by the elements X with E_*X in even degrees. Then E_0X is free of rank 1 over E_0. If we choose a generator $a \in E_0X$ then we can define a crossed homomorphism $\phi : \mathbb{G}_n \to E_0^\times$ by the formula

$$ga = \phi(g)a, \qquad g \in \mathbb{G}_n.$$

This defines a homomorphism

$$e : \mathrm{Pic}^0_{K(n)} \longrightarrow H^1(\mathbb{G}_n, E_0^\times) \tag{1.2.1}$$

to the algebraic Picard group of invertible Morava modules. We write κ_n for the kernel of e; this is the subgroup of *exotic elements* in the Picard group.

Notation: Both $\mathrm{Pic}_{K(n)}$ and $H^1(\mathbb{G}_n, E_0^\times)$ are abelian groups where the group operation is written as multiplicatively; thus $e(X \wedge Y) = e(X)e(Y)$.

The following result explains the hypothesis on the prime in the equivalence of dualities (1.1). This appears in the literature in various guises; the exact criterion on the prime depends on the setting, but the proof is always the same as in Theorem 5.4 of [23].

Proposition 1.9   *Suppose $2p > \max\{n^2 + 1, 2n + 2\}$. Then $\kappa_n = 0$ and the map e is an injection.*

Proof Suppose X and Y are two invertible spectra so $E_0(X) \cong E_0(Y)$ as
Morava modules. Let $D_n Y = F(Y, L_{K(n)} S^0)$ be the $K(n)$-local Spanier-
Whitehead dual of Y. Then $D_n Y$ is the inverse of Y in $\mathrm{Pic}^0_{K(n)}$; hence,
$E_0 D_n Y$ is the inverse of $E_0 Y$ as an invertible Morava module. It follows
that $E_*(X \wedge D_n Y) \cong E_* S^0$ as Morava modules and we need only show
that the class

$$\iota \in H^0(\mathbb{G}_n, E_0(X \wedge D_n Y))$$

determined by this isomorphism is a permanent cycle. The differentials
will lie in subquotients of

$$H^{s+1}(\mathbb{G}_n, E_s), \quad s \geq 1.$$

Under the hypotheses here, we can now apply the sparseness result of part
(1) of Remark 1.4 and the horizontal vanishing line of Proposition 1.6. The
second of these requires $p - 1 > n$ and the first requires $2(p-1) + 1 > n^2$.
Combined they imply that all differentials on ι land in zero groups. $\square$

Remark 1.10 A more sophisticated variation of the argument used to
prove Proposition 1.9 will also show that e is surjective under the same
hypotheses on p and n. See [29] for details. Here is an outline.

Let M be an invertible graded Morava module and let

$$M^\vee = \mathrm{Hom}_{E_0}(M, E_0)$$

with conjugation $\mathbb{G}_n$-action; see Remark 1.4.1 on why this action arises. The
essential idea is to use a Toda-style obstruction theory with successively
defined obstructions

$$\theta_s \in H^{s+2}(\mathbb{G}_n, E_s \otimes_{E_0} M \otimes_{E_0} M^\vee) \cong H^{s+2}(\mathbb{G}_n, E_s), \quad s \geq 1,$$

to finding such an X with $E_0 X \cong M$. Such an obstruction theory can
be constructed using Toda's techniques [35] or a linearized version of the
vastly more complex obstruction theory of [14]. These obstruction groups
will vanish if $2p > \max\{n^2 + 1, 2n + 2\}$.

The basic example of an element in $\mathrm{Pic}^0_{K(n)}$ is the localized 2-sphere
$L_{K(n)} S^2$. Since $E_0 S^2 \cong E_{-2}$ we can choose $u \in E_{-2}$ as the generator
and the associated crossed homomorphism is $t_0 : \mathbb{G}_n \to E_0^\times$. See (1.1.4).

We next explore the underlying algebraic structure of the Picard group
$\mathrm{Pic}^0_{K(n)}$; in particular, it is a profinite abelian group and continuous
module over the rather unusual completion of the integers

$$\mathfrak{Z}_n \overset{\mathrm{def}}{=} \lim_k \mathbb{Z}/p^k(p^n - 1) . \tag{1.2.2}$$

The canonical isomorphisms $\mathbb{Z}/p^k(p^n - 1) \cong \mathbb{Z}/p^k \times \mathbb{Z}/(p^n - 1)$ assemble to give a continuous isomorphism of rings

$$\mathfrak{Z}_{\mathfrak{n}} \cong \mathbb{Z}_p \times \mathbb{Z}/(p^n - 1).$$

See Remark 1.23 for more thoughts on the ring $\mathfrak{Z}_{\mathfrak{n}}$.

The number $p^n - 1$ appears in a number of ways in $K(n)$-local homotopy theory; for example, the element $v_n = u^{-(p^n-1)} \in E_{2(p^n-1)}$. We explore that observation more in Remark 1.24 below. In this context, however, the ring $\mathfrak{Z}_{\mathfrak{n}}$ arises for a much more basic reason.

Lemma 1.11 *Let $(S, \mathfrak{m}_S)$ be a complete local ring with residue field $\mathbb{F}_{p^n} \cong S/\mathfrak{m}_s$. The abelian group structure on the group of units $S^\times$ extends to a continuous $\mathfrak{Z}_{\mathfrak{n}}$-module structure in the topology given by the isomorphism $S^\times \cong \lim(S/\mathfrak{m}^k)^\times$.*

Proof For any $a \in \mathfrak{Z}_{\mathfrak{n}}$ and any $x \in S^\times$ we must define an element $x^a \in S^\times$. Furthermore if $a = n \in \mathbb{Z}$, then we need $x^a = x^n$.

Since $(S/\mathfrak{m})^\times \cong \mathbb{F}_{p^n}^\times$, any $x \in S^\times$ has the property that $x^{p^n-1} \equiv 1$ modulo $\mathfrak{m}_S$ and, hence, that

$$x^{p^k(p^n-1)} \equiv 1 \quad \mathrm{mod} \quad \mathfrak{m}^{k+1}.$$

Let $a \in \lim_k \mathbb{Z}/p^k(p^n - 1) \in \mathfrak{Z}_{\mathfrak{n}}$. For each integer $k \geq 0$ choose an integer a_k so that $a_k \equiv a \in \mathbb{Z}/p^k(p^n - 1)$. Then the elements

$$x^{a_k} \in (S/\mathfrak{m}^{k+1})^\times$$

define an element $x^a \in S^\times \cong \lim(S/\mathfrak{m}^{k+1})^\times$ as needed. $\qquad\square$

The basic application of Lemma 1.11 is to the ring $S = E_0$. This further implies that the continuous cohomology group $H^1(\mathbb{G}_n, E_0^\times)$ is a continuous module over $\mathfrak{Z}_{\mathfrak{n}}$.

It turns out we can show $\mathrm{Pic}^0_{K(n)}$ is also a profinite module over $\mathfrak{Z}_{\mathfrak{n}}$ and the evaluation map

$$e : \mathrm{Pic}^0_{K(n)} \longrightarrow H^1(\mathbb{G}_n, E_0^\times)$$

is a continuous map of $\mathfrak{Z}_{\mathfrak{n}}$ modules. This can be deduced from Proposition 14.3.d of [24], which in turn depends heavily on [21]. The argument given here is essentially the same, but packaged to emphasize the role of the group κ_n of exotic elements in the Picard group and the cohomology group $H^1(\mathbb{G}_n, E_0^\times)$. We'll give a proof in Proposition 1.18 below.

Remark 1.12 Using nilpotence technology derived from [22] and working as in [24] §4 we can choose a sequence of ideals $J(i) \subseteq \mathfrak{m} \subseteq E_0$ and spectra $S/J(i)$ with the following properties:

1 $J(i+1) \subseteq J(i)$ and $\cap\, J(i) = 0$;

2 $E_0/J(i)$ is finite;

3 $E_0(S/J(i)) \cong E_0/J(i)$ and there are maps $q : S/J(i+1) \to S/J(i)$ realizing the quotient map $E_0/J(i+1) \to E_0/J(i)$;

4 there are maps $\eta = \eta_i : S^0 \to S/J(i)$ inducing the quotient map $E_0 \to E_0/J(i)$ and $q\eta_{i+1} = \eta_i : S/J(i) \to S/J(i)$;

5 if X a finite type n-spectrum, then the map $X \to$ holim $X \wedge S/J(i)$ induced by the maps η is an equivalence; and,

6 the $S/J(i)$ are μ-spectra; that is, there are maps

$$\mu : S/J(i) \wedge S/J(i) \to S/J(i)$$

so that $\mu(\eta \wedge 1) = 1 : S/J(i) \to S/J(i)$.

They also prove that items (1)-(5) characterize the tower $\{S/J(i)\}$ up to equivalence in the pro-category of towers under S^0. See Proposition 4.22 of [24].

Hovey and Strickland choose the $J(i)$ with the property that there are positive integers $a_0, a_1, \ldots, a_{n-1}$ (depending on i) so that

$$J(i) = (p^{a_0}, u_1^{a_1}, \ldots, u_{n-1}^{a_{n-1}}).$$

They don't quite say it explicitly, but in their construction they choose the a_i, $i \geq 1$, to be powers of p.

Remark 1.13 Let $G(i) \subseteq \mathrm{Pic}^0_{K(n)}$ be the set of equivalence classes X which can be given a $K(n)$-local equivalence

$$X \wedge S/J(i) \simeq S/J(i).$$

Item (6) of Remark 1.12 is used to show $G(i+1) \subseteq G(i)$. By Proposition 14.2 of [24] $G(i)$ is a finite index subgroup and $\cap\, G(i) = \{L_{K(n)}S^0\}$; thus the subgroups $G(i)$ define a separated profinite topology on $\mathrm{Pic}^0_{K(n)}$.

Lemma 1.14 *The evaluation map*

$$e : \mathrm{Pic}^0_{K(n)} \longrightarrow H^1(\mathbb{G}_n, E_0^\times)$$

is a continuous homomorphism of profinite abelian groups.

Proof Parts (1) and (2) of Remark 1.12 imply the ideals $\{J(i)\}$ define the same topology as $\{\mathfrak{m}^k\}$ on E_0. By Part (2) of Remark 1.12 we have $H^0(\mathbb{G}_n, E_0/J(i)^\times)$ is finite, hence $\lim^1 H^0(\mathbb{G}_n, E_0/J(i)^\times) = 0$ and the lim-lim^1 short exact sequence in cohomology gives

$$H^1(\mathbb{G}_n, E_0^\times) \cong \lim H^1(\mathbb{G}_n, E_0/J(i)^\times).$$

By definition the evaluation map e factors

$$e : \mathrm{Pic}^0_{K(n)}/G(i) \longrightarrow H^1(\mathbb{G}_n, E_0/J(i)^\times) .$$

The result follows. $\qquad\qquad\qquad\qquad\qquad\qquad\qquad\qquad\qquad\qquad$ □

Remark 1.15 Let $A = \lim A_i$ be a profinite abelian group. Let $U_i \subseteq A$ be the open and closed subgroups with $A/U_i \cong A_i$. If $B \subseteq A$ is a subgroup, let $V_i = B \cap U_i$ and $B_i = B/V_i$. Then $B_i \subseteq A_i$ is a subgroup and $B \to \lim B_i$ is the closure of B in A; thus B is closed if and only if $B = \lim B_i$. Note that B is then profinite. If B is closed then $A/B = \lim A_i/B_i$ since $\lim^1 B_i = 0$; hence A/B is also profinite.

From this we see that if $f : A \to B$ is a continuous map of profinite abelian groups then both the kernel and image of f are closed.

Remark 1.16 We now write down a criterion for identifying $\mathfrak{Z}_n$-modules. Let $A = \lim A_i$ be a profinite abelian group. Since the groups A_i are discrete, A is a continuous $\mathfrak{Z}_n$-module if and only if the following criterion holds:

(Criterion 1) Let $x = (x_i) \in \lim A_i = A$. Then for all i there is an integer N_i so that $p^{N_i}(p^n - 1)x_i = 0 \in A_i$.

This can be deduced directly from the definitions.

From Criterion 1 it follows that if $B \subseteq A$ is a closed subgroup and A is $\mathfrak{Z}_n$-module, then both B and A/B are $\mathfrak{Z}_n$-modules. There is a partial converse as well. We will say a profinite abelian group $A = \lim A_i$ is a pro-p-group if for all i the group A_i is a finite abelian p-torsion group. Note that by Criterion 1 any abelian pro-p-group is a $\mathfrak{Z}_n$-module. Suppose

$$0 \to A \to B \to C \to 0.$$

is a short exact sequence of profinite abelian groups, C is an $\mathfrak{Z}_n$-module and A is pro-p-group. Then B is a $\mathfrak{Z}_n$-module.

To prove this converse statement, first note that by reindexing if necessary we can write the short exact sequence as the inverse limit of a tower of short exact sequences

$$0 \to A_i \to B_i \to C_i \to 0.$$

Let $b = (b_i) \in B$ and (c_i) its image in C. Then for all i there is an integer N_i so that $p^{N_i}(p^n - 1)c_i = 0$, whence $p^{N_i}(p^n - 1)b_i \in A_i$. By hypothesis there is an integer M_i so that $p^{M_i}A_i = 0$, or $p^{(M_i+N_i)}(p^n - 1)b_i = 0$ as needed.

Lemma 1.17 *Suppose $A \to B \to C$ is a sequence of profinite abelian groups so that A is a pro-p-group and C is a 3_n-module. Then B is a 3_n-module and the maps are all 3_n-module maps.*

Proof By Remark 1.15, we may assume that the sequence is short exact and then we can apply the previous remark.   □

We now get to our core observation.

Proposition 1.18 *The profinite abelian group $\mathrm{Pic}^0_{K(n)}$ is a 3_n-module and the evaluation map*

$$e : \mathrm{Pic}^0_{K(n)} \longrightarrow H^1(\mathbb{G}_n, E_0^\times)$$

is a continuous map of 3_n-modules.

Proof By Lemma 1.14 we need only show $\mathrm{Pic}^0_{K(n)}$ is a continuous 3_n-module. We apply Lemma 1.17 to the exact sequence

$$0 \to \kappa_n \to \mathrm{Pic}^0_{K(n)} \to H^1(\mathbb{G}_n, E_0^\times).$$

For this to work we need to know that κ_n is a pro-p-group. Since e is a continuous map between profinite groups, its kernel κ_n is a profinite group. So it suffices to show that there is an integer K so that $p^K \kappa_n = 0$. The argument is standard; it can be deduced from [21] and see also, for example, [16]. Here are the details.

If $X \in \kappa_n$, then $E_0 X \cong E_0$ as a Morava module. A choice of isomorphism defines a class

$$\iota_X \in H^0(\mathbb{G}_n, E_0 X).$$

The choice of ι_X is unique up to the group of automorphisms $\mathrm{Aut}(E_0) \cong \mathbb{Z}_p^\times$ of E_0 as a Morava module; therefore, we have that $X \simeq L_{K(n)}S^0$ if and only if ι_X is a permanent cycle in the Adams-Novikov Spectral Sequence for X. Define a filtration of κ_n by setting

$$F_s = \{X \in \kappa_n \mid d_r(\iota_X) = 0, \ r < s \ \}.$$

Then $F_{s+1} \subset F_s$ and d_s defines an injection

$$F_s/F_{s+1} \longrightarrow E_s^{s,s-1}$$

By part (3) of Remark 1.4 we have a uniform vanishing line for the

Adams-Novikov Spectral Sequence; hence, this filtration is finite. To finish the proof we need to show F_s/F_{s-1} is of bounded p-torsion for all s.

The group $E_s^{s,s-1}$ is a subquotient of $H^s(\mathbb{G}_n, E_{s-1})$. We now use parts (1) and (2) of Remark 1.4 to conclude $H^s(\mathbb{G}_n, E_{s-1})$ is bounded p-torsion and the result follows. $\qquad\square$

Remark 1.19 It is possible to give a crude upper bound on the integer K so that $p^K \kappa_n = 0$ by summing the upper bounds of the integers k_s so that p^{k_s} annihilates $H^s(\mathbb{G}_n, E_{s-1})$.

Furthermore, in our admittedly very few known examples, κ_n is a finite group. For $n = 1$ and $p = 1$, see [21]. For $n = 2$ and $p = 3$, see [13]. For $n = 2$ and $p = 2$, the cohomology calculations of [3] show κ_2 must be finite.

Remark 1.20 The homomorphism $\mathbb{Z} \to \mathrm{Pic}_{K(n)}^0$ sending n to S^{2n} extends to a homomorphism of profinite $\mathfrak{z}_n$-modules $\mathfrak{z}_n \to \mathrm{Pic}_{K(n)}^0$, which we write as $a \mapsto S^{2a}$. This first appeared in [21]; the argument there can be distilled from the one given here; in particular it uses the tower $\{S/J(i)\}$.

Each of the spectra $S/J(i)$ is a type n complex and, hence, there is an integer N_i so that $S/J(i)$ has $v_n^{p^{N_i}}$-self map. Then $S^{2p^{N_i}(p^n-1)} \in G(i)$. We can arrange for $N_{i+1} \geq N_i$ and $N_i \to \infty$ as $i \to \infty$. Then the map $\mathbb{Z} \to \mathrm{Pic}_{K(n)}^0$ sending n to S^{2n} extends to map

$$\mathbb{Z} \to \mathfrak{z}_n \cong \lim \mathbb{Z}/p^{N_i}(p^n - 1) \to \lim \mathrm{Pic}_{K(n)}^0/G(i) \cong \mathrm{Pic}_{K(n)}^0.$$

This procedure writes S^{2a} as a homotopy inverse limit in the $K(n)$-local category

$$S^{2a} \simeq \mathrm{holim}\, S^{2a_i} \wedge S/J(i)$$

where $\{a_i\}$ is a sequence of integers so that $a_i \equiv a$ modulo $2p^{N_i}(p^n - 1)$. The transition maps in this tower are given by the maps

$$S^{2a_{i+1}} \wedge S/J(i+1) \xrightarrow{\quad q \quad} S^{2a_{i+1}} \wedge S/J(i) \xrightarrow{\quad v \quad} S^{2a_i} \wedge S/J(i)$$
$$(1.2.3)$$

where we have write $a_{i+1} = a_i + 2m_i p^{N_i}(p^n - 1)$ and v is a $v_n^{p^{N_i} m_i}$-self map.

We can now record the following result.

Proposition 1.21 *Let X be a type n complex with $v_n^{p^k}$-self map. Let*

$a \in \mathfrak{Z}_{\mathfrak{n}}$ *and let* a_k *be any integer so that* $a_k \equiv a$ *modulo* $p^k(p^n - 1)$. *Then there is* $K(n)$-*local equivalence*

$$S^{2a} \wedge X \simeq \Sigma^{2a_k} X.$$

Proof By Proposition 14.3.c of [24] the subgroup

$$G_X = \{P \in \mathrm{Pic}^0_{K(n)} \mid P \wedge X \simeq X \} \subseteq \mathrm{Pic}^0_{K(n)}$$

is of finite index, so there is an i so that $G(i) \subseteq G_X$. In particular, if $S/J(i)$ has $v_n^{p^{N_i}}$ self map, then $S^{2p^{N_i}(p^n-1)} \in G_X$. It follows that if a_i is an integer so that $a \equiv a_i$ modulo $2p^{N_i}(p^n - 1)$ then there is $K(n)$-local equivalence

$$S^{2a} \wedge X \simeq S^{2a_i} \wedge X.$$

If we assume $N_i \geq k$, we can then use the $v_n^{p^k}$-self map of X to write $S^{2a_i} \wedge X \simeq S^{2a_k} \wedge X$.   □

Remark 1.22 Proposition 1.21 has the following algebraic analog. Let $a \in \mathfrak{Z}_{\mathfrak{n}}$ and let a_k be any integer so that $a \equiv a_k$ modulo $p^k(p^n - 1)$. Then there is an isomorphism of Morava modules

$$E_0(S^a)/\mathfrak{m}^k \cong E_0(S^{a_k})/\mathfrak{m}^k$$

where $\mathfrak{m} \subseteq E_0$ is the maximal ideal. This can be also be proved by a calculation with crossed homomorphisms.

Note that

$$E_* S^{2a} \cong \lim \Sigma^{2a_i} E_*/J(i)$$

is a free E_*-module generated by an element x_a in degree 0 which reduces to the canonical generator $u^{a_k} \in \Sigma^{2a_k} E_0/J(I)$. We conclude that for $g \in \mathbb{G}_n$

$$g_* x_a = t_0(g)^a x_a.$$

The function t_0 was defined in Remark 1.3.

Remark 1.23 The ring $\mathfrak{Z}_{\mathfrak{n}} = \lim_k \mathbb{Z}/p^k(p^n - 1)$ is a somewhat unusual completion of the integers; thus, it might be worthwhile to analyze it and various of the elements therein. The split short exact sequences

$$0 \longrightarrow \mathbb{Z}/p^k \xrightarrow{\times (p^n - 1)} \mathbb{Z}/p^k(p^n - 1) \longrightarrow \mathbb{Z}/(p^n - 1) \longrightarrow 0$$

assemble to give a split short exact sequence

$$0 \longrightarrow \mathbb{Z}_p \xrightarrow{\times (p^n - 1)} \mathfrak{Z}_{\mathfrak{n}} \longrightarrow \mathbb{Z}/(p^n - 1) \longrightarrow 0 .$$

To give a specific element of order $p^n - 1$ in 3_n, notice that

$$p^{nk} = p^{n(k-1)}(p^n - 1) + p^{n(k-1)}.$$

and, hence, $p^{nk} \equiv p^{n(k-1)}$ modulo $p^{n(k-1)}(p^n - 1)$. Thus the elements

$$p^{nk} \in \mathbb{Z}/p^{nk}(p^n - 1)$$

assemble to give an element α of order $p^n - 1$ in the inverse limit. We will use the notation

$$\alpha = \lim p^{nk} \in 3_n.$$

In using this limit notation, we must be wary of the different completions. We have $0 = \lim p^{nk} \in \mathbb{Z}_p$, but $0 \neq \alpha = \lim p^{nk} \in 3_n$.

We can also analyze an element of 3_n which will appear in the main algebraic result, Proposition 1.30. Recall that $r(n) = (p^n - 1)/(p - 1)$. Define

$$\lambda = \lim_k \; p^{nk} r(n) \in \lim_k \mathbb{Z}/p^{nk}(p^n - 1) = 3_n.$$

Then

$$\lambda = \alpha r(n) \in 3_n.$$

Since $r(n)(p - 1) = p^n - 1$ we have that λ is torsion of exact order $p - 1$.

Remark 1.24   In the spirit of [23], we could begin with the Johnson-Wilson theory $E(n)$ with

$$E(n)_* \cong \mathbb{Z}_{(p)}[v_1, \ldots, v_{n-1}, v_n^{\pm 1}].$$

This is a complex oriented theory with obvious p-typical formal group law; compare (1.1.1). There is a map $E(n) \to E_n$ of complex oriented theories inducing an equivalence

$$L_{K(n)}E(n) \simeq E_n^{hF}$$

where E^{hF} is the Devinatz-Hopkins fixed points [9] with respect to

$$F = \mathbb{F}_{p^n}^\times \rtimes \mathrm{Gal}(\mathbb{F}_{p^n}/\mathbb{F}_p) \subseteq \mathbb{G}_n.$$

Since $E(n)$ and E^{hF} are exactly v_n-periodic, we have a degree function

$$\deg : \mathrm{Pic}_{K(n)} \longrightarrow \mathbb{Z}/2(p^n - 1)$$

sending X to the degree of generator of $\pi_* L_{K(n)}(E(n) \wedge X) \cong E_*^{hF} X = (E_* X)^F$. Proposition 1.18 now implies the kernel of this map is an abelian pro-p-group and, in particular, a continuous module over $\mathbb{Z}_p$. This is exactly the statement of Proposition 14.3.d of [24].

If X has degree zero, then $(E^{hF})_0 X$ is an invertible $(E^{hF})_0 E^{hF}$-comodule in some appropriate category of completed $(E^{hF})_0$-modules, but since F is not normal in $\mathbb{G}_n$ this category of comodules is not obviously a category of modules over some group. Thus the analog of the evaluation map (1.2.1) is a bit harder to define and analyze.

1.3 The determinant and the reduction modulo p

In (1.1.4) we introduced the crossed homomorphism $t_0 : \mathbb{G}_n \to E_0^{\times}$ defining the invertible Morava module $E_0 S^2$. The determinant homomorphism defines another invertible module, this time not arising from the unlocalized stable category. The purpose of the section is to write down exactly how these two modules are related. See Proposition 1.30, Theorem 1.32, and Proposition 1.34.

Remark 1.25 (The determinant)   Let $\operatorname{End}(H_n/\mathbb{F}_{p^n})$ be the endomorphism ring of the Honda formal group law over $\mathbb{F}_{p^n}$. By Theorem A2.2.18 of [30], there is an isomorphism

$$\mathbb{W}\langle S \rangle / (S^n - p, Sa = a^{\phi}S) \cong \operatorname{End}(H_n/\mathbb{F}_{p^n})$$

where $S = x^p \in \operatorname{End}(H_n/\mathbb{F}_{p^n})$ and $a \mapsto a^{\phi}$ is the lift of the Frobenius of $\mathbb{F}_{p^n}$ to $\mathbb{W} = W(\mathbb{F}_{p^n})$.

The right action of

$$\mathbb{S}_n = \operatorname{Aut}(H_n/\mathbb{F}_{p^n}) \subseteq \operatorname{End}(H_n/\mathbb{F}_{p^n})$$

on the endomorphism ring determines a composition

$$\mathbb{S}_n \xrightarrow{\ A\ } \operatorname{Gl}_n(\mathbb{W}) \xrightarrow{\ \det\ } \mathbb{W}^{\times}$$

where the last map is the determinant. The matrix representation of a typical element

$$g = a_0 + a_1 S + \cdots a_{n-1} S^{n-1} \in \mathbb{S}_n$$

with respect to the $\mathbb{W}$-basis $\{S^i \mid 0 \le i \le n-1\}$ of the endomorphism ring is

$$A(g) = \begin{bmatrix} a_0 & pa_{n-1}^{\phi} & \cdots & pa_1^{\phi^{n-1}} \\ a_1 & a_0^{\phi} & \cdots & pa_2^{\phi^{n-1}} \\ \vdots & \vdots & \ddots & \vdots \\ a_{n-1} & a_{n-2}^{\phi} & \cdots & a_0^{\phi^{n-1}} \end{bmatrix}.$$

From this we conclude that

$$\det A(g) \equiv a_0^{r(n)} \qquad \text{modulo } p \qquad (1.3.1)$$

with $r(n) = (p^n - 1)/(p - 1)$. In addition, we have that $\det A(g) \in \mathbb{Z}_p^\times$; hence, we get a homomorphism

$$\mathbb{S}_n \rtimes \text{Gal}(\mathbb{F}_{p^n}/\mathbb{F}_p) \longrightarrow \mathbb{Z}_p^\times \times \text{Gal}(\mathbb{F}_{p^n}/\mathbb{F}_p) \xrightarrow{p_1} \mathbb{Z}_p^\times$$

which we call det. This defines an action of $\mathbb{G}_n$ on $\mathbb{Z}_p$ which we write $\mathbb{Z}_p[\det]$ and if M is any Morava module we define a new Morava module

$$M[\det] = M \otimes \mathbb{Z}_p[\det]$$

with the diagonal action. The invertible module $E_0[\det]$ determines the element of $H^1(\mathbb{G}_n, E_0^\times)$ defined by the homomorphism given by the composition

$$\mathbb{G}_n \xrightarrow{\det} \mathbb{Z}_p^\times \xrightarrow{\subseteq} E_0^\times.$$

Remark 1.26 (Realizing the determinant) There is a canonical determinant sphere $S[\det] \in \text{Pic}_{K(n)}$ with $E_0 S[\det] \cong E_0[\det]$ as Morava modules. Complete details are in [2], but the argument is essentially due to the second author here. Assume $p > 2$. Here is an outline. Since the classifying space $B\mathbb{Z}_p$ is a model for the p-completion of S^1, the evident action of $\mathbb{Z}_p^\times$ on $\mathbb{Z}_p$ and suspension define an action of $\mathbb{Z}_p^\times$ on the completed sphere spectrum S_p^0. The determinant $\det : \mathbb{G}_n \to \mathbb{Z}_p^\times$ then gives an action of $\mathbb{G}_n$ on S_p^0. We then can define

$$S[\det] = (E \wedge S_p^0)^{h\mathbb{G}_n}$$

where we use the diagonal action. The prime 2 requires a more delicate argument.

Remark 1.27 Recall from Remark 1.23 we have defined numbers $\alpha = \lim p^{nk}$ and $\lambda = \lim p^{nk} r(n)$ in $\mathfrak{z}_n$. Let $(S, \mathfrak{m}_k)$ be a complete local ring with $S/\mathfrak{m}_k \cong \mathbb{F}_{p^n}$. For any $x \in S^\times$, $x^{p^n} \equiv x$ modulo $\mathfrak{m}_S$, so

$$x^{p^{nk}} \equiv x^{p^{n(k-1)}} \qquad \text{modulo } \mathfrak{m}_S^k.$$

Thus

$$T_1(x) = \lim_k x^{p^{nk}} = x^\alpha \in S$$

exists and $T_1(x) \equiv x$ modulo $\mathfrak{m}_S$. If $x \equiv y$ modulo $\mathfrak{m}_S$, then $x^{p^{nk}} \equiv y^{p^{nk}}$

modulo $\mathfrak{m}_S^k$ so we have that T_1 factors

$$S^\times \longrightarrow \mathbb{F}_{p^n}^\times \overset{T}{\longrightarrow} S^\times$$

and $T(xy) = T(x)T(y)$. In particular, $T(x)^{p^n-1} = 1$. The element $T(x)$ is the Teichmüller lift of x. In a radical but standard abuse of notation, we often simply write x for $T(x)$.

We can apply these constructions to $H^1(\mathbb{G}_n, E_0^\times)$ as well. If $g \in \mathbb{S}_n$ then $g = g(x) \in \mathbb{F}_{p^n}[[x]]$ with leading coefficient $g'(0) \in \mathbb{F}_{p^n}^\times$. Since $t_0(g) \equiv g'(0)$ modulo $\mathfrak{m}$ we have

$$t_0(g)^\alpha = T_1(t_0(g)) = g'(0)^\alpha = g'(0) \in E_0^\times \qquad (1.3.2)$$

and since $\det(g) = g'(0)^{r(n)}$ modulo $\mathfrak{m}$ we have

$$t_0(g)^\lambda = (t_0(g)^\alpha)^{r(n)} = g'(0)^{r(n)} = \det(g)^\alpha \in E_0^\times.$$

for all g, or $t_0^\lambda = \det^\alpha$ as a crossed homomorphisms on $\mathbb{S}_n$. Since both functions are constant on the Galois group, we also have $t_0^\lambda = \det^\alpha$ as crossed homomorphisms on $\mathbb{G}_n$.

We now turn to an examination of $X \wedge V(0)$ where $X \in \mathrm{Pic}_{K(n)}$. Here $V(0)$ is the mod p Moore spectrum.

Lemma 1.28 *Let S be a torsion-free complete local ring with maximal ideal $\mathfrak{m}_S$ and suppose $S/\mathfrak{m}_S$ is of characteristic $p > 2$. Then there is a short exact sequence*

$$0 \longrightarrow S \overset{\exp(p-)}{\longrightarrow} S^\times \longrightarrow (S/p)^\times \longrightarrow 1.$$

Proof An element $x \in S$ is a unit if and only if $0 \neq x \in S/\mathfrak{m}_S$. Thus the map $S^\times \to (S/p)^\times$ is onto with kernel K consisting of all elements of the form $1+py$. Note that if $x \in S$, the power series $\exp(px) = \sum_{n=0}^\infty p^n x^n/n!$ converges since $p > 2$. This defines a homomorphism

$$\exp(p-) : S \longrightarrow K$$

with inverse sending $1 + py$ to $(1/p)\log(1 + py)$. $\qquad\square$

If we set $S = E_0$ we get an exact sequence in cohomology

$$H^1(\mathbb{G}_n, E_0) \overset{\exp(p-)}{\longrightarrow} H^1(\mathbb{G}_n, E_0^\times) \longrightarrow H^1(\mathbb{G}_n, (E_0/p)^\times). \quad (1.3.3)$$

Remark 1.29 (Definition of $\zeta = \zeta_n$) Let $p > 2$. The Teichmüller homomorphism (see Remark 1.27) $T : \mathbb{F}_p^\times \to \mathbb{Z}_p^\times$ defines a splitting of the projection $\mathbb{Z}_p^\times \to \mathbb{F}_p^\times$ and determines an isomorphism

$$1 + p\mathbb{Z}_p \times \mathbb{F}_p^\times \xrightarrow{\cong} \mathbb{Z}_p^\times.$$

We'll confuse elements in $\mathbb{F}_p^\times$ with their Teichmüller lifts.

Let $g \in \mathbb{S}_n$. Since $\det(g) \equiv g'(0)^{r(n)} \bmod p$, we have

$$\theta(g) \overset{\text{def}}{=} g'(0)^{-r(n)} \det(g) \in 1 + p\mathbb{Z}_p.$$

We now define $\zeta = \zeta_n : \mathbb{S}_n \to \mathbb{Z}_p$ to be the composite

$$\mathbb{S}_n \xrightarrow{\theta} 1 + p\mathbb{Z}_p \xrightarrow[\cong]{\ell} \mathbb{Z}_p$$

where $\ell(1 + py) = (1/p)\log(1 + py)$. As with the determinant, ζ extends to a homomorphism $\zeta : \mathbb{G}_n \to \mathbb{Z}_p$ using the projection:

$$\mathbb{S}_n \rtimes \mathrm{Gal}(\mathbb{F}_{p^n}/\mathbb{F}_p) \xrightarrow{\zeta \times 1} \mathbb{Z}_p \times \mathrm{Gal}(\mathbb{F}_{p^n}/\mathbb{F}_p) \xrightarrow{p_1} \mathbb{Z}_p$$

The homomorphism ζ defines a cohomology class also called $\zeta = \zeta_n \in H^1(\mathbb{G}_n, \mathbb{Z}_p)$. In a further abuse of notation we also write ζ for image of this cohomology class in $H^1(\mathbb{G}_n, E_0)$ induced by the inclusion $\mathbb{Z}_p \subseteq E_0$.

There is a class $\zeta \in H^1(\mathbb{G}_n, E_0)$ when $p = 2$ as well; it takes a few more words to define.

The following result is implicit in various forms in various places. At $n = 2$ and $p = 3$ the exact sequence of (1.3.3) is used in the work of Karamanov [25]; see also the work on the Picard group in [26] §5.

Recall that

$$r(n) = \frac{p^n - 1}{p - 1} = p^{n-1} + \cdots + p + 1.$$

Recall also that we are writing the group operation in $H^1(\mathbb{G}_n, E_0)$ multiplicatively. In the following result we will confuse a cohomology class with a crossed homomorphism representative.

Proposition 1.30 *Let $p > 2$ and let $\zeta \in H^1(\mathbb{G}_n, E_0)$ be the class defined in Remark 1.29 . Then*

$$\exp(p\zeta) = t_0^{-\lambda} \det \in H^1(\mathbb{G}_n, E_0^\times)$$

where

$$\lambda = \lim_k p^{nk} r(n) \in \lim_k \mathbb{Z}/p^{nk}(p^n - 1) = 3_n.$$

Remark 1.31 We saw in Remark 1.23 that the element $t_0^\lambda \in H^1(\mathbb{G}_n, E_0^\times)$ is torsion of exact order $(p-1)$.

Proof The class ζ is defined in Remark 1.29 as a class in $H^1(\mathbb{G}_n, \mathbb{Z}_p)$, so we examine the diagram

$$
\begin{array}{ccc}
H^1(\mathbb{G}_n, \mathbb{Z}_p) & \xrightarrow{\exp(p-)} & H^1(\mathbb{G}_n, \mathbb{Z}_p^\times) \\
\downarrow & & \downarrow \\
H^1(\mathbb{G}_n, E_0) & \xrightarrow{\exp(p-)} & H^1(\mathbb{G}_n, E_0^\times) \, .
\end{array}
$$

By construction $\exp(p\zeta) \in H^1(\mathbb{G}_n, \mathbb{Z}_p^\times)$ is determined by the function on $\mathbb{G}_n \cong \mathbb{S}_n \rtimes \mathrm{Gal}(\mathbb{F}_{p^n}/\mathbb{F}_p)$ given by

$$(g, \phi) \mapsto g'(0)^{-r(n)} \det(g).$$

Note that we have written $g'(0) \in \mathbb{Z}_p^\times$ for its image under the Teichmüller splitting $T : \mathbb{F}_p^\times \to \mathbb{Z}_p^\times$. From (1.3.2) we have that

$$t_0(g)^\alpha = g'(0)^\alpha = g'(0) \in E_0^\times.$$

where $\alpha = \lim p^{nk} \in \mathfrak{Z}_n$. Then for $(g, \phi) \in \mathbb{S}_n \rtimes \mathrm{Gal}(\mathbb{F}_{p^n}/\mathbb{F}_p)$ we conclude

$$t_0(g, \phi)^\lambda = ((t_0(g)^\alpha)^{r(n)} = g'(0)^{r(n)}.$$

The result follows. $\qquad\qquad\qquad\qquad\qquad\qquad\qquad\qquad\qquad\qquad\square$

Proposition 1.30, the long exact sequence of (1.3.3), and the identification of cohomology classes with invertible modules yield a canonical isomorphism of graded Morava modules

$$E_*(S^{2\lambda})/p \cong E_*[\det]/p$$

and, thus, by arguments similar to those in Remark 1.8, we have the following corollary.

Theorem 1.32 *Let* $2p > \max\{n^2 + 1, 2n + 2\}$ *and* $\lambda = \lim_k p^{nk} r(n) \in \mathfrak{Z}_n$. *Then there is a* $K(n)$-*local equivalence*

$$S^{2\lambda} \wedge V(0) \simeq S[\det] \wedge V(0).$$

Remark 1.33 (The case $n = 1$) The calculations at $n = 1$ are somewhat anomalous because they are so simple, partly because $\mathbb{G}_1 = \mathbb{Z}_p^\times$ and partly because $\mathfrak{m} = (p) \subseteq E_0 = \mathbb{Z}_p$, so setting $p = 0$ is a strong requirement. But more than all that at $n = 1$ we have $\det = t_0$ or, on invertible modules, we have $E[\det] = E_{-2} = E_0 S^2$. If $p > 2$, (1.3.3) becomes a short exact sequence in cohomology

$$0 \longrightarrow H^1(\mathbb{G}_1, \mathbb{Z}_p) \xrightarrow{\exp(p-)} H^1(\mathbb{G}_1, \mathbb{Z}_p^\times) \longrightarrow H^1(\mathbb{G}_1, \mathbb{F}_p^\times) \longrightarrow 0$$

is isomorphic to

$$0 \to \mathbb{Z}_p \to \mathbb{Z}_p^\times \to \mathbb{F}_p^\times \to 0.$$

The cohomology group $H^1(\mathbb{G}_1, \mathbb{Z}_p^\times)$ is free of rank one over

$$\mathfrak{z}_1 \cong \lim \mathbb{Z}/p^k(p-1)$$

with generator t_0. The function $t_0(g)^\alpha = g'(0)$ is an element of order $p-1$, where we have confused $g'(0)$ with it Teichmüller lift to $\mathbb{Z}_p$. Finally, by Proposition 1.30, the function $g \mapsto g'(0)^{-1} t_0(g)$ topologically generates the copy of the p-adic integers.

For $n = 2$ and primes $p > 2$, both the algebraic and the topological Picard group have been calculated. The algebraic result is as follows. If $p > 3$, this result is proved explicitly in [26] §5, but it was known much earlier to the second author. At the prime 3, the algebraic Picard group is the subject of [25].

Proposition 1.34 (The case $n = 2$) *Let $n = 2$ and $p > 2$. The long exact sequence*

$$\cdots \to H^1(\mathbb{G}_2, E_0) \xrightarrow{\exp(p-)} H^1(\mathbb{G}_2, E_0^\times) \longrightarrow H^1(\mathbb{G}_2, (E_0/p)^\times) \to \cdots$$

becomes a short exact sequence

$$0 \longrightarrow \mathbb{Z}_p \longrightarrow H^1(\mathbb{G}_2, E_0^\times) \longrightarrow \mathfrak{z}_2 \longrightarrow 0 \ .$$

The group $H^1(\mathbb{G}_2, E_0)$ is topologically generated by ζ and the group $H^1(\mathbb{G}_2, E_0^\times)$ is topologically generated by cohomology classes defined by t_0 and $\det$. Furthermore, the image of ζ in $H^1(\mathbb{G}_2, E_0^\times)$ is

$$t_0^{-\lambda} \det$$

where $\lambda = \lim_k p^{2k}(p+1) \in \mathfrak{z}_2$.

Remark 1.35 If $p > 2$, the map

$$e : \mathrm{Pic}^0_{K(2)} \longrightarrow H^1(\mathbb{G}_2, E_0^\times)$$

is a surjection as the target is topologically generated by $E_0 S^2$ and $E_0 S[\det]$. If $p > 3$, it then follows from Proposition 1.9 that e is an isomorphism. At the prime 3, the map e is only onto and the kernel was computed in [13].

1.4 Two dualities

In the next section – see Theorem 1.42 – we give a result relating Brown-Comenetz and Spanier-Whitehead duality. We go through the two dualities in this section and how they work with the Adams-Novikov Spectral Sequence (1.1.2). The ring R in this section will be shorthand for E_0.

1.4.1 Spanier-Whitehead duality

We define the $K(n)$-local Spanier-Whitehead dual of a spectrum X by

$$D_n X = F(X, L_{K(n)} S^0).$$

A spectrum X in the $K(n)$-local category is dualizable if for all $K(n)$-local spectra Z the natural map

$$D_n X \wedge Z \to F(X, Z)$$

is a $K(n)$-local equivalence; that is, if $L_{K(n)}(D_n X \wedge Z) \simeq F(X, Z)$. By Theorem 8.6 of [24] X is dualizable if and only if $E_* X$ is finitely generated as an E_*-module.

Remark 1.36   If X is dualizable, there is an isomorphism

$$E_t D_n X = E^{-t} X.$$

To compute $E^* X$ we use a Universal Coefficient Spectral Sequence. The constructions is §III.13 of [1], although an easy adaptation to the $K(n)$-local category is needed. Combining this with the algebra of Appendix A of [24] we get a natural spectral sequence

$$\mathrm{Ext}^p_{E_*}(E_* X, \Sigma^q E_*) \Longrightarrow E^{p+q} X$$

where Ext_{E_*} denotes the derived functors of the functor Hom_{E_*} of continuous homomorphisms in the category of graded $L_0^{\mathfrak{m}}$-complete E_*-modules. See Remark 1.5. By definition $\Sigma^q E_* = \pi_* \Sigma^q E$. Since

$$\mathrm{Hom}_{E_*}(E_* X, \Sigma^q E_*) \cong \mathrm{Hom}_{E_0}(E_q X, E_0)$$

we can rewrite the spectral sequence as

$$\text{Ext}^p_{E_0}(E_q X, E_0) \Longrightarrow E^{p+q} X. \tag{1.4.1}$$

The abutment has a continuous action of $\mathbb{G}_n$ through the action on E; this spectral sequence becomes a $\mathbb{G}_n$-equivariant spectral sequence if we give the E_2-term the conjugation action. Concretely, if $\varphi \in \text{Hom}_{E_0}(M, N)$ and $g \in \mathbb{G}_n$ we have $(g\varphi)(x) = g\varphi(g^{-1}x)$.

If $E_* X$ is $\mathfrak{m}$-torsion – for example if X is a type-n complex – then (1.1.7) and (1.4.1) imply that

$$E_t D_n X \cong \text{Hom}_R(E_{-t-n} X, R/\mathfrak{m}^\infty).$$

This is an isomorphism of Morava modules if we use the conjugation action on the module of homomorphisms. We then define a functor $D_S(-)$ on $\mathfrak{m}$-torsion Morava modules to Morava modules by

$$D_S(M) = \text{Hom}_{E_0}(M, R/\mathfrak{m}^\infty)$$

with the conjugation action, so that we have

$$E_t D_n X \cong D_S(E_{-t-n} X). \tag{1.4.2}$$

Remark 1.37 (SW duality and the ANSS spectral sequence)
Suppose $E_* X$ is finite and, hence, $\mathfrak{m}$-torsion. The isomorphism of Morava modules (1.4.2) allows us to rewrite the Adams-Novikov Spectral Sequence for $D_n X$ as

$$H^s(\mathbb{G}_n, D_S(E_{-t-n} X)) \Longrightarrow \pi_{t-s} D_n X. \tag{1.4.3}$$

Example 1.38   For self-dual X, the self-duality extends to the Adams-Novikov Spectral Sequence. To be specific, we work at chromatic height 2 and at $p > 2$. Let $V(0)$ be the mod p Moore spectrum and let $v_1 : \Sigma^{2(p-1)} V(0) \to V(0)$ be a v_1-self map. We define $M(1, n)$ by the cofiber sequence

$$\Sigma^{2n(p-1)} V(0) \xrightarrow{v_1^n} V(0) \longrightarrow M(1, n).$$

Since $v_1 = u_1 u^{1-p} \in E_* V(0) = E_*/p$, we have that $E_* M(1, n)$ is concentrated in even degrees and $E_0 M(1, n) = \mathbb{F}_{p^2}[u_1]/(u_1^n)$

Because $p > 2$, the v_1-self map of $V(0)$ is unique up to a unit in $\mathbb{Z}/p$, so we have a $K(n)$-equivalence $D_2 M(1, n) \simeq \Sigma^{-2n(p-1)-2} M(1, n)$. This fact and (1.4.2) give isomorphisms of Morava modules

$$E_{2t+2n(p-1)+2} M(1, n) \cong E_{2t} D_2 M(1, n) \cong \text{Hom}_{E_0}(E_{2t-2} M(1, n), E_0/\mathfrak{m}^\infty)$$

which sends the generator $u^{-t-n(p-1)}$ of $E_{2t+2n(p-1)+2}M(1,n)$ to the function

$$u^{-t+1} \mapsto 1/pu_1^n = u^{-n(p-1)}/pv_1^n.$$

Making these substitutions for $M(1,n)$, the Adams-Novikov Spectral Sequence (1.4.3) becomes

$$H^s(\mathbb{G}_n, E_{2t+2n(p-1)+2}M(1,n)) \Longrightarrow \pi_{2t-s}\Sigma^{-2n(p-1)-2}L_{K(2)}M(1,n).$$

This is, up to reindexing, isomorphic to the original Adams-Novikov Spectral Sequence for $M(1,n)$.

1.4.2 Brown-Comentez duality

Let $\mathbb{Z}/p^\infty = \mathbb{Q}/\mathbb{Z}_{(p)}$ and let I be the Brown-Comenetz dual of the sphere; this spectrum is defined by the natural isomorphism

$$\mathrm{Hom}(\pi_0 X, \mathbb{Z}/p^\infty) \cong [X, I].$$

If X is $K(n)$-local and if $M_n X$ is the fiber of

$$X \to L_{n-1}X = L_{K(0)\vee\ldots\vee K(n-1)}X,$$

then we define the $K(n)$-local Brown-Comenetz dual[2] of X to be

$$I_n X = F(M_n X, I) \simeq F(X, I_n).$$

Here $I_n = I_n L_{K(n)}S^0$.

By the work of Gross and the second author, we know the Morava module $E_0 I_n$. Let's recall the outline of the argument from [33]. The fact that $\mathbb{G}_n$ is a virtual Poincaré duality group of dimension n^2 and the fact that the top local cohomology group of $R = E_0$ is in degree n (see (1.1.6)) yield an isomorphism of Morava modules

$$E_0 I_n = E_{-n^2-n} \otimes_R \Omega^{n-1}$$

where

$$\Omega^{n-1} = \bigwedge^{n-1} \Omega_{R/\mathbb{W}} = R\{du_1 \wedge \ldots \wedge du_{n-1}\}$$

is the top exterior power of the differentials on $R = E_0$ over $\mathbb{W} = W(\mathbb{F}_{p^n})$. It follows that I_n is dualizable in the $K(n)$-local category, and hence

$$I_n X \simeq D_n X \wedge I_n. \tag{1.4.4}$$

---

[2] The nomenclature shifted some time after the appearance of [24] and $K(n)$-local Brown-Comenetz duality became to be called Gross-Hopkins duality. We adhere to the older version.

One of the main results of [19] is that there is an isomorphism of Morava modules

$$\Omega^{n-1} \cong E_{2n}[det] \tag{1.4.5}$$

and hence

$$E_0 I_n \cong E_{-n^2+n}[\det] \cong E_0 S^{n^2-n}[\det]. \tag{1.4.6}$$

From this it follows that for X finite and $\mathfrak{m}$-torsion

$$E_t I_n X \cong \operatorname{Hom}_R(E_{-t-2n+n^2} X, E_0[\det]/\mathfrak{m}^\infty) \tag{1.4.7}$$

Remark 1.39 (BC duality and the ANSS)   We continue to write $R = E_0$. If M is finite and $\mathfrak{m}$-torsion, define $D_I M = \operatorname{Hom}_R(M, \Omega^{n-1}/\mathfrak{m}^\infty)$ with the conjugation action. Then we have from 1.4.5) and (1.4.7) that

$$E_t I_n X = D_I(E_{-t+n^2} X) \tag{1.4.8}$$

if X is finite and $E_* X$ is $\mathfrak{m}$-torsion. In [33], Strickland defined an equivariant residue map

$$H^{n^2}(\mathbb{G}_n, \Omega^{n-1}/\mathfrak{m}^\infty) \to \mathbb{Z}/p^\infty.$$

The residue map then defines a pairing

$$H^s(\mathbb{G}_n, M) \otimes H^{n^2-s}(\mathbb{G}_n, D_I M) \to \mathbb{Z}/p^\infty .$$

Under the hypothesis that $p - 1 > n$ (compare Proposition 1.6), this is a perfect pairing. The Adams-Novikov Spectral Sequence for X

$$H^s(\mathbb{G}_n, D_I(E_{-t+n^2} X)) \Longrightarrow \pi_{t-s} I_n X. \tag{1.4.9}$$

can then be rewritten as

$$H^{n^2-s}(\mathbb{G}_n, E_{-t+n^2} X)^\vee \Longrightarrow (\pi_{s-t} X)^\vee$$

where $A^\vee = \operatorname{Hom}(A, \mathbb{Z}/p^\infty)$. This dualized spectral sequence is presumably dual to the original spectral sequence.

1.5 The duality theorem

Here we come to the algebraic and topological consequence of Proposition 1.30. We have defined an element

$$\lambda = \lim_k p^{nk} r(n) \in \lim_k \mathbb{Z}/p^{nk}(p^n - 1) = \mathfrak{Z}_\mathfrak{n}.$$

Recall from Remark 1.5 that a graded Morava module M_* is finite if M_0 and M_1 are finite. Such a Morava module is automatically $\mathfrak{m}$-torsion, where $\mathfrak{m} \subseteq E_0$ is the maximal ideal. The results of the previous sections now can be combined to give the following.

Proposition 1.40   *1.) Let M be a finite Morava module and suppose $pM = 0$. Then there is a natural isomorphism of Morava modules*

$$D_I M \cong E_0(S^{2\lambda - 2n}) \otimes_{E_0} D_S M.$$

2.) Suppose $E_* X$ is a finite Morava module and suppose $pE_* X = 0$. Then there is a natural isomorphism

$$E_* I_n X \cong E_0(S^{2\lambda + n^2 - n}) \otimes_{E_0} E_* D_n X.$$

Proof   This all follows from Proposition 1.30, which gives an isomorphism of Morava modules

$$E_0[\det]/p \cong E_0(S^{2\lambda})/p.$$

For part (1), we combine the following isomorphisms

$$
\begin{aligned}
D_I M &\cong \mathrm{Hom}_R(M, \Omega^{n-1}/\mathfrak{m}^\infty) \\
&\cong \mathrm{Hom}_R(M, R/m^\infty) \otimes_R E_{2n}[\det] \\
&\cong D_S M \otimes_R E_0(S^{2\lambda - 2n}).
\end{aligned}
$$

We use Proposition 1.30 in the last line. Part (2) follows from Part (1) and the isomorphisms

$$
\begin{aligned}
E_t I_n X &\cong D_I(E_{-t+n^2} X) \\
E_t D_n X &\cong D_S(E_{-t-n} X)
\end{aligned}
$$

of (1.4.2) and (1.4.8).   □

Proposition 1.41   *Let $2p > \max\{n^2 + 1, 2n + 2\}$ and let X be a finite type n spectrum with $p1_X = 0 : X \to X$. Then there is a weak equivalence in the $K(n)$-local category*

$$I_n X \cong S^{2\lambda + n^2 - n} \wedge D_n X.$$

Proof   We have that

$$I_n X \simeq D_n X \wedge I_n \simeq D_n X \wedge S^{n^2 - n}[\det].$$

We also have

$$S^{n^2 - n}[\det] \wedge V(0) \simeq S^{2\lambda + n^2 - n} \wedge V(0).$$

Given our assumptions on n and p, this is equivalent to the statement that

$$E_0 I_n/p = E_{-n+n^2}[\det]/p \cong E_{-n+n^2}(S^{2\lambda})/p.$$

See Proposition 1.9. Since $pE_*X = 0$ we have natural isomorphisms of Morava modules

$$E_* I_n X \cong E_* D_n X \otimes_{E_0} E_0 S^{n^2-n}[\det]$$
$$\cong E_* D_n X \otimes_{E_0} E_0 S^{2\lambda+n^2-n} \qquad (1.5.1)$$
$$\cong E_* (D_n X \wedge S^{2\lambda+n^2-n}).$$

which we are trying to extend to an equivalence of spectra.

Because $p1_X = 0$ and $p > 2$, we have a split cofiber sequence

$$D_n X \xrightarrow{\ i\ } D_n X \wedge V(0) \xrightarrow{\ q\ } \Sigma D_n X \ .$$

We choose a section $s : \Sigma D_n X \to D_n X \wedge V(0)$ of q. Because the isomorphisms of (1.5.1) are natural, we have a commutative diagram of split short exact sequences of Morava modules

$$
\begin{array}{ccc}
0 & & 0 \\
\downarrow & & \downarrow \\
E_* D_n X \wedge S^{n^2-n}[\det] & \xrightarrow{\ \cong\ } & E_* D_n X \wedge S^{2\lambda+n^2-n} \\
\downarrow{\scriptstyle i_*} & & \downarrow{\scriptstyle i_*} \\
E_*(D_n X \wedge S^{n^2-n}[\det] \wedge V(0)) & \xrightarrow{\ \cong\ } & E_*(D_n X \wedge S^{2\lambda+n^2-n} \wedge V(0)) \\
\downarrow{\scriptstyle q_*} & & \downarrow{\scriptstyle q_*} \\
E_* \Sigma D_n X \wedge S^{n^2-n}[\det] & \xrightarrow{\ \cong\ } & E_* \Sigma D_n X \wedge S^{2\lambda+n^2-n} \\
\downarrow & & \downarrow \\
0 & & 0.
\end{array}
$$

$$(1.5.2)$$

We will be done if we can find a map $f : D_n X \wedge S^{n^2-n}[\det] \to D_n X \wedge S^{2\lambda+n^2-n}$ realizing the topmost horizontal map of Morava modules. By Theorem 1.32 we have an equivalence $S^{n^2-n}[\det] \wedge V(0) \to S^{2\lambda+n^2-n} \wedge V(0)$, so we do have a realization of the middle horizontal map; that is, we have a diagram in which we'd like to complete the dotted arrow:

$$D_n X \wedge S^{n^2-n}[\det] \xrightarrow{i} D_n X \wedge S^{n^2-n}[\det] \wedge V(0)$$

$$\big\downarrow f \qquad\qquad\qquad\qquad\qquad \big\downarrow g$$

$$D_n X \wedge S^{2\lambda+n^2-n} \xrightarrow{i} D_n X \wedge S^{2\lambda+n^2-n} \wedge V(0) \xrightarrow{q} D_n X \wedge S^{2\lambda+n^2-n}.$$

For the composite $qgi : D_n X \wedge S^{n^2-n}[\det] \to D_n X \wedge S^{2\lambda+n^2-n}$ we have

$$E_*(qgi) = 0$$

as the algebraic diagram (1.5.2) above commutes and the columns are exact sequences. Using the section $s : \Sigma D_n X \to D_n X \wedge V(0)$ chosen above we form

$$g' = g - sqg.$$

Then $qg'i = qgi - qsqgi = 0$ and

$$E_*(g'i) = E_*(gi) - E_*(s)E_*(qgi) = E_*(gi).$$

Thus, we can let f be the unique map so that $if = g'i$. □

We finish by interpreting the number λ as an integer in some special cases. We will say a Morava module has a $v_n^{p^k}$-self map if $v_n^{p^k} = u^{p^k(1-p^n)}$ induces an isomorphism of Morava modules.

$$M \longrightarrow E_{2p^k(p^n-1)} \otimes_{E_0} M$$

Theorem 1.42   *Let M be finite Morava module. If $\mathfrak{m}^k M = 0$, then M has a $v_n^{p^k}$-self map. If, in addition, $pM = 0$ then*

$$D_I M = \Sigma^{2p^{nk}r(n)-2n} D_S M.$$

Proof   To prove the first statement, note that the class $v_n = u^{1-p^n} \in E_{2(p^n-1)}$ is $\mathbb{G}_n$-invariant modulo $\mathfrak{m}$. Since $\mathbb{G}_n$ acts through graded ring maps, $v_n^{p^k}$ is invariant modulo $\mathfrak{m}^{k+1}$. The second statement follows from Remark 1.22 and Proposition 1.40. □

Theorem 1.43   *Let $2p > \max\{n^2 + 1, 2n + 2\}$ and let X be a finite type n spectrum with a $v_n^{p^k}$-self map. Suppose $p1_X = 0 : X \to X$. Then*

$$I_n X = \Sigma^{2p^{nk}r(n)+n^2-n} D_n X.$$

Proof   This follows from Propositions 1.21 and 1.41. □

Remark 1.44   Theorem 1.43 has strong implications for the Adams-Novikov spectral sequence of a Spanier-Whitehead self-dual complex. For example, let $n = 2$ and $p > 3$. Let $M(1, s)$ be the cofiber of $v_1^s : \Sigma^{2s(p-1)} V(0) \to V(0)$; see Example 1.38. Then, $M(1, s)$ is Spanier-Whitehead self dual. We check $M(1, s)$ has a $v_2^{p^k}$-self map as long as $s \leq p^k$. Furthermore $D_2 M(1, s) = \Sigma^{-qn-2} M(1, s)$ where $q = 2(p-1)$. Then

$$I_2 M(1, p^k) = \Sigma^{2p^{2k}(p+1)+2} D_2 M(1, p^k) = \Sigma^{2p^{2k}(p+1)-p^k q} L_{K(2)} M(1, p^k).$$

In combination with Remark 1.37 and Remark 1.39 this forces a very strong duality onto the Adams-Novikov Spectral Sequence for $M(1, s)$. This is often what people mean when they point to charts to discuss Gross-Hopkins duality.

The spectrum $M(1, 1)$ is the Smith-Toda complex $V(1)$ and we have

$$I_2 V(1) \simeq \Sigma^{2(p+1)+2} D_2 V(1) \simeq \Sigma^4 V(1).$$

For this example, the duality in the Adams-Novikov Spectral Sequence discussed in Remark 1.39 can be checked by hand. In fact, $E_* V(1) = \mathbb{F}_{p^2}[u^{\pm 1}]$ and there is an isomorphism

$$\mathbb{F}_p[v_2^{\pm 1}] \otimes \Lambda(\zeta_2) \otimes A \cong H^*(\mathbb{G}_2, E_* V(1))$$

where A is the 3-dimensional Poincaré duality algebra on classes

$$h_0 \in H^1(\mathbb{G}_2, E_{2(p-1)} V(1)) \qquad h_1 \in H^1(\mathbb{G}_2, E_{-2(p-1)} V(1))$$
$$g_0 \in H^2(\mathbb{G}_2, E_{2(p-1)} V(1)) \qquad g_1 \in H^2(\mathbb{G}_2, E_{-2(p-1)} V(1))$$
$$h_0 g_1 = g_0 h_1 \in h_0 \in H^3(\mathbb{G}_2, E_0 V(1)).$$

See Theorem 6.3.22 of [30].

Remark 1.45   The hypothesis that p be large with respect to n is needed in Theorem 1.42. At the prime 2 there is no non-trivial finite CW spectrum X such that the identity $1_X : X \to X$ has order 2. To see this let n be the largest integer so that there is a class $x \in H^* X = H^*(X, \mathbb{F}_2)$ with $\mathrm{Sq}^n(x) \neq 0$. Then if $a \in H^0 V(0)$ is the bottom class, we have $\mathrm{Sq}^{n+1}(x \times a) \neq 0$ in $H^*(X \wedge V(0))$, so $X \wedge V(0)$ cannot split as $X \vee \Sigma X$. This simple argument is due to Mark Mahowald.

The prime 2 may always be an outlier, but at $n = 2$ and $p = 3$, the Smith-Toda complex $V(1)$ is a counterexample to extending Theorem 1.42. The identity map $V(1) \to V(1)$ does have order 3. By [6], $V(1)$ has a v_2^9-self map, and hence $L_{K(2)} V(1)$ is 144-periodic. In [12] we showed that the homotopy groups of $L_{K(2)} V(1)$ are exactly 144-periodic; see also the charts at the end of [11]. In particular, there is no v_2^3-self map. We also noted in §7 of [11] that $I_2 \wedge V(1) \simeq \Sigma^{-22} V(1)$; see also [4]. This implies

$$I_2 V(1) \simeq I_2 \wedge D_2 V(1) \simeq \Sigma^{-28} L_{K(2)} V(1) \simeq \Sigma^{116} L_{K(2)} V(1)$$

since $D_2 V(1) \simeq \Sigma^{-6} L_{K(2)} V(1)$. By contrast

$$\Sigma^{2p^{nk} r(n) + n^2 - n} D_n V(1) \simeq \Sigma^{650} D_2 V(1) \simeq \Sigma^{644} L_{K(2)} V(1).$$

Since $644 \equiv 68$ modulo 144, the equivalence of Theorem 1.43 can't hold. The map

$$e : \mathrm{Pic}^0_{K(2)} \longrightarrow H^1(\mathbb{G}_2, E_0^\times)$$

is not injective at $n = 2$ and $p = 3$. The difference $116 - 68 = 48$ is related to the fact that there is an element P in the kernel of the homomorphism of e (1.2.1) so that $P \wedge V(1) \simeq \Sigma^{48} L_{K(2)} V(1)$. All of this is covered in excruciating detail in [13] and [11]. Scholars in this field will recognize the number 48 as having near-cabalistic significance; this is only one of the many places it appears.

References

[1] J. F. Adams. *Stable homotopy and generalised homology.* Chicago Lectures in Mathematics. University of Chicago Press, Chicago, IL, 1974. Reprint of the 1974 original.

[2] T. Barthel, A. Beaudry, P. G. Goerss, and V. Stojanoska. Constructing the determinant sphere using a Tate twist. *arXiv e-prints*, page arXiv:1810.06651, 2018.

[3] A. Beaudry, P. G. Goerss, and H.-W. Henn. Chromatic splitting for the $K(2)$-local spheres at $p = 2$. *arXiv e-prints*, page arXiv:1712.08182, 2017.

[4] M. Behrens. A modular description of the $K(2)$-local sphere at the prime 3. *Topology*, 45(2):343–402, 2006.

[5] M. Behrens. The homotopy groups of $S_{E(2)}$ at $p \geq 5$ revisited. *Adv. Math.*, 230(2):458–492, 2012.

[6] M. Behrens and S. Pemmaraju. On the existence of the self map v_2^9 on the Smith-Toda complex $V(1)$ at the prime 3. In *Homotopy theory: relations with algebraic geometry, group cohomology, and algebraic K-theory*, volume 346 of *Contemp. Math.*, pages 9–49. Amer. Math. Soc., Providence, RI, 2004.

[7] M. P. Brodmann and R. Y. Sharp. *Local cohomology: an algebraic introduction with geometric applications*, volume 60 of *Cambridge Studies in Advanced Mathematics*. Cambridge University Press, Cambridge, 1998.

[8] E. S. Devinatz and M. J. Hopkins. The action of the Morava stabilizer group on the Lubin-Tate moduli space of lifts. *Amer. J. Math.*, 117(3):669–710, 1995.

[9] E. S. Devinatz and M. J. Hopkins. Homotopy fixed point spectra for closed subgroups of the Morava stabilizer groups. *Topology*, 43(1):1–47, 2004.

[10] W. G. Dwyer, J. P. C. Greenlees, and S. B. Iyengar. Gross-Hopkins duality and the Gorenstein condition. *J. K-Theory*, 8(1):107–133, 2011.

[11] P. G. Goerss and H.-W. Henn. The Brown-Comenetz dual of the $K(2)$-local sphere at the prime 3. *Adv. Math.*, 288:648–678, 2016.

[12] P. G. Goerss, H.-W. Henn, and M. Mahowald. The homotopy of $L_2 V(1)$ for the prime 3. In *Categorical decomposition techniques in algebraic topology (Isle of Skye, 2001)*, volume 215 of *Progr. Math.*, pages 125–151. Birkhäuser, Basel, 2004.

[13] P. G. Goerss, H.-W. Henn, M. Mahowald, and C. Rezk. On Hopkins' Picard groups for the prime 3 and chromatic level 2. *J. Topol.*, 8(1):267–294, 2015.

[14] P. G. Goerss and M. J. Hopkins. Moduli spaces of commutative ring spectra. In *Structured ring spectra*, volume 315 of *London Math. Soc. Lecture Note Ser.*, pages 151–200. Cambridge Univ. Press, Cambridge, 2004.

[15] J. P. C. Greenlees and J. P. May. Derived functors of I-adic completion and local homology. *J. Algebra*, 149(2):438–453, 1992.

[16] D. Heard. *Morava modules and the K(n)-local Picard group*. University of Melbourne, 2014. Thesis (Ph.D.) https://drew-heard.github.io/Thesis.pdf.

[17] H.-W. Henn. Centralizers of elementary abelian p-subgroups and mod-p cohomology of profinite groups. *Duke Math. J.*, 91(3):561–585, 1998.

[18] H.-W. Henn, N. Karamanov, and M. Mahowald. The homotopy of the $K(2)$-local Moore spectrum at the prime 3 revisited. *Math. Z.*, 275(3-4):953–1004, 2013.

[19] M. J. Hopkins and B. H. Gross. Equivariant vector bundles on the Lubin-Tate moduli space. In *Topology and representation theory (Evanston, IL, 1992)*, volume 158 of *Contemp. Math.*, pages 23–88. Amer. Math. Soc., Providence, RI, 1994.

[20] M. J. Hopkins and B. H. Gross. The rigid analytic period mapping, Lubin-Tate space, and stable homotopy theory. *Bull. Amer. Math. Soc. (N.S.)*, 30(1):76–86, 1994.

[21] M. J. Hopkins, M. Mahowald, and H. Sadofsky. Constructions of elements in Picard groups. In *Topology and representation theory (Evanston, IL, 1992)*, volume 158 of *Contemp. Math.*, pages 89–126. Amer. Math. Soc., Providence, RI, 1994.

[22] M. J. Hopkins and J. H. Smith. Nilpotence and stable homotopy theory. II. *Ann. of Math. (2)*, 148(1):1–49, 1998.

[23] M. Hovey and H. Sadofsky. Invertible spectra in the $E(n)$-local stable homotopy category. *J. London Math. Soc. (2)*, 60(1):284–302, 1999.

[24] M. Hovey and N. P. Strickland. Morava K-theories and localisation. *Mem. Amer. Math. Soc.*, 139(666):viii+100, 1999.

[25] N. Karamanov. On Hopkins' Picard group Pic_2 at the prime 3. *Algebr. Geom. Topol.*, 10(1):275–292, 2010.

[26] O. Lader. Une résolution projective pour le second groupe de Morava pour $p \geq 5$ et applications. Available from `https://tel.archives-ouvertes.fr/tel-00875761/document`, 2013.

[27] G. Li. *Picard Groups of $K(n)$-Local Categories*. ProQuest LLC, Ann Arbor, MI, 2019. Thesis (Ph.D.)–Northwestern University.

[28] J. Lubin and J. Tate. Formal moduli for one-parameter formal Lie groups. *Bull. Soc. Math. France*, 94:49–59, 1966.

[29] P. Pstragowski. Chromatic Picard groups at large primes. *arXiv e-prints*, page arXiv:1811.05415, 2018.

[30] D. C. Ravenel. *Complex cobordism and stable homotopy groups of spheres*, volume 121 of *Pure and Applied Mathematics*. Academic Press, Inc., Orlando, FL, 1986.

[31] H. Sadofsky. Hopkins' and Mahowald's picture of Shimomura's v_1-Bockstein spectral sequence calculation. In *Algebraic topology (Oaxtepec, 1991)*, volume 146 of *Contemp. Math.*, pages 407–418. Amer. Math. Soc., Providence, RI, 1993.

[32] K. Shimomura and A. Yabe. The homotopy groups $\pi_*(L_2S^0)$. *Topology*, 34(2):261–289, 1995.

[33] N. P. Strickland. Gross-Hopkins duality. *Topology*, 39(5):1021–1033, 2000.

[34] P. Symonds and T. Weigel. Cohomology of p-adic analytic groups. In *New horizons in pro-p groups*, volume 184 of *Progr. Math.*, pages 349–410. Birkhäuser Boston, Boston, MA, 2000.

[35] H. Toda. On spectra realizing exterior parts of the Steenrod algebra. *Topology*, 10:53–65, 1971.

2

Axiomatic Representation Theory of Finite Groups by way of Groupoids

Ivo Dell'Ambrogio[a]

Abstract

We survey several notions of Mackey functors and biset functors found in the literature and prove some old and new theorems comparing them. While little here will surprise the experts, we draw a conceptual and unified picture by making systematic use of finite groupoids. This provides a 'road map' for the various approaches to the axiomatic representation theory of finite groups, as well as some details which are hard to find in writing.

2.1 Introduction and results

This is a survey of several variants of Mackey functors and biset functors for finite groups appearing in the literature. (Beware: we survey *abstract formalisms*; the reader interested in *concrete examples* is referred to [31].) Our goal is to show that it becomes quite easy to relate these variants to one another and to rigorously prove comparison theorems, provided one embraces *finite groupoids* (categories with finitely many arrows all of which are invertible) and *2-categories* (categories equipped with 2-morphisms, *i.e.* arrows between arrows). The main reason for using finite groupoids is because they include all finite groups as well as all finite G-sets for each group G; the main reason for using the language of 2-categories is in order to exploit the fact that finite groupoids, together with functors and natural transformations, form a very nicely behaved 2-category.

[a] Department of Mathematics, University of Lille

Let us begin with a quick review of what is often referred to as "axiomatic representation theory". Roughly speaking, both Mackey functors and biset functors provide ways of encoding the various homomorphisms that arise in the representation theory of finite groups when one allows the group to vary. One typically encounters *restriction* maps and *induction* (also called *transfer* or *trace*) maps associated with inclusions $H \hookrightarrow G$ of subgroups, and possibly also *inflation* and *deflation* maps, associated with quotients $G \twoheadrightarrow G/N$ by normal subgroups. There may also be *isomorphism* maps coming from abstract isomorphisms of groups $G \cong G'$, or at least the special case of *conjugation* maps induced by conjugations $G \cong gGg^{-1}$ by an element g of some fixed 'ambient' group of which G is a subgroup. These families of maps interact by various (long) lists of basic relations, which are then promoted to the role of axioms of a formal algebraic theory.

Note that the above families of homomorphisms come in pairs of opposite variance: restriction/induction and inflation/deflation, with isomorphisms and conjugations having both variances since they are invertible. A classical idea, due to Lindner [20], is to simultaneously encode both variances in some category of 'spans', *i.e.* diagrams of the form $X \leftarrow S \rightarrow Y$ where the 'wrong-way' maps will induce the 'wrong-way' functoriality (variations on this idea abound in mathematics, see *e.g.* the many uses of *correspondences* in geometry). Another way to impose this symmetry is by using bisets, as proposed by Bouc [4]. Then a Mackey functor, resp. a biset functor, can be simply defined to be a linear representation of the category of spans, resp. of bisets.

Warning 2.1 Our usage of 'Mackey functors' for functors defined on any kind of span category originates in [20] and is now quite widespread. However, it is at odds with the tradition in representation theory, where the qualifier 'Mackey' is typically reserved for functors with restriction and induction maps but no inflations or deflations (if they have inflation maps, for instance, they will be called 'inflation functors', without the qualifier 'Mackey'). The two uses seem hard to reconcile, in particular with respect to the global variants surveyed below.

Let us first review Mackey functors, using groupoids.

Mackey functors

The unifying point of view we adopt here is that a *Mackey functor M* should be defined to be an abelian-groups-valued (or later, more generally, taking values in modules over some commutative ring $\Bbbk$) additive functor on the category of spans formed in a suitable 2-category of finite groupoids (see Section 2.3). Let us insist straight away that the additivity condition, $M(G_1 \sqcup G_2) \cong M(G_1) \oplus M(G_2)$, implies that the *data* of a Mackey functor can always be reduced to what happens to indecomposable groupoids, *i.e.* good old finite groups. Nevertheless, it is convenient to work with groupoids because they make definitions more conceptual and results easier to see and to prove.

We allow two parameters in this definition of Mackey functor. Firstly, the above-mentioned 'suitable' 2-category of groupoids, denoted below by $\mathbb{G}$, can be adjusted as needed: it will typically be either a (2-full) sub-2-category of the 2-category of all finite groupoids, or a comma 2-category over a fixed group(oid). Secondly, we further choose a distinguished (wide) sub-2-category $\mathbb{J} \subseteq \mathbb{G}$ which determines which functors of groupoids are allowed to induce 'wrong way' maps (*e.g.* inductions or deflations). By choosing the pair $(\mathbb{G}; \mathbb{J})$ adequately, the resulting notion of Mackey functor can be specialized to those in common use. By way of illustration, we will explicitly consider five of them (leaving further variations to the interested reader):

(1) The original *Mackey functors for a fixed group G* [12] [10]. They are equipped with: restriction maps, induction maps and conjugations in G. They appear all over equivariant mathematics, perhaps most notably in equivariant stable homotopy theory, as the algebraic structure with which the homotopy groups of 'genuine' G-spectra are naturally endowed ([19] [8]).

(2) *Global Mackey functors* in the sense of [11] and [25] [24]. These are defined on all finite group(oid)s and have maps of each kind (that is: restrictions, inductions, all isomorphisms, inflations and deflations).

(3) Global Mackey functors as above, but without deflation maps. These have been given various names: *functors with regular Mackey structure* [29], *inflation functors* [30], and *global $(\emptyset, \infty)$-Mackey functors* [18]. They appear, for instance, as the natural algebraic invariant of Schwede's global equivariant spectra (see [28, Thm. 4.2.6] and the discussion after it).

(4) Global Mackey functors as above, but without inflation and deflation maps. These are simply called *global Mackey functors* in [30] and [4] (*cf.* Warning 2.1).

(5) The *fused Mackey functors for G* of [5], also called *conjugation invariant Mackey functors* in [13], namely those Mackey functors for G

as in (1) which admit a reformulation as a kind of biset functors (*cf.* Remark 2.73).

As we will prove, the above five types of Mackey functors can be obtained by specializing our general definition to the following choices of the parameters $(\mathbb{G}; \mathbb{J})$, respectively:

For (1) : use $(\mathbb{G}; \mathbb{J}) = (\mathsf{gpd}^{\mathsf{f}}_{/G}; \mathsf{all})$, where $\mathsf{gpd}^{\mathsf{f}}_{/G}$ is the comma 2-category of groupoids faithfully embedded in G and $\mathbb{J} = \mathsf{all}$ just means that $\mathbb{J} = \mathbb{G}$. See Section 2.5.

For (2) : use $(\mathbb{G}; \mathbb{J}) = (\mathsf{gpd}; \mathsf{all})$, where we take the whole 2-category gpd of all finite groupoids and functors between them and where we allow the formation of all spans. See Definition 2.51.

For (3) : use $(\mathbb{G}; \mathbb{J}) = (\mathsf{gpd}; \mathsf{gpd}^{\mathsf{f}})$, where we consider the whole category of groupoids but we only allow spans $X \leftarrow S \rightarrow Y$ whose right leg $S \rightarrow Y$ is a faithful functor. See Example 2.46.

For (4) : use $(\mathbb{G}; \mathbb{J}) = (\mathsf{gpd}^{\mathsf{f}}; \mathsf{all})$, where both legs of all spans must be faithful functors. See Example 2.45.

For (5) : use $(\mathbb{G}; \mathbb{J}) = (\mathsf{gpd}^{\mathsf{f},\mathsf{fus}}_{/G}; \mathsf{all})$, where $\mathsf{gpd}^{\mathsf{f},\mathsf{fus}}_{/G}$ is the 'fused' variant of the comma 2-category of groupoids faithfully embedded in G. See Definition 2.65.

As the reader may guess, the above 2-categories are all related by evident inclusion and forgetful 2-functors. We will exploit this fact in order to easily establish comparison results.

In order to provide a uniform and conceptual construction of the span category for all of the above examples (and many more), we introduce the general notion of a *spannable pair* $(\mathbb{G}; \mathbb{J})$ (see Definition 2.31). A spannable pair consists of an extensive (2,1)-category $\mathcal{E}$ equipped with a suitably closed 2-subcategory $\mathbb{J}$ and sufficiently many *Mackey squares* (*i.e.* pseudo-pullbacks of groupoids). This abstract approach is developed in Section 2.3. In Section 2.4, we look in full details at the span category for the basic example (2), in order to dispel the (possibly intimidating) categorical abstractions of the general definition by reducing it to some classical combinatorics. In particular, we describe a presentation of the linear category of spans of groupoids (see Theorem 2.53).

Of course, we also need to explain how to connect the above definitions with the more familiar ones found in the literature. We will now briefly explain how to do this, beginning with (1) and (5) and the associated comma 2-categories.

From G-sets to groupoids: the transport groupoid

Mackey functors for a fixed group G, as in type (1) above, are typically expressed in terms of G-sets. The key tool for comparing Mackey functors for a fixed G with global Mackey functors is the *transport groupoid functor* $G \ltimes -$, which sends a G-set X to its *transport groupoid* (a.k.a. *action groupoid, homotopy quotient* or *Grothendieck construction*) $G \ltimes X$. The latter groupoid is canonically equipped with a faithful functor $G \ltimes X \rightarrowtail G$, which turns the transport groupoid construction into a functor

$$G \ltimes - : G\text{-set} \longrightarrow \mathsf{gpd}^{\mathsf{f}}_{/G}$$

from the category of G-sets into the comma 2-category of groupoids 'faithfully embedded' in G (Definition 2.60). This functor is a nice inclusion, in fact it is a biequivalence (an equivalence of 2-categories). As a consequence, Mackey functors for G, which can be defined to be linear representations of the category of spans in G-sets, turn out to be equivalent to representations of the category of spans in $\mathsf{gpd}^{\mathsf{f}}_{/G}$. Therefore they are the result of specializing our general notion of Mackey functors to the pair $(\mathbb{G}; \mathbb{J}) = (\mathsf{gpd}^{\mathsf{f}}_{/G}; \text{all})$ (see Corollary 2.64). All of this is already contained in [1, § B.1] but is briefly recalled at the beginning of Section 2.5 for the reader's convenience.

The 'G-local' and the 'global' settings are now compared by the forgetful 2-functor $\mathsf{gpd}^{\mathsf{f}}_{/G} \to \mathsf{gpd}$ which simply forgets the embedding $H \rightarrowtail G$ into G. This 2-functor is not 2-full, because 'being over G' puts a constraint on the natural isomorphisms between (faithful) functors that can be used in the comma 2-category, while in gpd we can use all of them. If we pull back these extra 2-cells and add them to $\mathsf{gpd}^{\mathsf{f}}_{/G}$, we obtain a variant of the comma 2-category which we denote $\mathsf{gpd}^{\mathsf{f,fus}}_{/G}$ and call the 2-category of *fused groupoids embedded into* G (see Definition 2.65). The 2-cells can be pulled further back onto the category of G-sets, which results in a 2-category $G\text{-set}^{\mathsf{fus}}$ consisting of finite G-sets, G-maps and *twisting maps* relating parallel G-maps (see Definition 2.66). If we truncate the 2-category $G\text{-set}^{\mathsf{fus}}$, the result is precisely Bouc's category $G\text{-}\underline{\mathsf{set}}$ of fused Mackey functors for G. By definition, fused Mackey functors (type (5) above) are representations of spans in $G\text{-}\underline{\mathsf{set}}$.

It follows that fused Mackey functors can be recovered as the Mackey functors for the 2-category $\mathbb{G} = \mathbb{J} = \mathsf{gpd}^{\mathsf{f,fus}}_{/G} \simeq G\text{-set}^{\mathsf{fus}}$ (see Corollary 2.72). The above arguments make it also easy to identify fused Mackey functors as those Mackey functors M for G which are *conjuga-*

tion invariant, i.e. such that the centralizer $C_G(H)$ acts trivially on the value $M(H)$ for every $H \leq G$.

This is all explained in Section 2.5.

Biset functors

As already mentioned, an alternative way to force symmetry on finite group(oid)s is to use bisets rather than spans. By definition, a (finite) *biset* (also called a *profunctor* or *bimodule*) $U \colon H \to G$ between two groupoids is a functor $U \colon H^{\mathrm{op}} \times G \to \mathsf{set}$ to (finite) sets[1]. Taken up to isomorphism and composed by tensor products (coends), bisets are the morphisms of a category with arbitrary finite direct sums. Similarly to Mackey functors, and following Bouc [4], we define here a *biset functor* to be an additive functor on some suitable (sub)category of bisets of groupoids. Here too additivity allows us to reduce everything to finite groups, hence in particular our definition of biset functors is equivalent to Bouc's definition, which only uses groups; but again, we want to keep all groupoids as they provide direct sums and allow us to define the realisation of spans (see below) in a natural way.

Just as with spans, also with bisets there are parameters we can twiddle: we can restrict the allowed class of objects (groupoids), or the allowed class of morphisms (bisets). We leave variations of the former kind to the interested reader. For the latter, we will study the following common three choices:

(2)′ Allow all bisets.

(3)′ Only allow right-free bisets: the resulting biset functors (called *inflation functors* in [4]) will have all types of maps except for deflations.

(4)′ Only allow bi-free bisets: the resulting biset functors (called *global Mackey functors* in [4]) will have neither deflations nor inflations.

Note that Webb [31, §8] provides a more combinatorial definition of biset functors, directly in terms of the maps induced by morphisms of groups and their relations (as in the classical definition of Mackey functor for G), without mentioning bisets, and under the name *globally defined Mackey functors* (see Remark 2.91.)

This is all explained in Section 2.6.

---

[1] Here the symmetry appears because every (G, H)-biset can be turned into an (H, G)-biset by precomposing with the inverse-arrows isomorphisms
$(-)^{-1} \colon G^{\mathrm{op}} \overset{\sim}{\to} G$ and $H \overset{\sim}{\to} H^{\mathrm{op}}$.

The realization of spans as bisets

The key tool for comparing Mackey functors and biset functors is the *realization functor* $\mathcal{R}$ from spans to bisets

$$\mathcal{R}: \mathsf{Span}(\mathsf{gpd}) \longrightarrow \mathsf{biset}$$

which sends a groupoid to itself and 'realizes' each (abstract) span of functors as a (concrete) biset. This construction, which already exists as a pseudo-functor between the *bi*categories of spans and bisets, was conjectured by Hoffnung [14, Claim 13], was foreshadowed by Nakaoka [25] [24], and was studied in more details in Huglo's thesis [15]. We recall its relevant features in Theorem 2.80. The way spans are realized as bisets is actually rather obvious and has been known to category theorists for a long time. What is apparently less known, but crucial for us, is the *(pseudo-) functoriality* of the construction, which holds provided one composes spans using Mackey squares, as we do (see § 2.16).

As a consequence of the mere existence of the realization functor $\mathcal{R}$, we can take any biset functor and pre-compose it with $\mathcal{R}$ in order to obtain a global Mackey functor. By matching the parameter choices for spans and bisets, we then obtain various comparison results involving *e.g.* the global Mackey functors of kind (2), (3) or (4) as above.

What is interesting here is that, *unless deflation maps are included*, this comparison yields an *equivalence* between the corresponding categories of global Mackey functors and of biset functors (that is, we have (3) = (3)′ and (4) = (4)′ but (2) ≠ (2)′); see Corollary 2.93. A version of this result was proved by Miller [23]; *cf.* Remark 2.94. If deflations are included in the package, then the two notions diverge and precomposition with $\mathcal{R}$ only yields an inclusion of biset functors as a full reflective subcategory of the corresponding category of global Mackey functors. The image of the latter inclusion consists precisely of those global Mackey functors that satisfy an extra identity called the *deflativity relation*. This applies *e.g.* to Mackey and biset functors with all maps, as in kind (2) above; see Corollary 2.83. This result is due independently to Ganter [11, App. A] (whose proof uses Webb's description of biset functors) and [25] (whose proof uses, instead of groupoids, a biequivalent 2-category $\mathbb{S}$ of 'variable group actions').

A road map of the formal representation theory of finite groups

To sum up, let us collect all of the above in a single picture:

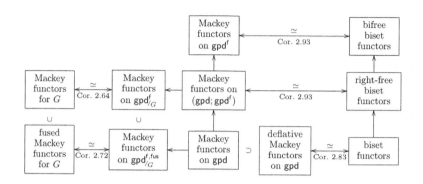

Each box is an abelian category of some sort of Mackey or biset functors, where we have indicated the parameter pair $(\mathbb{G}; \mathbb{J})$ where necessary, with $(\mathbb{G}; \mathbb{G}) =: \mathbb{G}$ for short. The arrows represent exact functors, with equivalences marked by "$\simeq$" and fully faithful inclusions by "$\subset$". We recall that gpd^f is the 2-category of finite groupoids with only faithful functors, $\mathsf{gpd}^f_{/G}$ is the comma 2-category of groupoids faithfully embedded in G (Definition 2.60), while $\mathsf{gpd}^{f,\mathsf{fus}}_{/G}$ denotes its fused variant (Definition 2.65).

As already partly evoked, and as will become clear in the course of the proofs, the above diagram of abelian categories is actually the result of taking representation categories on a diagram of bicategories and pseudo-functors, as follows (all notations will be explained in the chapter):

$$
\begin{array}{c}
\mathsf{Span}(\mathsf{gpd}^f) \xrightarrow{\mathcal{R}} \mathsf{biset}^{\mathsf{bif}}(\mathsf{gpd}) \xleftarrow[\sim_+]{} \mathsf{biset}^{\mathsf{bif}}(\mathsf{gr}) \\[2mm]
\mathsf{Span}(G\text{-set}) \xrightarrow[\sim]{\text{Prop. 2.62}} \mathsf{Span}(\mathsf{gpd}^f_{/G}) \longrightarrow \mathsf{Span}(\mathsf{gpd};\mathsf{gpd}^f) \xrightarrow{\mathcal{R}} \mathsf{biset}^{\mathsf{rf}}(\mathsf{gpd}) \xleftarrow[\sim_+]{} \mathsf{biset}^{\mathsf{rf}}(\mathsf{gr}) \\[2mm]
\mathsf{Span}(G\text{-set}^{\mathsf{fus}}) \xrightarrow[\text{Thm. 2.69}]{\sim} \mathsf{Span}(\mathsf{gpd}^{f,\mathsf{fus}}_{/G}) \longrightarrow \mathsf{Span}(\mathsf{gpd}) \xrightarrow[\text{Thm. 2.80}]{\mathcal{R}} \mathsf{biset}(\mathsf{gpd}) \xleftarrow[\sim_+]{} \mathsf{biset}(\mathsf{gr})
\end{array}
$$

More precisely, in order to obtain the first diagram from the second one we must: First, cut down the latter diagram to one of (usual) categories and functors by identifying isomorphic 1-morphisms (*i.e.* by applying the 1-truncation τ_1 of § 2.13). The result is a diagram of pre-additive categories, *i.e.* categories enriched over abelian monoids, and additive functors between them (see §2.23). Second, we must apply $\mathrm{Fun}_+(-, \mathrm{Ab})$ throughout, that is we take categories of additive functors into abelian groups. (The arrows marked $\sim_+$ induce equivalences only after this second operation, as they are inclusions in the additive hull; in fact, the resulting equivalences were omitted from the above diagram of abelian categories.) Note also that the portion of the diagram of bicategories lying to the left of the realization $\mathcal{R}$ results from applying the span-bicategory construction $\mathrm{Span}(-)$ (see Remark 2.36) to a suitable diagram of spannable pairs $(\mathbb{G}; \mathbb{J})$.

In conclusion, there exists a rich layer of underlying 2-categorical information behind these well-known categories of Mackey and biset functors. The exploration and mining of this stratum was begun in [1] and deserves to be taken further.

Remark 2.2 The literature on Mackey and biset functors for finite groups is rather vast and we have not attempted to list all variations on the theme, nor have we tried to assign historical precedence. Some history of the subject can be found in [31] and [4, § 1.4].

Remark 2.3 This chapter is an offshoot of [1], which developed the basic theory of *Mackey 2-functors*, a categorified version of Mackey functors whose values are additive categories instead of abelian groups. We have nonetheless strived to make this chapter self-contained. Indeed, while the use of groupoids and 2-categories arose quite naturally in the categorified context of [1], it is our hope that the present chapter will show – even to readers who do not particularly care for Mackey 2-functors – how the groupoidal viewpoint offers a useful organizing principle for the *usual*, merely abelian-group-valued Mackey functors. Conversely, we also hope that this survey may function as a guide and point of entry into the Mackey literature for those who are already fluent with 2-categories.

Notation 2.4 We will work over an arbitrary commutative ring with unit, denoted by $\Bbbk$. Common choices are the ring of integers, a field, or a nice local ring.

To be consistent with [1], a span $X \leftarrow S \rightarrow Y$ will be always visually understood as going from left to right, *i.e.* as a morphism $X \rightarrow Y$; to be consistent with [4], a biset $_G X_H$ will always be understood as going from right to left, *i.e.* as a morphism $H \rightarrow G$ (both are mere conventions).

This will make it slightly awkward in Section 2.6 where we compare spans and bisets.

Acknowledgements

We are grateful to Paul Balmer and Serge Bouc for their interest and for many useful comments on previous versions of this chapter. We would also like to thank an anonymous referee for their careful reading. Many thanks are due to Scott Balchin and Markus Szymik for their friendly and efficient editorial work.

2.2 Categorical preliminaries

We collect here some generalities, mostly to fix our terminology (which is standard and consistent with [1]). Categorically confident readers, or those familiar with [1], should skip ahead to Section 2.3 and refer back only if necessary.

2.5 Groupoids. A *groupoid* is a category where all morphisms are invertible. We will identify a group G with the groupoid having a single object $\bullet$ whose endomorphism monoid is $\mathrm{End}(\bullet) = G$. Under this identification, a homomorphism $f\colon G \to G'$ between groups is the same thing as a functor.

We will only consider groups and groupoids which are *finite*, that is, which only have finitely many objects and arrows.

Recall that a groupoid is *connected* if all its objects are isomorphic, in which case the groupoid is equivalent (as a category) to the full subcategory on any one of its objects, which is just a group. Thus every (finite) groupoid G is equivalent to a (finite) disjoint union of (finite) groups, with the equivalence depending on a chosen set of representative objects for all connected components.

2.6 Bicategories and 2-categories. (See *e.g.* [1, § A.1].) A *2-category* $\mathcal{C}$ is a category enriched in categories, *i.e.* it consists of a collection $\mathrm{Obj}\,\mathcal{C}$ of objects (or *0-morphisms*, *0-cells*), together with 'Hom' categories $\mathcal{C}(X, Y)$ for all pairs of objects X, Y (whose objects are the *1-morphisms* or *1-cells* $u\colon X \to Y$ of $\mathcal{C}$ and whose arrows are its *2-morphisms* or *2-cells* $\alpha\colon u \Rightarrow v$), and composition functors

$$\circ = \circ_{X,Y,Z}\colon \mathcal{C}(Y, Z) \times \mathcal{C}(X, Y) \to \mathcal{C}(X, Z)$$

subject to the usual (strict) unit and associativity equations. In particular, each object has a left-and-right identity 1-morphism Id_X. Our first example is the 2-category

gpd

consisting of finite groupoids, functors between them, and (necessarily invertible) natural transformations.

Notation 2.7 Given two parallel homomorphisms $f_1, f_2 \colon G \to H$ between groups, considered as functors between one-object groupoids (§ 2.5), a natural transformation $f_1 \Rightarrow f_2$ (a 2-cell in gpd) is completely determined by its unique component $x \colon \bullet \overset{\sim}{\to} \bullet$, which is an element $x \in H$ such that $^x f_1 = f_2$ (that is: $x f_1(g) x^{-1} = f_2(g)$ for all $g \in G$). We will write $\gamma_x \colon f_1 \Rightarrow f_2$ for this 2-cell.

A *bicategory* $\mathcal{B}$ is a 'relaxed' version of a 2-category, where the unit and associativity axioms only hold up to given coherent natural isomorphisms $\mathrm{Id}_X \circ u \cong u \cong u \circ \mathrm{Id}_X$ and $(w \circ v) \circ u \cong w \circ (v \circ u)$. Here 'coherent' means that all reasonable diagrams involving these isomorphisms must commute; as a consequence, each bicategory can effectively be replaced by a 2-category which is biequivalent (see below) to it.

In any bicategory, the composition of 2-cells within each Hom category is called *vertical* composition, while the effect of applying the composition functors $\circ_{X,Y,Z}$ to 1- or 2-morphisms is called *horizontal* composition. This is reflected by the usual layout in the 'cellular' diagram notation for a 2-morphism:

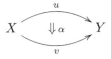

A 2-category is the same thing as a bicategory which is *strict*, that is whose unit and associativity isomorphisms are identity maps. An (ordinary) category, or *1-category*, can be seen as a *discrete* 2-category, that is one whose Hom categories only have identity arrows (hence 'are' just sets).

Remark 2.8 It is often straightforward to directly check that some given data defines a 2-category, *cf.* [22, § XII.3]. For a non-strict *bi*category, on the other hand, some more work may be required.

Example 2.9 (See e.g. [3, §7.8]) Let G and H be finite groupoids. A *finite biset (or bimodule, distributor, profunctor) of groupoids*, written

$U = {}_GU_H \colon H \to G$, is a functor $U \colon H^{\mathrm{op}} \times G \to \mathsf{set}$ to the category of finite sets. There is a bicategory $\mathsf{biset}(\mathsf{gpd})$ with finite groupoids as objects, finite bisets $H \to G$ as 1-morphisms from H to G, and natural transformations between them as 2-morphisms. The horizontal composition of two bisets ${}_GU_H$ and ${}_KV_G$ is given by their tensor product (*i.e.* coend)

$$V \otimes U = {}_KV \underset{G}{\otimes} U_H := \int^{g \in G} V(g,-) \times U(-,g) \colon H^{\mathrm{op}} \times K \longrightarrow \mathsf{set}.$$

Concretely, the image under $V \otimes U$ of $(h,k) \in \mathrm{Obj}(H^{\mathrm{op}} \times K)$ is the quotient set

$$(V \otimes U)(h,k) = \frac{\coprod_{g \in \mathrm{Obj}\, G} V(g,k) \times U(h,g)}{(\beta\varphi, \alpha) \sim (\beta, \varphi\alpha) \ \text{ if } \varphi \in G(g_1, g_2)} \in \mathrm{Obj}(\mathsf{set}).$$

Natural transformations between bisets induce maps on these quotients, and this defines the functoriality on maps of horizontal composition.

Example 2.10 If the groupoids G and H are just groups, to give a biset ${}_GU_H$ as in Definition 2.9 is the same thing as to give a G, H-biset in the sense of [4], that is, a set U together with a left action by G and a right action by H which commute: $(g \cdot u) \cdot h = g \cdot (u \cdot h)$. By restricting attention to bisets between groups, we obtain a 1-full and 2-full sub-bicategory of $\mathsf{biset}(\mathsf{gpd})$ that we denote by $\mathsf{biset}(\mathsf{gr})$.

2.11   Internal adjunctions and equivalences. Two 1-morphisms $\ell \colon X \to Y$ and $r \colon Y \to X$ in a bicategory $\mathcal{B}$ are *adjoint* if there exist 2-morphisms $\eta \colon \mathrm{Id}_X \Rightarrow r \circ \ell$ and $\varepsilon \colon \ell \circ r \Rightarrow \mathrm{Id}_Y$ such that $(\ell \circ \varepsilon) \circ (\eta \circ \ell) = \mathrm{id}_\ell$ and $(\varepsilon \circ r) \circ (r \circ \eta) = \mathrm{id}_r$. An *adjunction*, sometimes written $\ell \dashv r$, is the data of such a quadruple $(\ell, r, \eta, \varepsilon)$. We say the adjunction is an *adjoint equivalence* if η and ε are (necessarily mutually inverse) isomorphisms. More generally, a 1-morphism $\ell \colon X \to Y$ is an *equivalence* if there exist a 1-morphism $r \colon Y \to X$ and two invertible 2-morphisms $r\ell \cong \mathrm{id}_X$ and $\ell r \cong \mathrm{id}_Y$. Every equivalence can be completed to an adjoint equivalence.

Inside the 2-category of all categories, functors and natural transformations, these reduce to the usual notions of adjoint functors and equivalence of categories.

2.12   Pseudo-functors and biequivalences. A useful notion of morphism between two bicategories (or even 2-categories) is that of a pseudo-functor. A *pseudo-functor* $\mathcal{F} \colon \mathcal{B} \to \mathcal{C}$ consists of an assignment $X \mapsto \mathcal{F}X$ between the objects of $\mathcal{B}$ and $\mathcal{C}$, functors $\mathcal{F} = \mathcal{F}_{X,Y} \colon \mathcal{B}(X,Y) \to$

$\mathcal{B}(\mathcal{F}X, \mathcal{F}Y)$ between their Hom categories, and specified natural iso-morphisms $\mathrm{Id}_{\mathcal{F}X} \cong \mathcal{F}(\mathrm{Id}_X)$ and $\mathcal{F}(v) \circ \mathcal{F}(u) \cong \mathcal{F}(v \circ u)$ subject to suitable coherence axioms. The correct notion of an equivalence between bicategories is that of a *biequivalence*: a pseudo-functor $\mathcal{F}: \mathcal{B} \to \mathcal{C}$ such that there exists another pseudo-functor $\mathcal{G}: \mathcal{C} \to \mathcal{B}$ and isomorphisms $\mathcal{F} \circ \mathcal{G} \cong \mathrm{Id}_{\mathcal{C}}$ and $\mathcal{G} \circ \mathcal{F} \cong \mathrm{Id}_{\mathcal{B}}$. Here by isomorphism we mean an invertible *modification*, which is the correct notion of a morphism of pseudo-functors (see [1, A.1.14]). Equivalently, a pseudo-functor $\mathcal{F}: \mathcal{B} \to \mathcal{C}$ is a biequiva-lence iff each functor $\mathcal{F}_{X,Y}$ is an equivalence $\mathcal{B}(X,Y) \xrightarrow{\sim} \mathcal{C}(\mathcal{F}X, \mathcal{F}Y)$ of Hom categories and moreover each object Y of $\mathcal{C}$ is equivalent (in the internal sense of § 2.11) to one of the form $\mathcal{F}X$.

2.13   The truncation of a bicategory. The *1-truncation* (also called *classifying category*) of a bicategory $\mathcal{B}$, denoted $\tau_1(\mathcal{B})$, is the ordinary category with the same objects as $\mathcal{B}$ and whose morphisms are the isomorphism classes of 1-morphisms of $\mathcal{B}$, with the induced composition. That is, we look at 1-morphisms up to invertible 2-morphisms. This operation is functorial, in that it sends pseudo-functors $\mathcal{F}$ to ordinary functors $\tau_1 \mathcal{F}$, and preserves composition and identities.

Example 2.14   Recall from Example 2.10 the bicategory biset(gr) of fi-nite groups, finite bisets and morphisms of bisets. Its truncation τ_1 biset(gr) is precisely Bouc's category of bisets $\mathcal{B}$ of [4], and our conventions for composition are consistent with his. The inclusion 2-functor biset(gr) $\hookrightarrow$ biset(gpd) induces a fully faithful functor τ_1 biset(gr) $\hookrightarrow$ τ_1 biset(gpd).

2.15   Finite coproducts in a bicategory. Let $\mathcal{B}$ be any bicategory. An object $\varnothing$ is *initial* if the unique functor

$$\mathcal{B}(\varnothing, T) \xrightarrow{\sim} 1$$

to the final category (one object and one identity arrow) is an equivalence for every object $T \in \mathcal{B}$; thus, up to isomorphism there is precisely one 1-morphism $\varnothing \to T$. A diagram $X \xrightarrow{i_X} X \sqcup Y \xleftarrow{i_Y} Y$ of 1-morphisms is a *coproduct* if

$$(i_X^*, i_Y^*): \mathcal{B}(X \sqcup Y, T) \xrightarrow{\sim} \mathcal{B}(X, T) \times \mathcal{B}(Y, T) \qquad (2.2.1)$$

is an equivalence; in particular, for all 1-morphisms $u: X \to T$ and $v: Y \to T$ there is up to isomorphism a unique $(u, v): X \sqcup Y \to T$ such that $(u, v) i_X = u$ and $(u, v) i_Y = v$. Coproducts can be iterated any finite number of times, with similar uniqueness statements, and an initial object can be understood as the empty coproduct. Finite coproducts are unique only up to equivalence, but in many bicategories there are canonical constructions. The coproducts of $\mathcal{B}$ yield coproducts in the truncation $\tau_1 \mathcal{B}$.

If we require the above two equivalences to be *isomorphisms* of categories, we obtain the more familiar *strict* initial object and coproducts. In all our examples strict (and canonical) versions will be available, but all constructions will work with the above more relaxed notion, which has the advantage of being stable under biequivalence. Mostly we will ignore the difference.

2.16 Iso-comma squares and Mackey squares. (See [1, § 2.1-2].) A central role in this chapter will be played by certain diagrams which will provide a canonical replacement for Mackey formulas, namely *iso-comma squares* and *Mackey squares*. They are, respectively, a strict and a pseudo version of *homotopy pullbacks*. The second has the added advantage of being stable under biequivalence.

In a 2-category (or even a bicategory) $\mathcal{C}$, an invertible 2-cell γ of the form

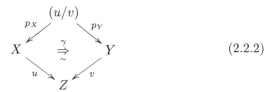

$$(2.2.2)$$

is an *iso-comma square* if it is the 2-universal invertible 2-cell sitting on top of $X \xrightarrow{u} Z \xleftarrow{v} Y$. More precisely, for any other invertible 2-morphism σ as on the left-hand side

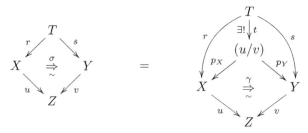

there is a unique 1-morphism t such that $r = p_X t$, $s = p_Y t$ and $\gamma t = \sigma$; we write

$$t = \langle r, s, \sigma \rangle$$

and call r, s and σ the *components* of t. Moreover, it is also required that for any two parallel 1-morphisms $t, t' : T \to (u/v)$, the 2-morphisms

$\alpha\colon t \Rightarrow t'$ be in bijection with pairs $(\beta_X\colon p_X t \Rightarrow p_X t', \beta_Y\colon p_Y t \Rightarrow p_Y t')$ of 2-morphisms between components, the bijection of course being given by $\alpha \mapsto (p_X\alpha, p_Y\alpha)$.

Similarly, we call a 2-cell as in (2.2.2) a *Mackey square*[2] if it satisfies almost the same 2-universal property as above, with the following difference: for each triple $\langle r, s, \sigma\rangle$, there exist a $t\colon T \to (u/v)$ and two 2-isomorphisms $\varphi\colon r \Rightarrow p_X t$ and $\psi\colon p_Y t \Rightarrow s$ such that $\sigma = (u\varphi)(\gamma t)(v\psi)$. (It follows that such a t is unique up to a non-unique isomorphism.)

Example 2.17 In the 2-category Cat of small categories, functors and natural transformations, iso-comma squares have the following canonical construction. The category (u/v) has for objects all triples (x, y, γ) with $x \in \mathrm{Obj}\, X$, $y \in \mathrm{Obj}\, Y$ and $\gamma\colon u(x) \overset{\sim}{\to} v(y)$ an isomorphism in Z; a morphism $(x, y, \gamma) \to (x', y', \gamma')$ is a pair (α, β) with $\alpha \in X(x, x')$, $\beta \in Y(y, y')$ and such that $v(\beta) \circ \gamma = \gamma' \circ u(\alpha)$ in Z. The two functors p_X and p_Y are the evident projections $(x, y, \gamma) \mapsto x$ and $(x, y, \gamma) \mapsto y$, and the natural isomorphism $\gamma\colon up_X \Rightarrow vp_Y$ has the 'tautological' component γ at the object (x, y, γ). If X, Y, Z happen to all be finite groupoids then so is (u/v), hence this construction also provides iso-comma squares for the 2-category gpd.

Remark 2.18 (See [1, 2.1.11-13]) Iso-comma squares and Mackey squares in any 2-category $\mathcal{C}$ can be nicely characterized in terms of iso-comma squares of their Hom categories, built in Cat as in Example 2.17. Namely, consider a 2-cell in $\mathcal{C}$:

$$
\begin{array}{ccc}
 & P & \\
{\scriptstyle p}\swarrow & & \searrow{\scriptstyle q} \\
X & \overset{\gamma}{\underset{\sim}{\Rightarrow}} & Y \\
{\scriptstyle u}\searrow & & \swarrow{\scriptstyle v} \\
 & Z &
\end{array}
\tag{2.2.3}
$$

For every object $T \in \mathrm{Obj}\,\mathcal{C}$, we can apply $\mathcal{C}(T, -)$ to it in order to obtain the following comparison functor:

$$
\mathcal{C}(T, P) \longrightarrow \big(\mathcal{C}(T, u)/\mathcal{C}(T, v)\big), \qquad t \longmapsto (pt, qt, \gamma t)
\tag{2.2.4}
$$

(this is the unique functor with components $\langle \mathcal{C}(T, p), \mathcal{C}(T, q), \mathcal{C}(T, \gamma)\rangle$). Then, as we see directly from the definitions, (2.2.3) is an iso-comma square (resp. a Mackey square) iff (2.2.4) is an isomorphism of categories (resp. an equivalence).

---

[2] Beware that in the literature both iso-comma squares and Mackey squares are sometimes called iso-comma squares, or pseudo-pullbacks, or even just pullbacks.

Remark 2.19 Iso-comma squares are, in particular, Mackey squares. If the iso-comma square over $X \xrightarrow{u} Z \xleftarrow{v} Y$ exists, then a square (2.2.3) is a Mackey square iff the comparison functor $\langle p, q, \gamma \rangle$ into the iso-comma square is an equivalence.

2.20 Comma 2-categories. Given an object C in a 2-category $\mathcal{C}$, the *comma 2-category* $\mathcal{C}_{/C}$ is the 2-category defined as follows (this also works for general bicategories, but we will not need it). Its objects are pairs (X, p) of an object X and a specified 1-cell $p \colon X \to C$ of $\mathcal{C}$. A 1-cell $(X, p) \to (Y, q)$ is a pair (u, θ) where $u \colon X \to Y$ is a 1-cell and $\theta \colon qu \overset{\sim}{\Rightarrow} p$ an invertible 2-cell of $\mathcal{C}$ (as θ is invertible, the choice of its direction is a matter of convention). Finally, a 2-cell $(u, \theta) \Rightarrow (u', \theta')$ is a 2-cell $\alpha \colon u \Rightarrow u'$ such that $\theta'(q\alpha) = \theta$ in $\mathcal{C}$:

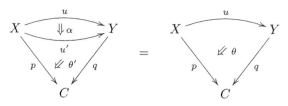

The vertical and horizontal compositions in $\mathcal{C}_{/C}$ are inherited from those of $\mathcal{C}$ in the evident way. There is a forgetful 2-functor $\mathcal{C}_{/C} \to \mathcal{C}$ which simply forgets the p and θ parts.

Remark 2.21 One verifies easily that $\mathcal{C}$ admits coproducts (§ 2.15) if and only if $\mathcal{C}_{/C}$ does so for every $C \in \mathcal{C}$. In this case, the forgetful 2-functor $\mathcal{C}_{/C} \to \mathcal{C}$ preserves and reflects them, in the sense that a diagram

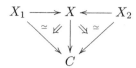

is a coproduct in $\mathcal{C}_{/C}$ iff the top row is a coproduct in $\mathcal{C}$.

Remark 2.22 If $\mathcal{C}$ admits coproducts $X \sqcup Y$ of any two objects, then any choice of adjoint quasi-inverses for all the functors (2.2.1) (or no choice at all if the coproducts are strict) defines a pseudo-functor

$$\mathcal{C}_{/X} \times \mathcal{C}_{/Y} \longrightarrow \mathcal{C}_{/X \sqcup Y}, \quad (A \xrightarrow{a} X, B \xrightarrow{b} Y) \mapsto (A \sqcup B \xrightarrow{a \sqcup b} X \sqcup Y)$$

in an evident way. Any two choices, of course, yield isomorphic pseudo-functors in a suitable sense.

2.23 Additive, semi-additive and $\Bbbk$-linear categories. If $\Bbbk$ is a commutative ring with unit, a $\Bbbk$-*linear category* $\mathcal{C}$ is a category enriched in $\Bbbk$-modules; this means that each Hom set $\mathcal{C}(X, Y)$ carries the structure of a $\Bbbk$-module and the composition maps are all $\Bbbk$-bilinear.

A $\Bbbk$-linear category is *additive* if moreover it has direct sums (a.k.a. biproducts) $X_1 \oplus \ldots \oplus X_n$ for every finite set of its objects, including an empty direct sum 0, a.k.a. a zero object. Direct sums are both categorical product and coproduct diagrams, and morphisms between them are determined by their matrix of components, and can be composed according to matrix multiplication. Direct sums can be used to recover the underlying additive monoid structure of each Hom set $\mathcal{C}(X, Y)$, since its zero element must be the unique map $x \to 0 \to y$ and the sum of $f, g \in \mathcal{C}(X, Y)$ must be the composite

$$f + g = \left(X \xrightarrow{\binom{1}{1}} X \oplus X \xrightarrow{\left(\begin{smallmatrix} f & 0 \\ 0 & g \end{smallmatrix}\right)} Y \oplus Y \xrightarrow{(1\ 1)} Y \right)$$

(see [22, VIII.2] or [1, A.5]).

We call *semi-additive* a category with finite direct sums which is enriched in abelian monoids (in the unique way possible, as above).

A functor $F \colon \mathcal{C} \to \mathcal{D}$ is $\Bbbk$-*linear* if each component $F_{X,Y} \colon \mathcal{C}(X, Y) \to \mathcal{D}(FX, FY)$ is a $\Bbbk$-linear map. Similarly, if $\mathcal{C}$ and $\mathcal{D}$ are (only) enriched in abelian monoids, a functor $F \colon \mathcal{C} \to \mathcal{D}$ is said to be *additive* if it preserves the addition and zero element of each Hom monoid. Note that additive functors (hence in particular $\Bbbk$-linear ones) always preserve direct sum diagrams when they exist.

We will denote by $\mathrm{Fun}_{\Bbbk}(\mathcal{C}, \mathcal{D})$ the category of $\Bbbk$-linear functors $\mathcal{C} \to \mathcal{D}$ and natural transformations. Similarly, if $\mathcal{C}, \mathcal{D}$ are (only) enriched in abelian monoids, we will denote by $\mathrm{Fun}_{+}(\mathcal{C}, \mathcal{D})$ the category of additive functors between them.

Construction 2.24 (The $\Bbbk$-linearization) Let $\mathcal{C}$ be a category enriched in additive monoids. Its $\Bbbk$-*linearization* is the $\Bbbk$-linear category

$$\Bbbk\mathcal{C}$$

defined as follows. Its objects are the same, $\mathrm{Obj}(\Bbbk\mathcal{C}) := \mathrm{Obj}(\mathcal{C})$. Its Hom $\Bbbk$-modules $\Bbbk\mathcal{C}(X, Y)$ are obtained by first building the Grothendieck group completion $\mathcal{C}(X, Y)^{\pm}$ (the abelian group of formal differences) and then extending scalars: $\Bbbk\mathcal{C}(X, Y) := \Bbbk \otimes_{\mathbb{Z}} \mathcal{C}(X, Y)^{\pm}$. The composition maps of $\Bbbk\mathcal{C}$ are the unique $\Bbbk$-bilinear maps extending the composition maps of $\mathcal{C}$ along the canonical maps $\mathcal{C}(X, Y) \to \Bbbk\mathcal{C}(X, Y)$, $f \mapsto 1 \otimes (f - 0)$.

The latter canonical maps also define an evident functor $\mathcal{C} \to \Bbbk\mathcal{C}$ having the universal property that it induces, by precomposition, an isomorphism of categories

$$\mathrm{Fun}_\Bbbk(\Bbbk\mathcal{C}, \mathcal{D}) \xrightarrow{\sim} \mathrm{Fun}_+(\mathcal{C}, \mathcal{D})$$

for any given $\Bbbk$-linear category $\mathcal{D}$.

Clearly, if $\mathcal{C}$ has direct sums (*i.e.* is semi-additive) then these remain direct sums in the $\Bbbk$-linearization $\Bbbk\mathcal{C}$, so that the latter is an *additive* $\Bbbk$-linear category.

Remark 2.25   Note that the above construciton is not the same as the *free* $\Bbbk$-linearization, which can be performed on any category by taking the free $\Bbbk$-module on each Hom set. It is in fact important for us that we remember the existing (semi-) additive structure already present on our categories of spans and bisets.

2.3 Mackey functors for (2,1)-categories

As announced in the introduction, we want to define a Mackey functor to be a linear functor defined on a suitable category of spans of groupoids. Moreover, we wish to allow variations with sub- and comma-2-categories. A natural setting for constructing categories of spans, and covering all such cases, is that of an *extensive (2,1)-category* $\mathbb{G}$ with enough Mackey squares and equipped with a distinguished sub-2-category $\mathbb{J}$ with suitable closure properties (what we call a *spannable pair*; see Definition 2.31). As the next few pages may appear a little abstract, we urge the reader to keep in mind the example $\mathbb{G} = \mathbb{J} = \mathsf{gpd}$ of all finite groupoids, where Mackey squares are provided by the concrete iso-comma construction of Example 2.17, which allow us to check all the following claims by direct computations. This particular example will also be revisited in the next section in much detail. But for now, let us bask in some glorious generality:

Definition 2.26   A *(2,1)-category* is[3] a strict 2-category, as in § 2.6, where moreover all 2-morphisms are invertible.

---

[3] Roughly speaking, the general pattern in higher category theory is that an (n, k)-*category* has morphisms between morphisms between morphisms etc. up to level n, but only those up to level k are allowed to be non-invertible. Moreover, n-*category* is short for (n, n)-category (with the notable exception that Lurie [21] calls ∞-*categories* his models of $(\infty, 1)$-categories).

Definition 2.27 ([6])   A 2-category $\mathcal{E}$ is *extensive* if it admits all finite coproducts (see §2.23) and if moreover, for any pair of objects X, Y the pseudo-functor

$$\mathcal{E}_{/X} \times \mathcal{E}_{/Y} \xrightarrow{\sim} \mathcal{E}_{/X \sqcup Y}, \quad (A \xrightarrow{a} X, B \xrightarrow{b} Y) \mapsto (A \sqcup B \xrightarrow{a \sqcup b} X \sqcup Y)$$
$$(2.3.1)$$

induced on comma 2-categories (see §2.20 and Remark 2.22) by taking coproducts is a biequivalence (see §2.12). By [6, Thm. 2.3], a 2-category with finite coproducts is extensive if and only if: 1) it admits Mackey squares along all coproduct inclusions, and 2) it has the property that in any diagram of the form

$$
\begin{array}{ccccc}
T_A & \longrightarrow & T & \longleftarrow & T_B \\
a \downarrow & \simeq & c \downarrow & \simeq & \downarrow b \\
A & \xrightarrow{i_A} & A \sqcup B & \xleftarrow{i_B} & B
\end{array}
\qquad (2.3.2)
$$

the top row is a coproduct (*i.e.* $T_A \sqcup T_B \to T$ is an equivalence) iff the two squares are Mackey squares. (Beware that in [6] Mackey squares are simply called 'pullbacks' and the invertible 2-cells are omitted from all such diagrams.)

Remark 2.28   The notion of extensive category was introduced in [7] to capture, at the categorical level, the intuition of coproducts behaving like set-theoretical disjoint unions. A 1-category is *extensive* if it admits all finite coproducts and if the 1-categorical analogues of (2.3.1) are equivalences (or equivalently: if it is extensive when seen as a locally discrete 2-category). Basic examples include elementary toposes such as set or G-set. This definition was extended to 2- (and bi-) categories in [6], with small categories Cat and (finite) groupoids gpd providing basic examples. The definition pins down a certain compatibility between coproducts and (weak) pullbacks, which is what we need here in order to form nice categories of spans (*cf.* also [26]).

More precisely, we need the following two lemmas:

Lemma 2.29   Let $\mathcal{E}$ be an extensive 2-category (Definition 2.27). Then squares of the form

$$(2.3.3)$$

are Mackey squares for all 1-morphisms u, v.

Proof This is immediate from the characterization of extensive 2-categories recalled in Definition 2.27. □

Lemma 2.30 *In an extensive 2-category $\mathcal{E}$, if the two squares on the left are Mackey squares*

$$(2.3.4)$$

then so is the induced square on the right.

Proof Consider two Mackey squares as in (2.3.4). By Remark 2.18, we must show that the functor into the iso-comma category

$$\mathcal{E}(T, P_1 \sqcup P_2) \longrightarrow \big(\mathcal{E}(T, (u_1, u_2)) / \mathcal{E}(T, w) \big) \qquad (2.3.5)$$

$$t \longmapsto \big((p_1 \sqcup p_2)t, (q_1, q_2)t, (\gamma_1, \gamma_2)t \big)$$

is an equivalence for every T. Let (x, y, φ) be any object in the target, that is $x \colon T \to X_1 \sqcup X_2$ and $y \colon T \to Y$ are 1-cells and $\varphi \colon (u_1, u_2)x \Rightarrow wy$ is a 2-cell in $\mathcal{E}$. By the extensivity of $\mathcal{E}$, there exists a decomposition $T_1 \sqcup T_2 \xrightarrow{\sim} T$ which identifies x with the coproduct of some $x_1 \colon T_1 \to X_1$ and $x_2 \colon T_2 \to X_2$. By precomposing with the canonical inclusions $T_\ell \to T$ ($\ell = 1, 2$), we can also write y and φ in their two components $y_\ell \colon T_\ell \to Y$ and

respectively. Using the two given Mackey squares on $\xrightarrow{u_\ell} \xleftarrow{w}$, the above

2-cells φ_ℓ can be written as pastings of the form

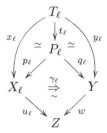

for some $t_\ell\colon T_\ell \to P_\ell$ (for $\ell = 1, 2$). Let $t := (T \simeq T_1 \sqcup T_2 \xrightarrow{t_1 \sqcup t_2} P_1 \sqcup P_2)$. By construction, the image of t under the functor (2.3.5) is isomorphic to the given triple (x, y, φ). This shows that (2.3.5) is essentially surjective.

A similar extensivity argument for 2-cells $\alpha\colon t \Rightarrow t'$ shows fully faithfulness, whence the desired equivalence. The remaining details are straightforward and are left to the reader. $\qquad\square$

Definition 2.31 (Spannable pair)   We call *spannable pair* a pair $(\mathbb{G}; \mathbb{J})$ where:

(1) $\mathbb{G}$ is a $(2, 1)$-category (Definition 2.26) which we assume essentially small, *i.e.* the equivalence classes of objects form a set and every Hom category is small;
(2) $\mathbb{G}$ is also an extensive 2-category (Definition 2.27); and
(3) $\mathbb{J}$ is a distinguished class of 1-cells of $\mathbb{G}$;

and the pair $(\mathbb{G}; \mathbb{J})$ satisfies the following three axioms:

(a) The class $\mathbb{J}$ contains the equivalences of $\mathbb{G}$ (see § 2.12) and is closed under horizontal composition and under taking isomorphic 1-cells. In particular, we may identify $\mathbb{J}$ with the corresponding *2-full* sub-2-category of $\mathbb{G}$ (*i.e.* if $\alpha\colon u \Rightarrow v$ is a 2-cell of $\mathbb{G}$ with u, v in $\mathbb{J}$ then α is also in the sub-2-category $\mathbb{J}$).
(b) For any 1-cells $X \xrightarrow{i} Z \xleftarrow{u} Y$ with $i \in \mathbb{J}$, there exists in $\mathbb{G}$ a Mackey square

(see § 2.16), and moreover $q \in \mathbb{J}$.

(c) For any finite set $\{u_\ell : X_\ell \to Y\}_\ell$ of 1-cells of $\mathbb{G}$ with common target, the 1-cell $(u_\ell)_\ell : \coprod_\ell X_\ell \to Y$ is in $\mathbb{J}$ iff each u_ℓ is.

In the special case where $\mathbb{G} = \mathbb{J}$ we write $\mathbb{G} := (\mathbb{G}; \mathbb{J})$ for short and call $\mathbb{G}$ a *spannable (2,1)-category*.

Remark 2.32   By (a) and (c), the canonical 1-cells

$$X \xrightarrow{i_X} X \sqcup Y \xleftarrow{i_Y} Y$$

of a coproduct belong to $\mathbb{J}$, because the identity $\mathrm{Id}_{X \sqcup Y} = (i_X, i_Y)$ does. Similarly, for every X the unique 1-cell $\varnothing \to X$ from the initial object (= empty coproduct) is in $\mathbb{J}$, because it factors as the composite

$$\varnothing \xrightarrow{i_\varnothing} \varnothing \sqcup X \simeq X$$

of two 1-cells in $\mathbb{J}$.

Remark 2.33   Notice that a spannable 2-category $\mathbb{G}$ (*i.e.* the case of a spannable pair with $\mathbb{J} = \mathbb{G}$, which will cover most of our explicit examples) is precisely the same thing as an essentially small extensive (2,1)-category admitting arbitrary Mackey squares. Indeed, the closure properties (a)–(c) are then automatically satisfied.

Construction 2.34 (The category of spans $\mathrm{Sp}(\mathbb{G}; \mathbb{J})$)   Let $(\mathbb{G}; \mathbb{J})$ be a spannable pair as in Definition 2.31. We construct a category

$$\mathrm{Sp}(\mathbb{G}; \mathbb{J})$$

whose objects are the same as those of $\mathbb{G}$, and where a morphism $X \to Y$ is the equivalence class of a span of 1-morphisms of $\mathbb{G}$

$$
\begin{array}{ccc}
 & S & \\
{}^{u}\nearrow & & \searrow{}^{i \in \mathbb{J}} \\
X & \cdots\cdots\cdots\cdots\!\!\!\!\longrightarrow & Y
\end{array}
$$

where the 'forward' one belongs to $\mathbb{J}$. Two spans $X \xleftarrow{u} S \xrightarrow{i} Y$ and $X \xleftarrow{u'} S' \xrightarrow{i'} Y$ are *equivalent* iff there exist in $\mathbb{G}$ a diagram

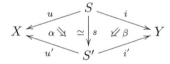

where $s : S \xrightarrow{\sim} S'$ is an equivalence and $\alpha : u \Rightarrow u's$ and $\beta : i \Rightarrow i's$ are 2-morphisms (which are invertible, this being a (2,1)-category). The composition of two (equivalence classes of) spans $[X \xleftarrow{u} S \xrightarrow{i} Y]$ and

$[Y \xleftarrow{v} T \xrightarrow{j} Z]$ is the span $[X \xleftarrow{up} P \xrightarrow{jq} Z]$ which is obtained by constructing a Mackey square in the middle:

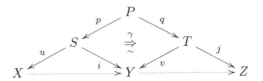

(Note that such a Mackey square exists and $jq \in \mathbb{J}$ by the hypotheses (a) and (b) on a spannable pair.) The identity morphism of an object X is given by the span $[X \xleftarrow{\mathrm{Id}} X \xrightarrow{\mathrm{Id}} X]$.

Proposition 2.35 *Construction 2.34 yields a well-defined category* $\mathrm{Sp}(\mathbb{G}; \mathbb{J})$, *which moreover is semi-additive (see § 2.23). Explicitly, the zero object is given by the initial object $\varnothing$ (empty coproduct) of $\mathbb{G}$ and the direct sum diagram for two objects X, Y is given by the four spans (i.e. two spans considered in both directions)*

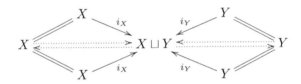

where $X \xrightarrow{i_X} X \sqcup Y \xleftarrow{i_Y} Y$ is the coproduct of X and Y in $\mathbb{G}$. The spans

and

provide the zero map $X \to Y$ and the sum of two maps $[X \xleftarrow{u_\ell} S_\ell \xrightarrow{i_\ell} Y]$, $\ell \in \{1, 2\}$.

Proof These are rather straightforward verifications, as follows.

To see that the composition of spans is well-defined, associative and unital, one must repeatedly employ, in a routine way, the 2-universal property of Mackey squares and the closure properties (a) and (b) of the spannable pair $(\mathbb{G}; \mathbb{J})$.

Thanks to Remark 2.32 and property (c), we may form the zero span and the sum of two spans as indicated. The universal property of coproducts ensures that this operation is associative and unital on each Hom set. To verify that the composition of $\mathrm{Sp}(\mathbb{G}; \mathbb{J})$ preserves zero spans, it suffices to notice that the squares of the form

are Mackey squares by Lemma 2.29 (set $A = X = \varnothing$ in the first one and $B = Y = \varnothing$ in the second one). Similarly, composition preserves sums of spans because the sum of two Mackey squares, as in (2.3.4) (and also its left-right mirror version), is again a Mackey square by Lemma 2.30. Thus $\mathrm{Sp}(\mathbb{G}; \mathbb{J})$ is enriched in abelian monoids.

Finally, to verify the four equations for the claimed biproducts we can use that

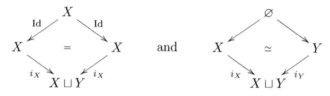

are Mackey squares, which again follows from Lemma 2.29, this time by specializing the two squares in (2.3.3) to $u = \mathrm{Id}_X$ and $B = \varnothing$. Note that the four spans comprising a biproduct diagram are all permissible by Remark 2.32.   □

Remark 2.36 The category $\mathrm{Sp}(\mathbb{G}; \mathbb{J})$ of Construction 2.34 is only the shadow of a richer structure. Indeed, it is precisely the 1-truncation (see § 2.13)

$$\mathrm{Sp}(\mathbb{G}; \mathbb{J}) = \tau_1(\mathsf{Span}(\mathbb{G}; \mathbb{J}))$$

of a *bicategory* of spans $\mathsf{Span}(\mathbb{G}; \mathbb{J})$. This is explained in [1, Ch. 4] in all details, for the situation where the necessary Mackey squares of $\mathbb{G}$ are actually iso-comma squares; the additive aspects are covered in [1, Ch. 7] but only for the example $\mathbb{G} = \mathsf{gpd}$ of groupoids. It is straightforward to generalize the construction of $\mathsf{Span}(\mathbb{G}; \mathbb{J})$ to any spannable pair $(\mathbb{G}; \mathbb{J})$ as in Definition 2.31, but it involves some choices (namely, in order to obtain specified composition functors one must choose a pseudo-inverse for each of the equivalences (2.2.4)).

Remark 2.37 There is an alternative way of describing the category $\mathrm{Sp}(\mathbb{G}; \mathbb{J})$, where the operations of forming spans and 1-truncating are permuted. In this other picture, we begin by forming the truncated (ordinary) category $\tau_1 G$, where a morphism is an isomorphism class $[u]$ of 1-morphisms in G. Then we consider spans in $\tau_1 G$, that is diagrams

$$X \xleftarrow{[u]} S \xrightarrow{[i]} Y$$

with $i \in J$, and we identify isomorphic spans, where two spans $([u], [i])$ and $([u'], [i'])$ are *isomorphic* if there exists a commutative diagram

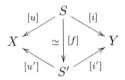

in $\tau_1 G$ with $[f]$ invertible. Note however that, in general, one cannot define the composition of $\mathrm{Sp}(G; J)$ purely in terms of the ordinary category $\tau_1 G$, because one would still need to remember the (images in $\tau_1 G$ of the) Mackey squares of G. (Sometimes said images admit an intrinsic characterization in $\tau_1 G$, but not always.)

Definition 2.38 (The $\Bbbk$-linear category of spans, $\mathrm{Sp}_\Bbbk(G; J)$) We define the $\Bbbk$-*linear category of spans in* G *with respect to* J to be the $\Bbbk$-linearization (Construction 2.24) of the semi-additive category $\mathrm{Sp}(G; J)$ of Proposition 2.35:

$$\mathrm{Sp}_\Bbbk(G; J) := \Bbbk\big(\mathrm{Sp}(G; J)\big).$$

Thus the objects remain the same and the morphisms of $\mathrm{Sp}_\Bbbk(G; J)$ are formal $\Bbbk$-linear combinations of morphisms of $\mathrm{Sp}(G; J)$, in a way that preserves sums.

The span construction is functorial in the following evident sense:

Proposition 2.39 *Let* $\mathcal{F}\colon (G; J) \to (G'; J')$ *be a* morphism of spannable pairs, *by which we mean a pseudo-functor (e.g. a 2-functor)* $\mathcal{F}\colon G \to G'$ *preserving (up to equivalence) finite coproducts and Mackey squares and such that* $\mathcal{F}(J) \subseteq J'$. *Then* $\mathcal{F}$ *induces a* $\Bbbk$-*linear functor on the span categories of Definition 2.38*

$$\mathrm{Sp}_\Bbbk(\mathcal{F})\colon \mathrm{Sp}_\Bbbk(G; J) \longrightarrow \mathrm{Sp}_\Bbbk(G'; J')$$

sending $[X \xleftarrow{u} S \xrightarrow{i} Y]$ *to* $[\mathcal{F}X \xleftarrow{\mathcal{F}u} \mathcal{F}S \xrightarrow{\mathcal{F}i} \mathcal{F}Y]$. *Moreover*

$$\mathrm{Sp}_\Bbbk(\mathcal{F}_2 \circ \mathcal{F}_1) = \mathrm{Sp}_\Bbbk(\mathcal{F}_2) \circ \mathrm{Sp}_\Bbbk(\mathcal{F}_1) \quad and \quad \mathrm{Sp}_\Bbbk(\mathrm{Id}_{(G;J)}) = \mathrm{Id}_{\mathrm{Sp}_\Bbbk(G;J)}$$

and if $\mathcal{F}$ is a biequivalence then $\mathrm{Sp}_{\Bbbk}(\mathcal{F})$ is an equivalence of $\Bbbk$-linear categories.

Proof All claims are immediate from the constructions.   □

Under a Krull-Schmidt-type finiteness hypothesis on $\mathbb{G}$, easily checked in all our examples, the Hom modules of $\mathrm{Sp}_{\Bbbk}(\mathbb{G}; \mathcal{J})$ become particularly nice:

Proposition 2.40 *Let $(\mathbb{G}; \mathcal{J})$ be a spannable pair and consider $\mathrm{Sp}_{\Bbbk}(\mathbb{G}; \mathcal{J})$, the associated $\Bbbk$-linear category of spans. Then:*

(1) The Hom $\Bbbk$-modules of $\mathrm{Sp}_{\Bbbk}(\mathbb{G}; \mathcal{J})$ are all free, provided this holds: For every object X of $\mathbb{G}$ there exists an equivalence $X \simeq X_1 \sqcup \ldots \sqcup X_n$ to a coproduct of finitely many objects X_i which are indecomposable with respect to coproducts; moreover, if $X \simeq X'_1 \sqcup \ldots \sqcup X'_m$ is another such decomposition, then $n = m$ and there exist a permutation $\sigma \in \Sigma_n$ and equivalences $X'_i \simeq X_{\sigma(i)}$.

(2) Assuming the hypothesis of (1) holds, the Hom $\Bbbk$-modules of $\mathrm{Sp}_{\Bbbk}(\mathbb{G}; \mathcal{J})$ are all finitely generated (and free), provided that for every object X the comma category $(\tau_1 \mathbb{G})/X = \tau_1(\mathbb{G}/X)$ only has finitely many $\sqcup$-indecomposable objects.

Proof Notice that it suffices to show the claims hold in the special case $\Bbbk = \mathbb{Z}$, as the general case will immediately follow by extending scalars.

Assume the hypothesis of (1), and fix two objects X, Y. Choose a full set of representatives

$$s_\ell := \left[X \xleftarrow{u_\ell} S_\ell \xrightarrow{i_\ell} Y \right], \quad \ell \in \Lambda$$

for those spans in $\mathrm{Sp}(\mathbb{G}; \mathcal{J})(X, Y)$ whose middle object S_ℓ is indecomposable (the latter property is invariant under equivalence of spans). We claim that $\{s_\ell\}_{\ell \in \Lambda}$ is a basis of $\mathrm{Sp}_{\mathbb{Z}}(\mathbb{G}; \mathcal{J})(X, Y)$.

By hypothesis (1), we may decompose the middle object of every span between X and Y into a coproduct of indecomposables, which induces a sum-decomposition of the span. Hence the s_ℓ generate the abelian monoid $\mathrm{Sp}(\mathbb{G}; \mathcal{J})(X, Y)$.

To show they are linearly independent, we first prove that the abelian monoid $\mathrm{Sp}(\mathbb{G}; \mathcal{J})(X, Y)$ is cancellative, that is: if s, s', t are elements such that $s + t = s' + t$ then $s = s'$. Thus consider three parallel spans

$$s = \left[X \xleftarrow{u} S \xrightarrow{i} Y \right],$$

$$s' = \left[X \xleftarrow{u'} S' \xrightarrow{i'} Y \right],$$

$$t = \left[X \xleftarrow{v} T \xrightarrow{j} Y \right]$$

and suppose we have an equivalence

$$X \xleftarrow[(u',v)]{(u,v)} \begin{matrix} S \sqcup T \\ \alpha \searrow \quad \simeq \downarrow f \quad \not\Leftarrow \beta \\ S' \sqcup T \end{matrix} \xrightarrow[(i',j)]{(i,j)} Y \; . \tag{2.3.6}$$

Now let us decompose the spans $s+t$ and $s'+t$ as sums of indecomposable spans, using the representatives chosen above. In particular, we get decompositions

$$S \simeq \coprod_\ell S_\ell^{\sqcup n_\ell}, \quad S' \simeq \coprod_\ell S_\ell^{\sqcup n'_\ell} \text{ and } \quad T \simeq \coprod_\ell S_\ell^{\sqcup m_\ell}$$

where the n_ℓ, n'_ℓ, m_ℓ are non-negative integers, almost all of which are zero. By the extensivity of $\mathbb{G}$, the equivalence f must be a coproduct of 1-cells between these factors. By the uniqueness part of hypothesis (1), and since we picked one S_ℓ per equivalence class, we see that f must be *diagonal* with respect to them.

Now, it is *a priori* possible that, for a fixed ℓ, the equivalence f matches some of the identical factors S_ℓ within S and T or S' and T, but after composing f if necessary with a self-equivalence of the target which permutes said factors, we may assume that it has the form $f \simeq f_1 \sqcup f_2$ for two equivalences $f_1 \colon S \overset{\sim}{\to} S'$ and $f_2 \colon T \overset{\sim}{\to} T$. (Note also that two different $s_\ell \neq s_k$ may have equivalent middle objects $S_\ell \simeq S_k$, but f cannot match two such middle objects because then, by (2.3.6), f together with (suitable components of) α and β would yield an equivalence $s_\ell = s_k$; but this can only hold if $\ell = k$.) Then $\alpha = (\alpha_1, \alpha_2)$ and $\beta = (\beta_1, \beta_2)$ must decompose accordingly. In particular, we obtain from (2.3.6) an equivalence

$$X \xleftarrow[u']{u} \begin{matrix} S \\ \alpha_1 \searrow \quad \simeq \downarrow f_1 \quad \not\Leftarrow \beta_1 \\ S' \end{matrix} \xrightarrow[i']{i} Y$$

showing that $s = s'$ in $\mathrm{Sp}(\mathbb{G}; \mathbb{J})(X, Y)$, as claimed.

Next, we show that the set $\{s_\ell\}_\ell$ is a basis of the abelian monoid $\mathrm{Sp}(\mathsf{G}; \mathsf{J})(X, Y)$ (*i.e.* every element can be written as a sum of generators in a unique way). It will then easily follow from cancellativity that it is also a basis for the $\mathbb{Z}$-module $\mathrm{Sp}(\mathsf{G}; \mathsf{J})(X, Y)$. Consider two finite sums yielding the same element:

$$\sum_\ell n_\ell s_\ell = \sum_\ell m_\ell s_\ell$$

with $n_\ell, m_\ell \in \mathbb{N}$ (and almost all zero). This means that there is an equivalence

$$
X \xleftarrow[(u_\ell)]{(u_\ell)} \underset{\coprod_\ell S_\ell^{\sqcup m_\ell}}{\overset{\coprod_\ell S_\ell^{\sqcup n_\ell}}{\underset{\simeq\,\Big\downarrow f}{\alpha\searrow\quad \nearrow\beta}}} \xrightarrow[(i_\ell)_\ell]{(i_\ell)_\ell} Y
$$

of spans. By extensivity and (1) as before, the equivalence f down the middle must decompose diagonally as a coproduct of equivalences $f_\ell\colon S_\ell^{\sqcup n_\ell} \overset{\sim}{\to} S_\ell^{\sqcup m_\ell}$ at each ℓ such that $\alpha\colon u_\ell \cong u_\ell f_\ell$ and $\beta\colon i_\ell \cong i_\ell f_\ell$. Moreover, each of these must match the factors one-on-one, since the S_ℓ are indecomposable. Hence we must have $n_\ell = m_\ell$ for all ℓ, as claimed.

For part (2), the finiteness of the basis $\{s_\ell\}_\ell$ in an easy consequence of the hypothesis that each comma category $(\tau_1 \mathsf{G})_{/X}$ (or equivalently each comma 2-category $\mathsf{G}_{/X}$) only has finitely many indecomposable objects.   □

Definition 2.41 (Mackey functor; *cf.* [1, Def. 2.5.4])   Let $\mathrm{Sp}_\Bbbk(\mathsf{G}; \mathsf{J})$ be the category of spans of Construction 2.34 for some spannable pair $(\mathsf{G}; \mathsf{J})$. A $\Bbbk$-*linear Mackey functor for* G *with respect to* J is defined to be a $\Bbbk$-linear (hence also additive, *i.e.* direct-sum preserving) functor

$$M\colon \mathrm{Sp}_\Bbbk(\mathsf{G}; \mathsf{J}) \longrightarrow \Bbbk\text{-Mod}$$

to the abelian category of $\Bbbk$-modules, and a morphism of Mackey functors is simply a natural transformation. We denote by

$$\mathrm{Mack}_\Bbbk(\mathsf{G}; \mathsf{J}) := \mathrm{Fun}_\Bbbk(\mathrm{Sp}_\Bbbk(\mathsf{G}; \mathsf{J}), \Bbbk\text{-Mod})$$

the resulting $\Bbbk$-linear category. By the general properties of $\Bbbk$-linearization, we have a canonical additive functor $\mathrm{Bis}(\mathbf{gpd}) \to \mathrm{Bis}_\Bbbk(\mathbf{gpd})$ which induces an isomorphism

$$\mathrm{Mack}_\Bbbk(\mathsf{G}; \mathsf{J}) \overset{\sim}{\longrightarrow} \mathrm{Fun}_+(\mathrm{Sp}(\mathsf{G}; \mathsf{J}), \Bbbk\text{-Mod}) \qquad (2.3.7)$$

of $\Bbbk$-linear categories, identifying additive functors on $\mathrm{Sp}(\mathbb{G}; \mathbb{J})$ and $\Bbbk$-linear functors on $\mathrm{Sp}_\Bbbk(\mathbb{G}; \mathbb{J})$.

Remark 2.42 Explicitly, a Mackey functor for $(\mathbb{G}; \mathbb{J})$ as in Definition 2.41 (and via (2.3.7)) consists of a $\Bbbk$-module $M(X)$ for every object X of $\mathbb{G}$, a $\Bbbk$-homomorphism $u^* \colon M(Y) \to M(X)$ for every 1-morphism $u \in \mathbb{G}(X, Y)$ (corresponding to the image of the span $[Y \overset{u}{\leftarrow} X = X]$) and a $\Bbbk$-homomorphism $u_* \colon M(X) \to M(Y)$ if furthermore $u \in \mathbb{J}$ (the image of the span $[X = X \overset{u}{\to} Y]$). This data is subject to the following rules:

(1) *Functoriality:* $(\mathrm{Id}_X)^* = (\mathrm{Id}_X)_* = \mathrm{id}_{M(X)}$ and $(uv)^* = v^* u^*$ for all $X \in \mathbb{G}$ and all composable 1-cells u, v; and also $(uv)_* = u_* v_*$ when u, v belong to $\mathbb{J}$.

(2) *Isomorphism invariance:* If two 1-cells $u \simeq v$ are isomorphic in the category $\mathbb{G}(X, Y)$ then $u^* = v^*$, and also $u_* = v_*$ whenever u, v belong to $\mathbb{J}$.

(3) *Additivity:* The canonical morphism $M(X_1 \sqcup X_2) \overset{\sim}{\to} M(X_1) \oplus M(X_2)$ is an isomorphism for all $X_1, X_2 \in \mathbb{G}$, as well as $M(\emptyset) \overset{\sim}{\to} 0$.

(4) *Mackey formula:* For every Mackey square in $\mathbb{G}$ with i (and thus also j) in $\mathbb{J}$

we have $u^* i_* = j_* v^* \colon M(X) \to M(Y)$.

Remark 2.43 Consistently with the previous remark, we may use the short-hand notation

$$u^* := [Y \overset{u}{\leftarrow} X = X], \qquad i_* := [Z = Z \overset{i}{\to} W]$$

for the 'contravariant' and 'covariant' spans associated to all 1-cells $u \in \mathbb{G}$ and $i \in \mathbb{J}$. Note that every span can be written as $[\overset{u}{\leftarrow} \overset{i}{\to}] = i_* u^*$. This defines two faithful functors

$$(-)^* \colon \tau_1 \mathbb{G}^{\mathrm{op}} \longrightarrow \mathrm{Sp}(\mathbb{G}; \mathbb{J}), \qquad (-)_* \colon \tau_1 \mathbb{J} \longrightarrow \mathrm{Sp}(\mathbb{G}; \mathbb{J})$$

which can be jointly used to formulate a simple universal property for the categories $\mathrm{Sp}(\mathbb{G}; \mathbb{J})$ and $\mathrm{Sp}_\Bbbk(\mathbb{G}; \mathbb{J})$; see *e.g.* [1, A.4].

Here is the list of all examples of Mackey functors over spannable pairs $(\mathbb{G}; \mathbb{J})$ that will be considered in this chapter.

Example 2.44 Our first example is $\mathbb{G} = \mathbb{J} = $ gpd, the 2-category of all finite groupoids, functors and natural isomorphisms. This is a spannable (2,1)-category, as one verifies easily (*cf.* Remark 2.33). In this case, Definition 2.41 specializes to the notion of (*global*) *Mackey functor* studied for example in [11] and [25]. We fully investigate the span category $\mathrm{Sp}_{\Bbbk}($gpd$)$ in Section 2.4.

The next two examples can also be understood as kinds of biset functors, as will be proved in Corollary 2.93.

Example 2.45 Similarly, we may consider $\mathbb{G} = \mathbb{J} = $ gpd^{f}, the 2-category of all finite groupoids, *faithful* functors, and natural isomorphisms. The resulting Mackey functors are also sometimes called *global Mackey functors* (*cf.* [4, Ex. 3.2.6]).

Example 2.46 As a proper example of a spannable *pair*, consider $\mathbb{G} = $ gpd and $\mathbb{J} = $ gpd^{f}. The closure properties (a)-(c) are easily checked. The resulting Mackey functors are sometimes called *inflation functors* (*cf.* [4, Ex. 3.2.5]).

Example 2.47 For a fixed finite group G, we may consider the comma 2-category $\mathbb{G} = \mathbb{J} = $ gpd$^{\mathsf{f}}_{/G}$ of gpd^{f} over G. The resulting Mackey functors are the classical *Mackey functors for* G. See Section 2.5.

Example 2.48 Again for a fixed G, we will consider a variant gpd$^{\mathsf{f},\mathsf{fus}}_{/G}$ of the comma 2-category of Example 2.47, where we add all 2-morphisms which exist after forgetting to gpd. This imposes extra relations on the associated notion of Mackey functors, which turn out to be Bouc's *fused Mackey functors for* G. See Section 2.5.

Remark 2.49 Many more variations are possible and useful. In particular, we may also want to restrict the class of *objects* of $\mathbb{G}$. For example, we may consider spans (of various kinds) only between groupoids consisting of coproducts of finite p-groups, or coproducts of subquotients of a single fixed group G (*cf.* [31, § 9] and [4, Part III]).

2.4 Presenting spans of groupoids

In this section we take a hands-on approach to the central object of this chapter, the category of spans of groupoids. In particular, we provide a presentation by generators and relations as a linear category. This presentation should look familiar to all practitioners of Mackey and biset functors, and indeed will be useful for the comparison results of the last section. The reader who is easily bored by long lists of relations may read Remark 2.55 and then skip ahead to Section 2.5. However our intention is not to be soporific, but rather to demonstrate how the usual 'combinatorial' approach to Mackey and biset functors works just as well with spans of groupoids.

Fix a commutative ring $\Bbbk$. The category we are interested in is

$$\mathrm{Sp}_{\Bbbk}(\mathsf{gpd}),$$

the $\Bbbk$-linear category of spans in finite groupoids, as defined in Definition 2.38.

Remark 2.50 Definition 2.38 can be applied with $\mathbb{G} := \mathsf{gpd}$ because the latter is a spannable (2,1)-category in the sense of Definition 2.31, as one can easily verify directly. In particular, gpd has well-behaved (strict) finite coproducts given by disjoint unions, and arbitrary (strict) Mackey squares given by the iso-comma construction of Example 2.17. Moreover gpd satisfies the finiteness hypotheses of Proposition 2.40, so that the Hom $\Bbbk$-modules of $\mathrm{Sp}_{\Bbbk}(\mathsf{gpd})$ are all finitely generated and free. Let us quickly recall from the previous section that the objects of $\mathrm{Sp}_{\Bbbk}(\mathsf{gpd})$ are the finite groupoids, and a morphism $\varphi \colon G \to H$ is a formal $\Bbbk$-linear combination of equivalence classes of spans

$$[G \xleftarrow{a} P \xrightarrow{b} H],$$

where a, b are any two functors whose common source is a *connected* groupoid P. In practice, we also use non-connected P but then we must identify $[G \xleftarrow{a} P \xrightarrow{b} H]$ with the sum $\sum_i [G \xleftarrow{a_i} P_i \xrightarrow{b_i} H]$, where $P = \coprod_i P_i$ is any disjoint union decomposition and a_i, b_i are the corresponding components of a, b. Here two spans $G \xleftarrow{a} P \xrightarrow{b} H$ and $G \xleftarrow{a'} P' \xrightarrow{b'} H$ are *equivalent* iff there exists an equivalence $f \colon P \xrightarrow{\sim} P'$ making the diagram

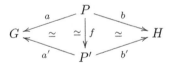

commute up to isomorphisms of functors $a'f \simeq a$, $b'f \simeq b$. Disjoint unions provide the direct sums of objects as well as, when taken at the middle object, the additive structure of each Hom module (see Proposition 2.35). Composition in $\mathrm{Sp}_{\Bbbk}(\mathsf{gpd})$ is induced $\Bbbk$-bilinearly from the composition of two spans $[G \xleftarrow{a} P \xrightarrow{b} H]$ and $[H \xleftarrow{c} Q \xrightarrow{d} K]$ obtained by constructing the iso-comma square (b/c). The groupoid (b/c) will rarely be connected, hence the result of composing two spans is typically a *sum* of classes of connected spans.

Definition 2.51 (Global Mackey functor)   The $\Bbbk$-linear category of *global Mackey functors* is obtained by specializing Definition 2.41 to $\mathbb{G} = \mathbb{J} = \mathsf{gpd}$:

$$\mathrm{Mack}_\Bbbk(\mathsf{gpd}) := \mathrm{Fun}_\Bbbk(\mathrm{Sp}_\Bbbk(\mathsf{gpd}), \Bbbk\text{-Mod}).$$

This, in the form of Remark 2.42, is precisely the same definition used in [11] and is equivalent to the definition of Mackey functor used in [25] (via a biequivalence between gpd and Nakaoka's 2-category $\mathbb{S}$ of 'sets with variable finite group action'; see [24]).

Notation 2.52 (Elementary spans)   We introduce five families of maps of $\mathrm{Sp}_\Bbbk(\mathsf{gpd})$ which we call *elementary spans*. These spans are all connected (*i.e.* their middle object is connected) by virtue of only involving groups. Let $i\colon H \hookrightarrow G$ denote a subgroup inclusion, $p\colon G \twoheadrightarrow G/N$ a quotient homomorphism, and $f\colon G \overset{\sim}{\to} G'$ an isomorphism between finite groups. Then we write

$$\mathrm{Res}^G_H := i^* = [G \overset{i}{\hookleftarrow} H = H] \qquad \mathrm{Ind}^G_H := i_* = [H = H \overset{i}{\hookrightarrow} G]$$

$$\mathrm{Inf}^G_{G/N} := p^* = [G/N \overset{p}{\leftarrow} G = G] \quad \mathrm{Def}^G_{G/N} := p_* = [G = G \overset{p}{\twoheadrightarrow} G/N]$$

$$\mathrm{Iso}(f) := (f^{-1})^* = f_* = [G \overset{f^{-1}}{\longleftarrow} G' = G'] = [G = G \overset{f}{\to} G']$$

for the corresponding equivalence classes of spans in $\mathrm{Sp}_\Bbbk(\mathsf{gpd})$ and call them, respectively, *restrictions*, *inductions*, *deflations*, *inflations* and *isomorphisms*.

Theorem 2.53 (A presentation of $\mathrm{Sp}_\Bbbk(\mathsf{gpd})$)   *As an additive $\Bbbk$-linear category, $\mathrm{Sp}_\Bbbk(\mathsf{gpd})$ is generated by the elementary spans of Notation 2.52; this means that every object is a direct sum of finite groups, and that every morphism is a matrix of maps between groups, each of which can be obtained as a $\Bbbk$-linear combination of composites of elementary spans. Moreover, the ideal of all relations between the maps of $\mathrm{Sp}_\Bbbk(\mathsf{gpd})$ is generated by the following three families:*

0. Triviality relations:

  (a) For every group G:

$$\mathrm{Res}^G_G = \mathrm{id}_G, \quad \mathrm{Ind}^G_G = \mathrm{id}_G, \quad \mathrm{Def}^G_{G/1} = \mathrm{id}_G, \quad \mathrm{Inf}^G_{G/1} = \mathrm{id}_G.$$

(b) For every inner *automorphism* $f \colon G \xrightarrow{\sim} G$:

$$\mathrm{Iso}(f) = \mathrm{id}_G \,.$$

1. Transitivity relations:

(a) For all subgroups $K \leq H \leq G$:

$$\mathrm{Res}_K^H \circ \mathrm{Res}_H^G = \mathrm{Res}_K^G, \qquad \mathrm{Ind}_H^G \circ \mathrm{Ind}_K^H = \mathrm{Ind}_K^G \,.$$

(b) For any two composable isomorphisms $G \xrightarrow{f} G' \xrightarrow{f'} G''$:

$$\mathrm{Iso}(f') \circ \mathrm{Iso}(f) = \mathrm{Iso}(f' \circ f) \,.$$

(c) For any two normal subgroups $N, M \trianglelefteq G$ *with* $N \leq M$:

$$\mathrm{Inf}_{G/N}^G \circ \mathrm{Inf}_{G/M}^{G/N} = \mathrm{Inf}_{G/M}^G \,, \qquad \mathrm{Def}_{G/M}^{G/N} \circ \mathrm{Def}_{G/N}^G = \mathrm{Def}_{G/M}^G \,.$$

2. Commutativity relations:

(a) For a subgroup $K \leq G$ *and an isomorphism* $f \colon G \xrightarrow{\sim} G'$, *where* $\tilde{f} \colon K \xrightarrow{\sim} f(K)$ *denotes the isomorphism induced by* f:

$$\mathrm{Iso}(\tilde{f}) \circ \mathrm{Res}_K^G = \mathrm{Res}_{f(K)}^{G'} \circ \mathrm{Iso}(f) \,,$$
$$\mathrm{Iso}(f) \circ \mathrm{Ind}_K^G = \mathrm{Ind}_{f(K)}^{G'} \circ \mathrm{Iso}(\tilde{f}) \,.$$

(b) For a normal subgroup $N \trianglelefteq G$ *and an isomorphism* $f \colon G \xrightarrow{\sim} G'$, *where* $\overline{f} \colon G/N \xrightarrow{\sim} G'/f(N)$ *is the isomorphism induced by* f:

$$\mathrm{Iso}(\overline{f}) \circ \mathrm{Def}_{G/N}^G = \mathrm{Def}_{G'/f(N)}^{G'} \circ \mathrm{Iso}(f) \,,$$
$$\mathrm{Iso}(f) \circ \mathrm{Inf}_{G/N}^G = \mathrm{Inf}_{G'/f(N)}^{G'} \circ \mathrm{Iso}(\overline{f}) \,.$$

(c) (The Mackey formula.) For any two subgroups $H, K \leq G$, *where* $[H\backslash G/K]$ *denotes any full set of representatives of the double cosets* $H\backslash G/K$ *and* $c_x \colon H^x \cap K \xrightarrow{\sim} H \cap {}^x K$ *is the conjugation isomorphism* $g \mapsto {}^x g = xgx^{-1}$:

$$\mathrm{Res}_H^G \circ \mathrm{Ind}_K^G = \sum_{x \in [H\backslash G/K]} \mathrm{Ind}_{H \cap {}^x K}^H \circ \mathrm{Iso}(c_x) \circ \mathrm{Res}_{H^x \cap K}^K \,.$$

(d) For any two normal subgroups $M, N \trianglelefteq G$ *such that* $M \cap N = 1$ *we have:*

$$\mathrm{Def}_{G/N}^G \circ \mathrm{Inf}_{G/M}^G = \mathrm{Inf}_{G/MN}^{G/N} \circ \mathrm{Def}_{G/MN}^{G/M} \,.$$

(e) For a subgroup $H \leq G$ and a normal subgroup $N \trianglelefteq G$, where $f \colon H/H \cap N \xrightarrow{\sim} HN/N$ denotes the canonical isomorphism:

$$\mathrm{Def}^G_{G/N} \circ \mathrm{Ind}^G_H = \mathrm{Ind}^{G/N}_{HN/N} \circ \mathrm{Iso}(f) \circ \mathrm{Def}^H_{H/H\cap N},$$

$$\mathrm{Res}^G_H \circ \mathrm{Inf}^G_{G/N} = \mathrm{Inf}^H_{H/H\cap N} \circ \mathrm{Iso}(f^{-1}) \circ \mathrm{Res}^{G/N}_{HN/N}.$$

(f) For a subgroup $H \leq G$ and a normal subgroup $N \trianglelefteq G$ such that $N \leq H$:

$$\mathrm{Res}^{G/N}_{H/N} \circ \mathrm{Def}^G_{G/N} = \mathrm{Def}^H_{H/N} \circ \mathrm{Res}^G_H,$$

$$\mathrm{Ind}^G_H \circ \mathrm{Inf}^H_{H/N} = \mathrm{Inf}^G_{G/N} \circ \mathrm{Ind}^{G/N}_{H/N}.$$

Remark 2.54 As commonly done, we have identified quotients such as $G/1 = G$ in 0.(a) or $(G/M)/MN = G/MN$ in 2.(d), *i.e.* we have omitted some (very canonical) isomorphism maps from the presentation. This should not cause any problems.

Remark 2.55 The presentation in Theorem 2.53 is almost identical to the presentation provided in [4, § 1.1.3] for the biset category; for ease of comparison, we have kept the same notations and numbering, except for changing '3.' to '0.'. Indeed, the *only difference* lies in the family 2.(d), where Bouc's 2.(d) is stronger than ours because it holds for any pair $M, N \trianglelefteq G$ of normal subgroups, with no condition on their intersection. Spans and bisets will be compared in Section 2.6, but the conclusion should already be clear: bisets can be obtained from spans simply by requiring 2.(d) to hold for all normal subgroups $M, N \trianglelefteq G$, not only those with $M \cap N = 1$. We will see that it actually suffices to replace the above 2.(d) by the following *deflativity relation*

$$\mathrm{Def}^G_{G/N} \circ \mathrm{Inf}^G_{G/N} = \mathrm{id}_{G/N}$$

for all normal subgroups $N \trianglelefteq G$.

Remark 2.56 The condition $M \cap N = 1$ in relation 2.(d) is equivalent to the square

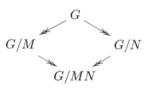

of quotient maps being a pull-back of groups. Indeed, the comparison map to the pull-back $f \colon G \to G/M \times_{G/MN} G/N =: P$, $g \mapsto f(g) = (gM, gN)$, is injective iff $M \cap N = 1$, and is always automatically surjective: Given $x = (g_1 M, g_2 N) \in P$, we can write $g_1 m_1 n_1 = g_2 m_2 n_2$ for some $m_1, m_2 \in M$ and $n_1, n_2 \in N$, hence $g_2 = g_1 m_1 n_1 n_2^{-1} m_2^{-1} \in g_1 m N$ for some $m \in M$ (recall N is normal), hence $g_2 N = g_1 m N$ and of course also $g_1 M = g_1 m M$, hence $x = f(g_1 m)$, showing surjectivity.

To prove the theorem, we can first reduce the problem from groupoids to groups.

Lemma 2.57 *Let* $\mathrm{Sp}_{\Bbbk}(\mathrm{gr}) \subset \mathrm{Sp}_{\Bbbk}(\mathrm{gpd})$ *denote (by a slight abuse of notation) the full subcategory whose objects are all finite groups. Then* $\mathrm{Sp}_{\Bbbk}(\mathrm{gpd})$ *is the additive hull of* $\mathrm{Sp}_{\Bbbk}(\mathrm{gr})$, *that is: every object of* $\mathrm{Sp}_{\Bbbk}(\mathrm{gpd})$ *is (isomorphic to) a direct sum of objects of* $\mathrm{Sp}_{\Bbbk}(\mathrm{gr})$, *and every map decomposes as a matrix of maps in* $\mathrm{Sp}_{\Bbbk}(\mathrm{gr})$.

Proof Every finite groupoid is equivalent to a finite disjoint union of groups, and disjoint unions provide the direct sums (= biproducts) in $\mathrm{Sp}_{\Bbbk}(\mathrm{gpd})$. The rest follows immediately from the basic properties of direct sums (§ 2.23). □

Proof of Theorem 2.53 Since the relations listed in Theorem 2.53 only involve maps between groups, it will suffice to show that $\mathrm{Sp}_{\Bbbk}(\mathrm{gr})$ is generated *as a $\Bbbk$-linear category* (without using direct sums) by the same generators and relations. The theorem will then follow from Lemma 2.57 by taking additive hulls.

Let $\mathcal{F}$ denote the free $\Bbbk$-linear category generated by the elementary spans and let $\mathcal{J}$ denote the $\Bbbk$-linear categorical ideal of maps in $\mathcal{F}$ generated by all the relations in the theorem (*i.e.* by the corresponding differences, of course). Thus by construction we have a $\Bbbk$-linear functor $\Phi \colon \mathcal{F} \to \mathrm{Sp}_{\Bbbk}(\mathrm{gr})$, and we must show that it induces an equivalence

$$\overline{\Phi} \colon \mathcal{F}/\mathcal{J} \xrightarrow{\sim} \mathrm{Sp}_{\Bbbk}(\mathrm{gr}) \,.$$

Let us prove that the factorization $\overline{\Phi}$ exists, that is, that the three families of relations are all satisfied in $\mathrm{Sp}_{\Bbbk}(\mathrm{gr})$. This is a straightforward verification, which involves constructing many iso-comma squares in order to compute the composites of various pairs of elementary spans $[a, b]$ and $[c, d]$:

$$(2.4.1)$$

Most relations are immediately obtained by composing spans where either b or c is an isomorphism (*e.g.* an identity map), in which case the iso-comma square can be replaced by an equivalent *commutative* square where two parallel sides are isomorphisms (*e.g.* identities); see [1, Rem. 2.1.9] if necessary. This takes care of all transitivity relations 1. and of the commutativity relations 2.(a),(b) as well as 2.(e); for the latter, we may also use the equality of maps $(H \hookrightarrow G \twoheadrightarrow G/N) = (H \twoheadrightarrow H/H \cap N \cong HN/N \hookrightarrow G/N)$.

The triviality relations 0.(a) are trivial and 0.(b) can be checked as follows. Recall that we are viewing groups as one-object groupoids and homomorphisms as functors (§ 2.5-2.6), so that if the isomorphism f is $c_x \colon G \xrightarrow{\sim} G$, $g \mapsto xgx^{-1}$, conjugation by an element $x \in G$, then there is a natural isomorphism $\gamma_x \colon \mathrm{Id}_G \Rightarrow c_x$ as in Notation 2.7. This yields an equivalence

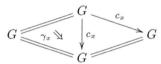

between the spans $\mathrm{Iso}(c_x) = [G = G \xrightarrow{c_x} G]$ and $\mathrm{id}_G = [G = G = G]$.

For the remaining relations, we will compose elementary spans where neither b nor c is an isomorphism, *i.e.* where both are non-trivial inclusions or quotient maps (and hence where a and d are identities, since these are elementary spans). We will deduce various commutativity relations of the form $c^* b_* = \tilde{b}_* \tilde{c}^*$ (in the notation $(-)^*$ and $(-)_*$ of Remark 2.43 applied to (2.4.1)).

Lemma 2.58 *If either b or c is a surjection of groups, then the iso-comma square and the pullback square over $\xrightarrow{b} \xleftarrow{c}$ are equivalent. More precisely, the canonical comparison functor w with components $\langle \mathrm{pr}_1, \mathrm{pr}_2, \mathrm{id} \rangle$ (see § 2.16)*

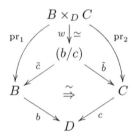

is an equivalence.

Proof Indeed, say that c is surjective (the case of b is similar). Then w sends the unique object $\bullet$ of the pullback group $B \times_D C$ to $(\bullet_B, \bullet_C, \mathrm{id}_{\bullet_D})$, and every object $(\bullet, \bullet, \delta) \in \mathrm{Obj}(b/c)$ is isomorphic to $(\bullet, \bullet, \mathrm{id})$ via a map (id, γ) for any $\gamma \in C$ such that $c(\gamma) = \delta$. It is immediate to see that w is also fully faithful.

(More generally, pull-backs and iso-commas along a functor f are equivalent if f has the invertible path lifting property, *i.e.* if it is a fibration in the canonical model structure; see [17].) □

This lemma takes care of most remaining cases. For instance, a pull-back as in Remark 2.56 is equivalent to the iso-comma square over $G/M \to G/MN \leftarrow G/N$ and the commutivity relation stated in 2.(d) then follows immediately. This was the case where b, c are both quotient maps; the cases where one is a quotient and the other an inclusion prove the relations 2.(f). (For the latter, use also that the square

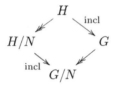

is automatically a pullback.)

The only remaining relation is the Mackey formula 2.(c), which corresponds to the case where both $b =: j$ and $c =: i$ are inclusions. This is proved by decomposing the relevant iso-comma square by the equivalence $v := \langle(\mathrm{incl}), (\mathrm{incl}\, c_x), (\gamma_x)\rangle$ below which, concretely, sends the object $\bullet_{H^x \cap K}$ to $(\bullet_K, \bullet_H, x) \in \mathrm{Obj}(j/i)$ and sends $h^x = k \in H^x \cap K$ of the component at $[x]$ to the map (k, h) of (j/i):

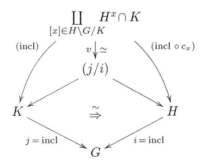

Here γ_x denotes the natural isomorphism

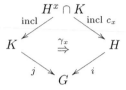

whose sole component is $x \in G$ (*cf.* [1, Rem. 2.2.7]).

Thus we have proved so far that the functor $\overline{\Phi}\colon \mathcal{F}/\mathcal{J} \to \mathrm{Sp}_\Bbbk(\mathrm{gr})$ is well-defined, and we must show that it is an equivalence. As $\overline{\Phi}$ is the identity on objects, it only remains to show that it is fully faithful. It is certainly full, because every span

$$
\begin{array}{ccc}
 & S & \\
{}^{b}\swarrow & & \searrow^{a} \\
H & \cdots\cdots\cdots\!\!\!\!> & G
\end{array}
\tag{2.4.2}
$$

of group homomorphisms admits the decomposition

$$
H \stackrel{}{\twoheadleftarrow} D \stackrel{\ell}{\underset{\sim}{\leftarrowtail}} S/N \stackrel{b}{\longleftarrow} S \stackrel{a}{\longrightarrow} S/M \stackrel{f}{\underset{\sim}{\longrightarrow}} B \rightarrowtail G
$$

where $B := \mathrm{Im}(a)$, $D := \mathrm{Im}(b)$, $M := \mathrm{Ker}(a)$ and $N := \mathrm{Ker}(b)$, and therefore can be written as the following composite of six elementary spans:

$$
\mathrm{Ind}_B^G \circ \mathrm{Iso}(f) \circ \mathrm{Def}_{S/M}^S \circ \mathrm{Inf}_{S/N}^S \circ \mathrm{Iso}(\ell^{-1}) \circ \mathrm{Res}_D^H \ .
\tag{2.4.3}
$$

That is, the image in $\mathrm{Sp}_\Bbbk(\mathrm{gr})$ of this formal composite is (2.4.2) by construction.

In order to prove faithfulness, let us first notice that the relations of $\mathcal{J}$ allow us to transform an arbitrary (composable) finite string of elementary spans in $\mathcal{F}/\mathcal{J}$ into a linear combination of strings of the form (2.4.3), so that the latter form a $\Bbbk$-linear generating set. Indeed, the relations 2.(a),(c),(e),(f) let us bring all induction maps to the left of all other elementary spans, where they can be combined in a single Ind by relation 1.(a). Similarly, 2.(d),(f) let us drag any deflations to the left of all inflations and restrictions, where they can be combined into a single Def by 1.(c). Inflations can be brought to the left of restrictions by 2.(e) and combined into one by 1.(c). Finally, by 2.(a),(b), isomorphisms end up cumulating in only two spots, as indicated, where they can be combined by 1.(b).

Consider now two such length-six formal strings, one as in (2.4.3) and one with primed notations $a', b', S', B', D', M', N', f', \ell'$. Assume that they have the same image in $\mathrm{Sp}_{\Bbbk}(\mathrm{gr})$, *i.e.* that there exists an equivalence of spans as follows:

$$
H \xleftarrow{\quad b \quad} \overset{S}{\underset{S'}{\underset{\cong}{\Big\downarrow s}}} \xrightarrow{\quad a \quad} G
$$

Here $\alpha\colon a \Rightarrow a's$ and $\beta\colon b \Rightarrow b's$ are some natural isomorphisms, which (since S is a group) are simply given by two elements $x \in G$ and $y \in H$ such that $a's = {}^x a$ and $b's = {}^y b$. Moreover, s is a group isomorphism. All this information can be reorganized into a commutative diagram of groups

$$
\begin{array}{ccccccccccccc}
H & \longleftarrow & D & \xleftarrow{\;f\;} & S/N & \longleftarrow & S & \longrightarrow & S/M & \xrightarrow{\;\ell\;} & B & \rightarrowtail & G \\
\cong \downarrow c_y & & \cong \downarrow c_y & & s_N \downarrow \cong & & s \downarrow \cong & & s_M \downarrow \cong & & c_x \downarrow \cong & & c_x \downarrow \cong \\
H & \longleftarrow & D' & \xleftarrow[f']{\cong} & S'/N' & \longleftarrow & S' & \longrightarrow & S'/M' & \xrightarrow[\ell']{\cong} & B' & \rightarrowtail & G
\end{array}
$$

$$(2.4.4)$$

where all maps are either inclusions, quotients or isomorphisms; here c_x and c_y denote conjugation isomorphisms by x and y, respectively, and s_M, s_N are the isomorphisms induced by s on the quotients (indeed $M' = s(M)$ and $N' = s(N)$). Each square corresponds to a relation of $\mathcal{J}$, with which we compute:

$$
\mathrm{Ind}_{B'}^G \, \mathrm{Iso}(\ell') \, \mathrm{Def}_{S'/M'}^{S'} \, \mathrm{Inf}_{S'/N'}^{S'} \, \mathrm{Iso}(f'^{-1}) \, \mathrm{Res}_{D'}^H
$$

$$
\overset{0.(b)}{=} \; \mathrm{Iso}(c_x^{-1}) \, \mathrm{Ind}_{B'}^G \, \mathrm{Iso}(\ell') \, \mathrm{Def}_{S'/M'}^{S'} \, \mathrm{Inf}_{S'/N'}^{S'} \, \mathrm{Iso}(f'^{-1}) \, \mathrm{Res}_{D'}^H \, \mathrm{Iso}(c_y)
$$

$$
\overset{2.(a)}{=} \; \mathrm{Ind}_B^G \, \mathrm{Iso}(c_x^{-1}) \, \mathrm{Iso}(\ell') \, \mathrm{Def}_{S'/M'}^{S'} \, \mathrm{Inf}_{S'/N'}^{S'} \, \mathrm{Iso}(f'^{-1}) \, \mathrm{Iso}(c_y) \, \mathrm{Res}_D^H
$$

$$
\overset{1.(b)}{=} \; \mathrm{Ind}_B^G \, \mathrm{Iso}(\ell) \, \mathrm{Iso}(s_M^{-1}) \, \mathrm{Def}_{S'/M'}^{S'} \, \mathrm{Inf}_{S'/N'}^{S'} \, \mathrm{Iso}(s_N) \, \mathrm{Iso}(f^{-1}) \, \mathrm{Res}_D^H
$$

$$
\overset{2.(b)}{=} \; \mathrm{Ind}_B^G \, \mathrm{Iso}(\ell) \, \mathrm{Def}_{S/M}^S \, \mathrm{Iso}(s^{-1}) \, \mathrm{Iso}(s) \, \mathrm{Inf}_{S/N}^S \, \mathrm{Iso}(f^{-1}) \, \mathrm{Res}_D^H
$$

$$
\overset{1.(b)}{=} \; \mathrm{Ind}_B^G \, \mathrm{Iso}(\ell) \, \mathrm{Def}_{S/M}^S \, \mathrm{Inf}_{S/N}^S \, \mathrm{Iso}(f^{-1}) \, \mathrm{Res}_D^H
$$

Hence the two given parallel length-six composites are already equal in $\mathcal{F}/\mathcal{J}$.

Thus the $\Bbbk$-linear map $\overline{\Phi}\colon \mathcal{F}/\mathcal{J}(H,G) \to \mathrm{Sp}_{\Bbbk}(\mathrm{gr})(H,G)$ can be restricted to a bijection between a generating set of $\mathcal{F}/\mathcal{J}(H,G)$ and a generating set of $\mathrm{Sp}_{\Bbbk}(\mathsf{gr})(H, G)$. As the latter $\Bbbk$-module is free (Remark 2.50), this map is an isomorphism. $\square$

2.5 G-sets vs groupoids

Let us fix a finite group G throughout this section. We begin by recalling from [1, App. B] how to use groupoids in order to capture the classical notion of Mackey functor for G.

First of all, recall that the category of $\Bbbk$-linear Mackey functors for G, which first (implicitly) appeared in [12], can be equivalently defined as the functor category over spans of G-sets:

$$\mathrm{Mack}_{\Bbbk}(G) := \mathrm{Fun}_{\Bbbk}(\mathrm{Sp}_{\Bbbk}(G\text{-set}), \Bbbk\text{-Mod}). \qquad (2.5.1)$$

This approach is due to Dress [10] and Lindner [20].

Remark 2.59   Here G-set denotes the ordinary category of finite left G-sets, seen as a discrete 2-category. It makes sense to apply to G-set the span category construction $\mathrm{Sp}_{\Bbbk}(-)$ of Definition 2.38 because it is a spannable 2-category as in Definition 2.31. Indeed, it is a (2,1)-category (like any discrete 2-category) which is extensive (like any elementary topos) and which moreover has arbitrary Mackey squares, because in an ordinary category the latter are the same thing as the usual pullbacks.

Definition 2.60 (Groupoids faithfully embedded in G)   Let $\mathsf{gpd}^{\mathsf{f}}$ denote the 2-category of finite groupoids, *faithful* functors between them, and natural transformations. We will consider the comma 2-category $\mathsf{gpd}^{\mathsf{f}}_{/G}$ of $\mathsf{gpd}^{\mathsf{f}}$ over G (as in § 2.20), and call its objects $(H, i_H : H \to G)$ *groupoids faithfully embedded in G.*

Definition 2.61 (Transport groupoid)   The *transport groupoid $G \ltimes X$* of a G-set X is the groupoid with set of objects $\mathrm{Obj}(G \ltimes X) := X$ and where an arrow $x \to y$ is a pair $(g, x) \in G \times X$ such that $gx = y$ (we will occasionally also write $g \colon x \to y$ for simplicity). Composition is induced by the multiplication in G via $(h, y)(g, x) := (hg, x)$. A G-equivariant map $f \colon X \to Y$ induces a faithful (!) functor $G \ltimes f \colon G \ltimes X \to G \ltimes Y$ which sends x to $f(x)$ and (g, x) to $(g, f(x))$. The transport groupoid comes equipped with a faithful functor $\pi_X \colon G \ltimes X \to G$ mapping $x \mapsto \bullet$ and $(g, x) \mapsto g$. (Note that the latter is just $G \ltimes (X \twoheadrightarrow G/G)$ followed by the obvious isomorphism $G \ltimes G/G = G$.)

Proposition 2.62 ([1, Prop. B.08]) *The transport groupoid of Definition 2.61 defines a 2-functor (strict pseudo-functor)*

$$G \ltimes - : G\text{-set} \to \mathsf{gpd}^{\mathsf{f}}_{/G},$$
$$X \mapsto (G \ltimes X, \pi_X),$$

which is a biequivalence between the discrete 2-category of G-sets and the comma 2-category of groupoids faithfully embedded in G. □

Like any biequivalence, this one preserves Mackey squares and generally shows that $\mathsf{gpd}^{\mathsf{f}}_{/G}$, just like G-set, is a spannable 2-category. One can also check the latter by hand, for instance the fact that the iso-comma squares of gpd induce iso-comma squares in $\mathsf{gpd}^{\mathsf{f}}_{/G}$.

By Proposition 2.39, if we apply the span construction of Definition 2.38 to the biequivalence of Proposition 2.62 we get:

Corollary 2.63 *The transport groupoid induces an equivalence*

$$\mathrm{Sp}_{\Bbbk}(G\text{-set}) \xrightarrow{\sim} \mathrm{Sp}_{\Bbbk}(\mathsf{gpd}^{\mathsf{f}}_{/G})$$

of $\Bbbk$-linear categories of spans. □

Finally, by taking categories of $\Bbbk$-linear functors:

Corollary 2.64 *The transport groupoid induces a $\Bbbk$-linear equivalence*

$$\mathrm{Mack}_{\Bbbk}(G) \xleftarrow{\sim} \mathrm{Mack}_{\Bbbk}(\mathsf{gpd}^{\mathsf{f}}_{/G})$$

between the classical category of Mackey functors for G and the category of generalized Mackey functors $\mathrm{Mack}_{\Bbbk}(\mathbb{G})$ (Definition 2.41) applied to the 2-category $\mathbb{G} = \mathsf{gpd}^{\mathsf{f}}_{/G}$ of groupoids faithfully embedded in G (Definition 2.60). □

* * *

For the remainder of this section, we adapt the above ideas in order to capture the *fused* Mackey functors of [5] (whose definition will be recalled later).

First, we 'correct' the fact that the forgetful functor $\mathsf{gpd}^{\mathsf{f}}_{/G} \to \mathsf{gpd}$ is not 2-full:

Definition 2.65 (The fused comma category) We consider the following variant $\mathsf{gpd}^{\mathsf{f,fus}}_{/G}$ of the comma 2-category $\mathsf{gpd}^{\mathsf{f}}_{/G}$ considered above, which we call the 2-category of *fused groupoids embedded into G*. The 2-category $\mathsf{gpd}^{\mathsf{f,fus}}_{/G}$ has the same 0-cells and 1-cells as $\mathsf{gpd}^{\mathsf{f}}_{/G}$, but has the larger class of 2-cells obtained by ignoring the compatibility requirement with the embeddings. Explicitly, the 0-cells of $\mathrm{gpd}_{/G}^{\mathrm{f,fus}}$ are finite groupoids embedded into G, i.e. pairs (H, i_H) with $i_H \colon H \rightarrowtail G$ faithful. The 1-cells $(H, i_H) \to (K, i_K)$ are pairs $(u \colon H \rightarrowtail K, \alpha_u \colon i_K u \overset{\sim}{\Rightarrow} i_H)$. The new 2-cells $(u \colon H \to K, \alpha_u) \Rightarrow (v \colon H \to K, \alpha_v)$ are simply 2-cells $\alpha \colon u \Rightarrow v$ in gpd, without further condition. There are obvious inclusion and forgetful 2-functors

$$\mathrm{gpd}_{/G}^{\mathrm{f}} \hookrightarrow \mathrm{gpd}_{/G}^{\mathrm{f,fus}} \to \mathrm{gpd}$$

which are the identity on objects.

We now introduce a 2-category which is to G-set what $\mathrm{gpd}_{/G}^{\mathrm{f,fus}}$ is to $\mathrm{gpd}_{/G}^{\mathrm{f}}$:

Definition 2.66 (The 2-category of fused G-sets) Let G^c denote the set G equipped with the conjugation left G-action $(g, x) \mapsto {}^g x = g x g^{-1}$. We define a 2-category $G\text{-}\mathsf{set}^{\mathsf{fus}}$ *of fused G-sets*, whose 0-cells and 1-cells are those of G-set, namely finite G-sets and G-equivariant maps, and whose 2-cells $\tau \colon f_1 \Rightarrow f_2 \colon X \to Y$ are given by G-maps $\tau \colon X \to G^c$ such that

$$\tau * f_1 = f_2,$$

where the notation means $\tau(x) \cdot f_1(x) = f_2(x)$ for all $x \in X$. (Such G-maps $\tau \colon X \to G^c$ are called *twisting maps*.) Vertical composition of 2-cells in $G\text{-}\mathsf{set}^{\mathsf{fus}}$ is defined by multiplication in G, that is, $(\tau' \cdot \tau)(x) = \tau'(x) \cdot \tau(x)$ for all $x \in X$, and horizontal composition of 2-cells

$$X \underset{f_2}{\overset{f_1}{\Downarrow \tau}} Y \underset{f_4}{\overset{f_3}{\Downarrow \sigma}} Z \quad = \quad X \underset{f_4 \circ f_2}{\overset{f_3 \circ f_1}{\Downarrow \sigma \circ \tau}} Y \qquad (2.5.2)$$

is given by $(\sigma \circ \tau)(x) = \tau(x) \cdot \sigma(f_1(x))$ for all $x \in X$. The identity 2-cell id_f of a 1-cell $f \colon X \to Y$ is given by the constant map $X \to G^c$, $\mathrm{id}_f(x) = e$ for all $x \in X$.

Proposition 2.67 *The construction in Definition 2.66 yields a well-defined (2,1)-category $G\text{-}\mathsf{set}^{\mathsf{fus}}$ whose 1-truncation $\tau_1(G\text{-}\mathsf{set}^{\mathsf{fus}})$ is equal to Bouc's ordinary category of fused G-sets, $G\text{-}\underline{\mathsf{set}}$, as defined in [5]. Moreover, this (2,1)-category has Mackey squares for all cospans, provided by the usual pullback squares of G-sets, and is in fact a spannable (2,1)-category (Definition 2.31).*

Proof The first claim is a direct verification from the definitions which we leave to the reader (*cf.* Remark 2.8). Bouc [5] defines G-$\underline{\text{set}}$ as the quotient category of G-set obtained by identifying any two parallel maps f_1, f_2 such that $\tau * f_1 = f_2$ for some twisting map τ (the resulting relation turns out to be a congruence); clearly, this is precisely the same as the truncated category $\tau_1(G\text{-set}^{\text{fus}})$.

Let us verify the claim about Mackey squares. Consider a pullback square of G-sets and view it inside G-set^{fus}:

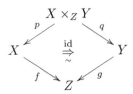

By Remark 2.18, we must show that for any G-set T the functor

$$(p, q, \text{id})_* : G\text{-set}^{\text{fus}}(T, X \times_Z Y) \longrightarrow G\text{-set}^{\text{fus}}(T, f)/G\text{-set}^{\text{fus}}(T, g)$$

$$(T \xrightarrow{u} X \times_Z Y) \longmapsto (pu, qu, fpu \overset{\text{id}}{\Rightarrow} gqu)$$

is an equivalence of categories.

Let $(t, s, \gamma \colon ft \Rightarrow gs)$ be any object of the target iso-comma category; thus $\gamma \colon T \to G^c$ is a G-map such that $\gamma * ft = gs$. Then for all $x \in T$ we compute

$$(f \circ (\gamma * t))(x) = f(\gamma(x) \cdot t(x)) = \gamma(x) \cdot f(t(x)) = (\gamma * (f \circ t))(x) \overset{\text{hyp.}}{=} (g \circ s)(x)$$

showing that $f(\gamma * t) = gs$, so that we may define a G-map $u \colon T \to X \times_Z Y$ into the pullback with components $(\gamma * t, s)$. The square of 2-morphisms of G-set^{fus}

$$
\begin{array}{ccc}
ft & \overset{\gamma}{=\!=\!\Rightarrow} & gs \\
f\gamma \Big\Vert\Big\downarrow & & \Big\Vert\Big\downarrow \text{id} \\
fpu & \underset{\text{id}}{=\!=\!\Rightarrow} & gqu
\end{array}
$$

commutes by the calculation $(x \in T)$

$$(f \circ \gamma)(x) = (\text{id}_f \circ \gamma)(x) = \gamma(x) \cdot \text{id}_f(t(x)) = \gamma(x)$$

showing that the pair (γ, id) is a well-defined isomorphism $(t, s, \gamma) \xrightarrow{\sim} (pu, qu, \text{id})$. This proves the essential surjectivity of $(p, q, \text{id})_*$.

To show it is fully faithful, consider two G-maps $u, v \colon T \to X \times_Z Y$. A morphism $(pu, qu, \text{id}) \to (pv, qv, \text{id})$ between their images is a pair (φ, ψ)

of 2-cells $\varphi\colon pu \Rightarrow pv$ and $\psi\colon qu \Rightarrow qv$ of G-set^{fus} such that $f\varphi = g\psi$, that is such that

$$
\begin{aligned}
\varphi(x) &= \varphi(x) \cdot \mathrm{id}_f(pu(x)) \\
&= (f \circ \varphi)(x) \\
&\overset{\mathrm{hyp.}}{=} (g \circ \psi)(x) \\
&= \psi(x) \cdot \mathrm{id}_g(qu(x)) \\
&= \psi(x)
\end{aligned}
$$

for all $x \in T$, so that in fact φ and ψ must be the same map $T \to G^c$. Moreover, by looking at the pullback components we obtain $\varphi * u = v$:

$$
\begin{aligned}
(\varphi * u)(x) = \varphi(x) \cdot u(x) &= \varphi(x) \cdot \big(pu(x), qu(x)\big) \\
&= \big(\varphi(x) \cdot pu(x), \varphi(x) \cdot qu(x)\big) = \big(pv(x), qv(x)\big) = v(x)
\end{aligned}
$$

In other words, φ also defines a 2-cell $u \Rightarrow v$ such that $(p, q, \mathrm{id})_*(\varphi) = (\varphi, \psi)$, and we see that $(p, q, \mathrm{id})_*$ induces a bijection on each Hom set as claimed.

By Remark 2.33, it remains to show that G-set^{fus} is an extensive bicategory (Definition 2.27). Indeed, the usual coproducts of G-sets are also (strict) coproducts in G-set^{fus}, and the 2-functor

$$
G\text{-set}^{\mathsf{fus}}_{/X} \times G\text{-set}^{\mathsf{fus}}_{/Y} \longrightarrow G\text{-set}^{\mathsf{fus}}_{/X \sqcup Y},
$$

$$
(A \overset{a}{\to} X, B \overset{b}{\to} Y) \mapsto (A \sqcup B \overset{a \sqcup b}{\longrightarrow} X \sqcup Y)
$$

is a biequivalence; both follow easily from the fact that the underlying ordinary category G-set is extensive (see Remark 2.28), together with the fact that a 2-cell $\varphi\colon (a \sqcup b) \Rightarrow (a' \sqcup b')$ in G-set^{fus}, that is a G-map $A \sqcup B \to G^c$ such that $\tau * (a \sqcup b) = a' \sqcup b'$, amounts to the same thing as a pair of 2-cells $\tau_A\colon a \Rightarrow a'$ and $\tau_B\colon b \Rightarrow b'$, that is G-maps $\tau_A\colon A \to G^c$ and $\tau_B\colon B \to B^c$ such that $\tau_A * a = a'$ and $\tau_B * b = b'$. $\qquad\square$

Remark 2.68 It is apparent from the proof that the Mackey squares of Proposition 2.67 are not strict in general, *i.e.* they are not iso-comma squares. Indeed, it seems that the 2-category G-set^{fus} does not admit any nontrivial iso-comma squares.

The two 2-categories we have just defined are actually equivalent:

Theorem 2.69 *For every finite group G, the transport groupoid 2-functor $G \ltimes -$ lifts to a biequivalence of spannable $(2,1)$-categories*

$$
\begin{array}{ccc}
G\text{-set} & \xrightarrow[\text{Prop. 2.62}]{\sim} & \mathsf{gpd}^{\mathsf{f}}_{/G} \\
\text{incl} \downarrow & & \downarrow \text{incl} \\
G\text{-set}^{\mathsf{fus}} & \xrightarrow[\sim]{\exists} & \mathsf{gpd}^{\mathsf{f,fus}}_{/G}
\end{array}
$$

extending the biequivalence of Proposition 2.62 along the two inclusions.

Proof The transport groupoid construction $X \mapsto (G \ltimes X, \pi_X)$ of Definition 2.61, which gave us the biequivalence $G\text{-set} \overset{\sim}{\to} \mathsf{gpd}^{\mathsf{f}}_{/G}$, can be extended to a well-defined 2-functor $G\text{-set}^{\mathsf{fus}} \to \mathsf{gpd}^{\mathsf{f,fus}}_{/G}$ by mapping a 2-cell $\tau \colon f_1 \Rightarrow f_2$, that is a G-map $\tau \colon X \to G^c$ such that $\tau * f_1 = f_2$, to the natural transformation

$$
G \ltimes \tau \colon G \ltimes f_1 \Rightarrow G \ltimes f_2
$$

of functors $G \ltimes X \to G \ltimes Y$ whose component at an object $x \in X$ of $G \ltimes X$ is given by the arrow $(\tau(x), f_1(x)) \in G \ltimes Y(f_1(x), f_2(x))$.

To see that this $G \ltimes -$ is a well-defined 2-functor it only remains to check that it preserve identity 2-cells and vertical and horizontal composition, all of which is straightforward from the definitions. For horizontal composition this may look a little counter-intuitive, so let us spell it out. Consider a horizontal composite (2.5.2) in $G\text{-set}^{\mathsf{fus}}$. By applying $G \ltimes -$ to the right-hand side, we get the natural transformation $G \ltimes (f_3 f_1) \Rightarrow G \ltimes (f_4 f_2)$ with component

$$
\tau(x) \cdot \sigma(f_1(x)) \colon f_3 f_1(x) \to f_4 f_2(x)
$$

at $x \in X$. After applying $G \ltimes -$ to the left-hand side, we may form the horizontal composite of $G \ltimes \tau$ and $G \ltimes \sigma$ which by definition is the diagonal of the following commutative square of natural transformations:

$$
\begin{array}{ccc}
(G \ltimes f_3)(G \ltimes f_1) & \overset{(G\ltimes\sigma)\,\mathrm{Id}}{=\!=\!=\!=\!\Longrightarrow} & (G \ltimes f_4)(G \ltimes f_1) \\
{\scriptstyle \mathrm{Id}\,(G\ltimes\tau)} \Big\Downarrow & & \Big\Downarrow {\scriptstyle \mathrm{Id}\,(G\ltimes\tau)} \\
(G \ltimes f_3)(G \ltimes f_2) & \underset{(G\ltimes\sigma)\,\mathrm{Id}}{=\!=\!=\!=\!\Longrightarrow} & (G \ltimes f_4)(G \ltimes f_2)
\end{array}
$$

By following the right-then-down path, we obtain the (vertical!) composite

$$
(G \ltimes f_4)(G \ltimes \tau) \circ (G \ltimes \sigma)(G \ltimes f_1)
$$

whose component at $x \in X$ is given by the following element of G:

$$\underbrace{(G \ltimes f_4)(G \ltimes \tau)(x)}_{\tau(x)} \cdot \underbrace{(G \ltimes \sigma)(G \ltimes f_1)(x)}_{\sigma(f_1(x))} \,.$$

We see that the two agree, hence $G \ltimes -$ preserves horizontal composites.

To verify that this 2-functor is a biequivalence as claimed, it suffices to prove that it yields a bijection on each set of 2-cells, which is an immediate consequence of Lemma 2.70 below.

By Proposition 2.67, the biequivalence of Theorem 2.69 is in fact a biequivalence of spannable (2,1)-categories. □

Lemma 2.70   *Let $f_i \colon X \to Y$ $(i = 1, 2)$ be two G-maps. For every natural transformation $\alpha = \{\alpha_x\}_{x \in X} \colon G \ltimes f_1 \Rightarrow G \ltimes f_2$ in* gpd *we have $\alpha = G \ltimes \tau$ for a unique twisting G-map $\tau \colon X \to G^c$ such that $\tau * f_1 = f_2$. Explicitly, τ is determined by setting $\alpha_x = (\tau(x), f_1(x))$ for all $x \in X$.*

Proof   This follows by inspecting the definitions, because a natural transformation $\alpha \colon G \ltimes f_1 \Rightarrow G \ltimes f_2$ is precisely a collection of pairs $\{\alpha_x = (\tau(x), f_1(x))\}_{x \in X}$ for elements $\tau(x) \in G$ satisfying $\tau(x)f_1(x) = f_2(x)$ and such that

$$
\begin{array}{ccc}
f_1(x) & \xrightarrow{\ (g, f_1(x))\ } & f_1(gx) \\
{\scriptstyle \alpha_x} \Big\downarrow & & \Big\downarrow {\scriptstyle \alpha_{gx}} \\
f_2(x) & \xrightarrow{\ (g, f_2(x))\ } & f_2(gx)
\end{array}
$$

commutes in $G \ltimes X$ for all $g \in G$. The latter means that $\tau(gx)g = g\tau(x)$ for all g, that is $x \mapsto \tau(x)$ is a G-map $X \to G^c$. Moreover, the requirement that $f_2(x) = \tau(x)f_1(x) = (\tau * f_1)(x)$ for all $x \in X$ means that $f_2 = \tau * f_1$. Hence $\tau \colon f_1 \Rightarrow f_2$ is a 2-cell in $G\text{-set}^{\mathsf{fus}}$ and $\alpha = G \ltimes \tau$ by the above. □

Corollary 2.71   *By applying the span construction $\mathrm{Sp}_{\Bbbk}(-)$ of Definition 2.38 to Theorem 2.69, we obtain a commutative square of $\Bbbk$-linear categories*

$$
\begin{array}{ccc}
\mathrm{Sp}_{\Bbbk}(G\text{-set}) & \xrightarrow{\ \simeq\ } & \mathrm{Sp}_{\Bbbk}(\mathsf{gpd}^{\mathsf{f}}_{/G}) \\
\Big\downarrow & {\scriptstyle \mathrm{Sp}_{\Bbbk}(G \ltimes -)} & \Big\downarrow \\
\mathrm{Sp}_{\Bbbk}(G\text{-set}^{\mathsf{fus}}) & \xrightarrow{\ \simeq\ } & \mathrm{Sp}_{\Bbbk}(\mathsf{gpd}^{\mathsf{f},\mathsf{fus}}_{/G})
\end{array}
$$

with horizontal equivalences. □

Corollary 2.72 *By taking functor categories in Corollary 2.71, we obtain the following commutative square of* $\Bbbk$*-linear categories of Mackey functors (Definition 2.41)*

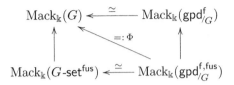

with horizontal equivalences. Moreover, the full image in $\mathrm{Mack}_\Bbbk(G)$ of the diagonal functor Φ coincides with the abelian subcategory $\mathrm{Mack}_\Bbbk^f(G)$ of fused Mackey functors of [5], also called conjugation-invariant Mackey functors by Hambleton-Taylor-Williams [13]. Concretely, they are precisely those $M \in \mathrm{Mack}_\Bbbk(G)$ such that for every subgroup $H \leq G$ the centralizer $C_G(H)$ acts trivially (via the conjugation maps) on $M(G/H)$.

In particular, this shows that fused Mackey functors too are a special case of our generalized Mackey functors (take $\mathbb{G} = \mathbb{J} = \mathrm{gpd}_{/G}^{\mathrm{f,fus}}$).

Proof of Corollary 2.72 The identification of the image of Φ with Bouc's $\mathrm{Mack}_\Bbbk^f(G)$ is now immediate from the definitions. Indeed, the latter is defined to be the (image in $\mathrm{Mack}_\Bbbk(G)$ of the) $\Bbbk$-linear category of spans on $G\text{-}\underline{\mathrm{set}} = \tau_1(G\text{-set}^{\mathrm{fus}})$ with composition induced by the pullbacks of G-set; by Proposition 2.67 and Remark 2.37 this is precisely the category $\mathrm{Sp}_\Bbbk(G\text{-set}^{\mathrm{fus}})$.

Finally, the explicit characterization of the Mackey functors in $\mathrm{Mack}_\Bbbk^f(G)$ follows, via Lemma 2.70, from the fact that an automorphism $\alpha \colon i \Rightarrow i$ in gpd of a subgroup inclusion homomorphism $i \colon H \hookrightarrow G$ is given by (*i.e.* has for its unique component) an element of G which centralizes H. More precisely, such an element $a \in C_G(H)$ of the centralizer defines a G-map $\tau_a \colon G/H \to G^c$ via $\tau_a(gH) = gag^{-1}$, hence a 2-cell $\tau_a \colon \mathrm{Id}_{G/H} \Rightarrow \tau_a * \mathrm{Id}_{G/H}$, hence an equivalence of spans in $G\text{-set}^{\mathrm{fus}}$:

The map $\tau_a * \mathrm{Id}_{G/H}$ sends gH to gaH, *i.e.* it is precisely the G-isomorphism $G/H \overset{\sim}{\to} G/H$ of conjugation by a (or its inverse, depending on conventions). This shows that any Mackey functor M factoring through the quotient $\mathrm{Sp}_\Bbbk(G) \to \mathrm{Sp}_\Bbbk(G\text{-set}^{\mathsf{fus}})$ has the property that $C_G(H)$ acts trivially on $M(G/H)$ for all $H \le G$.

The latter condition is also sufficient for a Mackey functor to factor via the quotient because, as one sees easily, *all 2-cells* of $G\text{-set}^{\mathsf{fus}}$ are generated by such τ_a by taking sums (of orbits G/H) and composites (of spans). (See [5, Thm. 2.11] for more details on this.) $\square$

Remark 2.73   The motivation for studying fused Mackey functors is that they are precisely the Mackey functors for G which can be formulated as biset functors (*cf.* Section 2.6). More precisely, the authors of [13] consider the (non-full) $\Bbbk$-linear subcategory $\mathrm{Bis}_\Bbbk(G) \subset \mathrm{Bis}_\Bbbk^{\mathsf{bif}}(\mathsf{gr})$ (see Notation 2.89) whose objects are the finite subgroups of G and whose maps are the *conjugation bisets* between them; the latter are all bisets which can be obtained by combining restriction and induction bisets (see Remark 2.85) as well as those isomorphism bisets given by conjugation by an element of G. Bouc [5] constructs a $\Bbbk$-linear functor $\mathrm{Sp}_\Bbbk(G\text{-set}) \to \mathrm{Bis}_\Bbbk^\oplus(G)$ to the additive hull $\mathrm{Bis}_\Bbbk^\oplus(G)$ of $\mathrm{Bis}_\Bbbk(G)$ which sends induction, restriction and conjugation spans to the homonymous bisets, and proves that it descends to an equivalence $\mathrm{Sp}_\Bbbk(G\text{-set}^{\mathsf{fus}}) \overset{\sim}{\to} \mathrm{Bis}_\Bbbk^\oplus(G)$ on spans of fused G-sets ([5, Thm. 2.11]). Hence, fused Mackey functors are indeed expressible as a kind of biset functors.

We suspect the latter equivalence is the 1-truncation of a biequivalence between $G\text{-set}^{\mathsf{fus}} \simeq \mathsf{gpd}_{/G}^{\mathsf{f},\mathsf{fus}}$ and a suitable comma bicategory of bisets over G, which can be obtained as a variant of the realization pseudofunctor $\mathcal{R}$ of Theorem 2.80 below. We do not pursue this idea here, as it would take us a little afield, but the next section should provide most of the ingredients to do so.

2.6 Spans vs bisets

In this section we investigate the relationship between Mackey functors and biset functors. Recall from Examples 2.9 and 2.14 the bicategory $\mathsf{biset}(\mathsf{gpd})$ of finite groupoids, bisets and biset morphisms, as well as its full sub-bicategory $\mathsf{biset}(\mathsf{gr})$ of finite groups. Consider their 1-truncations $\tau_1(\mathsf{biset}(\mathsf{gpd}))$ and $\tau_1(\mathsf{biset}(\mathsf{gr}))$.

Lemma 2.74   *The category $\tau_1\mathsf{biset}(\mathsf{gpd})$ is semi-additive (§ 2.23), with direct sums induced by the disjoint sums of groupoids. The sum of two bisets $H \to G$ and the zero biset $H \to G$ are given by the coproduct biset* *and the constantly empty biset*

$$[_G U_H] + [_G V_H] = [U \sqcup V] \quad and \quad 0_{H,G} = [\varnothing]$$

respectively. Moreover, $\tau_1\mathsf{biset}(\mathsf{gpd})$ is the semi-additive hull of its full subcategory $\tau_1\mathsf{biset}(\mathsf{gr})$, meaning that every object of $\tau_1\mathsf{biset}(\mathsf{gpd})$ is a direct sum of finite groups and every arrow is a matrix of arrows between groups.

Proof This is all straightforward. □

Notation 2.75 As usual, fix a ground commutative ring $\Bbbk$. We will write

$$\mathsf{Bis}(\mathsf{gpd}) := \tau_1\mathsf{biset}(\mathsf{gpd}) \quad and \quad \mathsf{Bis}(\mathsf{gr}) := \tau_1\mathsf{biset}(\mathsf{gr})$$

for the semi-additive categories of bisets of Lemma 2.74, and we will denote by

$$\mathsf{Bis}_\Bbbk(\mathsf{gpd}) := \Bbbk(\tau_1\mathsf{biset}(\mathsf{gpd})) \quad and \quad \mathsf{Bis}_\Bbbk(\mathsf{gr}) := \Bbbk(\tau_1\mathsf{biset}(\mathsf{gr}))$$

their $\Bbbk$-linearization as in Construction 2.24.

Remark 2.76 In the literature, the category $\mathsf{Bis}_\Bbbk(\mathsf{gr})$ is the one usually referred to as 'the biset category', rather than its additive hull $\mathsf{Bis}_\Bbbk(\mathsf{gpd})$. In [16], the authors provide an alternative explicit description of the additive completion, which avoids the use of groupoids and is related to the 2-category $\mathcal{S}$ of [25].

Definition 2.77 (Biset functor [4]) A ($\Bbbk$-*linear*) *biset functor* is a $\Bbbk$-linear (hence also additive, *i.e.* direct-sum preserving) functor

$$F\colon \mathsf{Bis}_\Bbbk(\mathsf{gpd}) \longrightarrow \Bbbk\text{-Mod}$$

to the abelian category of $\Bbbk$-modules. A morphism of biset functors is simply a natural transformation.

Remark 2.78 By the general properties of $\Bbbk$-linearization, we have a canonical additive functor $\mathsf{Bis}(\mathsf{gpd}) \to \mathsf{Bis}_\Bbbk(\mathsf{gpd})$ inducing a $\Bbbk$-linear isomorphism

$$\mathsf{Fun}_\Bbbk(\mathsf{Bis}_\Bbbk(\mathsf{gpd}), \Bbbk\text{-Mod}) \xrightarrow{\sim} \mathsf{Fun}_+(\mathsf{Bis}(\mathsf{gpd}), \Bbbk\text{-Mod})$$

of functor categories. Moreover, since additive functors extend essentially uniquely to the additive hull, we also have an equivalence

$$\mathsf{Fun}_\Bbbk(\mathsf{Bis}_\Bbbk(\mathsf{gpd}), \Bbbk\text{-Mod}) \xrightarrow{\sim} \mathsf{Fun}_\Bbbk(\mathsf{Bis}_\Bbbk(\mathsf{gr}), \Bbbk\text{-Mod}) =: \mathcal{F}$$

induced by the inclusion $\mathsf{Bis}_\Bbbk(\mathsf{gr}) \hookrightarrow \mathsf{Bis}_\Bbbk(\mathsf{gpd})$. This last functor category, under the notation $\mathcal{F}$, is the category of biset functors as defined in [4].

Remark 2.79 A more combinatorial description of biset functors, similar to Green's original axioms for Green and Mackey functors [12], is given in [31, §8] (under the name *globally defined Mackey functors*). See also Remark 2.91.

The following result provides a direct connection between bisets and spans:

Theorem 2.80 (Huglo [15]) *There is a well-defined pseudo-functor*

$$\mathcal{R}\colon \mathsf{Span}(\mathsf{gpd}) \longrightarrow \mathsf{biset}(\mathsf{gpd})$$

from the bicategory of spans in gpd *(Remark 2.36) to that of bisets (Definition 2.9), which is the identity on objects, and which sends a span* $H \xleftarrow{b} S \xrightarrow{a} G$ *of groupoids to the* G, H-*biset*

$$\mathcal{R}(H \xleftarrow{b} S \xrightarrow{a} G) := G(a-,-) \otimes_S H(-,b-)\colon H^{\mathrm{op}} \times G \longrightarrow \mathsf{Set}.$$

Moreover, the functor it induces on truncated 1-categories

$$\mathrm{Re} := \tau_1 \mathcal{R}\colon \mathrm{Sp}(\mathsf{gpd}) \longrightarrow \mathrm{Bis}(\mathsf{gpd})$$

is additive and full.

For the present purposes, we really only need the following corollary which is immediately obtained by $\Bbbk$-linearization:

Corollary 2.81 *There exists a full* $\Bbbk$-*linear functor*

$$\mathrm{Re}_\Bbbk := \Bbbk(\tau_1 \mathcal{R})\colon \mathrm{Sp}_\Bbbk(\mathsf{gpd}) \longrightarrow \mathrm{Bis}_\Bbbk(\mathsf{gpd})$$

which is the identity on objects and sends the class of a span $[H \xleftarrow{b} S \xrightarrow{a} G]$ *to the class of the tensor product* G, H-*biset* $[G(a-,-) \otimes_S H(-,b-)]$. $\quad\square$

Sketch of proof for Theorem 2.80 More details can be found in Huglo's PhD thesis [15], together with other properties of the pseudo-functor $\mathcal{R}$; see also [9, §5] for a compact account. The main observation is that for every functor $u\colon H \to G$ there is an internal adjunction (§2.11) in the bicategory $\mathsf{biset}(\mathsf{gpd})$ as follows:

$$\mathcal{R}_!(u) := G(u-,-) \left(\begin{array}{c} H \\ \Big\uparrow \\ \dashv \\ \Big\downarrow \\ G \end{array}\right) G(-,u-) =: \mathcal{R}^*(u)$$

Routine properties of adjunctions allow us to extend the collection of the left adjoints to a pseudo-functor $\mathcal{R}_!$: gpd → biset, and similarly the right adjoints to a pseudo-functor $\mathcal{R}^*$: gpd^{op} → biset, both pseudo-functors being the identity on objects. By an explicit computation, one verifies that these adjunctions satisfy the base-change formula with respect to iso-comma squares. It follows then by the universal property of the span bicategory (see [1, Thm. 5.2.1]) that the pseudo-functors $\mathcal{R}_!$ and $\mathcal{R}^*$ can be 'pasted together' in order to obtain a pseudo-functor $\mathcal{R}$ on Span(gpd) which sends a span $H \xleftarrow{b} S \xrightarrow{a} G$ to the horizontal composite $\mathcal{R}_!(a) \circ \mathcal{R}^*(b)$ in biset(gpd), as claimed.

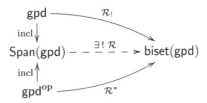

The additivity of $\tau_1\mathcal{R}$ boils down to the fact that $\mathcal{R}$ is the identity on objects and preserves disjoint unions.

In order to see that the functor $\tau_1\mathcal{R}$ is full, it can be shown that every biset ${}_GU_H$ is isomorphic to the image under $\mathcal{R}$ of a canonical span $H \leftarrow S_U \to G$, where the groupoid S_U is a suitable Grothendieck construction (category of elements). Alternatively, and more simply, it suffices to combine the additivity of $\tau_1\mathcal{R}$ with Remarks 2.85 and 2.86 below.   □

We are going to upgrade Corollary 2.81 to the following more precise result:

Theorem 2.82   *The realization pseudo-functor of Theorem 2.80 induces an isomorphism of* $\Bbbk$*-linear categories*

$$\text{Sp}_{\Bbbk}(\text{gpd}) \xrightarrow{\text{Re}_{\Bbbk}} \text{Bis}_{\Bbbk}(\text{gpd})$$
$$\searrow \qquad \nearrow_{\simeq}$$
$$\text{Sp}_{\Bbbk}(\text{gpd})/_\sim$$

which identifies the biset category with the quotient of the full span category of Section 2.4 obtained by factoring out the additive $\Bbbk$*-linear ideal generated by the relations*

$$[Q \xleftarrow{p} G \xrightarrow{p} Q] \sim \text{id}_Q$$

for all surjective group homomorphisms $p\colon G \to Q$*. Or equivalently* *(in terms of the elementary spans of Notation 2.52), generated by the 'deflativity relations'*

$$\mathrm{Def}^G_{G/N} \circ \mathrm{Inf}^G_{G/N} \sim \mathrm{id}_{G/N} \qquad (2.6.1)$$

for every normal subgroup $N \trianglelefteq G$.

Before proving the theorem, let us immediately record:

Corollary 2.83 *Precomposition with the functor $\mathrm{Re}_\Bbbk$ of Corollary 2.81 induces a fully faithful embedding of functor categories*

$$\mathrm{Re}^*_\Bbbk \colon \mathrm{Fun}_\Bbbk(\mathrm{Bis}_\Bbbk(\mathrm{gpd}), \Bbbk\text{-Mod}) \hookrightarrow \mathrm{Fun}_\Bbbk(\mathrm{Sp}_\Bbbk(\mathrm{gpd}), \Bbbk\text{-Mod})$$

which identifies biset functors (Definition 2.77) as a full reflexive $\Bbbk$-linear subcategory of global Mackey functors on gpd (Example 2.44). The essential image of the embedding consists precisely of the Mackey functors M with the property that $M([Q \leftarrow G \rightarrow Q]) = \mathrm{id}_{M(Q)}$ for every surjective group homomorphism $G \twoheadrightarrow Q$.

Proof The characterization of the image is immediate from Theorem 2.82.

The rest follows by standard arguments. Using that $\mathrm{Re}_\Bbbk$ is full and surjective on objects, one verifies immediately that it induces a fully faithful functor between functor categories. Its image is a reflective subcategory, *i.e.* the inclusion functor admits a left adjoint. In fact, this left adjoint (the 'reflection') is provided (after composing with $\mathrm{Re}^*_\Bbbk$) by the $\Bbbk$-linear left Kan extension along $\mathrm{Re}_\Bbbk$, *i.e.* by the unique colimit-preserving functor sending a representable Mackey functor $\mathrm{Sp}_\Bbbk(\mathrm{gpd})(G, -)$ to the corresponding representable biset functor $\mathrm{Bis}_\Bbbk(\mathrm{gpd})(G, -)$. $\square$

Remark 2.84 Corollary 2.83 contains the main result of [25]. Nakaoka calls (2.6.1) the *deflativity condition* and the Mackey functors arising this way from biset functors *deflative Mackey functors*.

In order to prove Theorem 2.82, it will suffice to compare suitable presentations of the two categories. In fact, the realization pseudo-functor $\mathcal{R}$ of Theorem 2.80 is not really necessary for the proof: once the two presentations are established, one can see that the functor of Corollary 2.81 must exist for formal reasons. However, it is nice to know that the comparison of spans and bisets comes from such a natural construction as $\mathcal{R}$, whose 2-categorical nature fits well in this chapter's philosophy.

Remark 2.85 (A presentation of the biset category) We recall from [4, §2.3] that every G, H-biset between finite groups decomposes as a coproduct of *transitive* G, H-bisets (*i.e.* those which are indecomposable with respect to coproducts), and every transitive biset $_GX_H : H \to G$ is isomorphic to a unique horizontal composite of the form:

$$G \xleftarrow{\text{Ind}_D^G} D \xleftarrow{\text{Inf}_{D/C}^D} D/C \xleftarrow{\text{Iso}(f)}_{\sim} B/A \xleftarrow{\text{Def}_{B/A}^B} B \xleftarrow{\text{Res}_B^H} H \qquad (2.6.2)$$

Here $A \trianglelefteq B \leq H$ and $C \trianglelefteq D \leq G$ are subgroups, with A and C normal in B and D, respectively, $f \colon B/A \overset{\sim}{\to} D/C$ is an isomorphism of groups, and the notations refer to the following five kinds of *elementary bisets* ([4, §2.3.9]):

- *Isomorphisms:* $\text{Iso}(f) := {}_GG_{G'}$, with G acting on itself on the left and G' acting on G by right multiplication via a group isomorphism $f \colon G' \overset{\sim}{\to} G$.
- *Restrictions:* $\text{Res}_H^G := {}_HG_G$, defined whenever H is a subgroup of G.
- *Inductions:* $\text{Ind}_H^G := {}_GG_H$, again for H a subgroup of G.
- *Deflations:* $\text{Def}_{G/N}^G := {}_{G/N}(G/N)_G$, for any normal subgroup N of G.
- *Inflations:* $\text{Inf}_{G/N}^G := {}_G(G/N)_{G/N}$, again for a normal subgroup N of G.

In particular, the morphisms of the $\Bbbk$-linear category $\text{Bis}_\Bbbk(\text{gr})$ are generated by the (isomorphism classes of) elementary bisets. As already mentioned in Remark 2.55, Bouc also provides a full list of relations for the elementary bisets, and therefore a presentation of $\text{Bis}_\Bbbk(\text{gr})$ as a $\Bbbk$-linear category (cf. [4, § 3.1]). Bouc's list is nearly identical with that in Theorem 2.53, the only difference being that Bouc's relation set 2.(d) does not require the condition "$M \cap N = 1$".

Remark 2.86 It is immediate to verify that the functor $\text{Re}_\Bbbk$ maps each elementary span of Notation 2.52 to the homonymous elementary biset of Remark 2.85. Just remember that, in our conventions, a span $[G \leftarrow S \to H]$ is read left-to-right while a biset $_HU_G$ is read right-to-left, *i.e.* they both stand for a map from G to H.

Proof of Theorem 2.82 In view of Corollary 2.81, it only remains to identify the kernel on morphisms of the $\Bbbk$-linear functor

$$\text{Re}_\Bbbk \colon \text{Sp}_\Bbbk(\text{gpd}) \to \text{Bis}_\Bbbk(\text{gpd}).$$

In fact, since the two inclusions $\text{Sp}_\Bbbk(\text{gr}) \subset \text{Sp}_\Bbbk(\text{gpd})$ and $\text{Bis}_\Bbbk(\text{gr}) \subset \text{Bis}_\Bbbk(\text{gpd})$ are additive hulls (by Lemma 2.57 and Lemma 2.74 respectively), we only need to look at the restriction of $\text{Re}_\Bbbk$ to a functor $\text{Sp}_\Bbbk(\text{gr}) \to \text{Bis}_\Bbbk(\text{gr})$ between groups.

Let us begin by noting that, since in $\mathrm{Sp}_{\Bbbk}(\mathrm{gr})$ we have the equality

$$[G = G \xrightarrow{p} Q] \circ [Q \xleftarrow{p} G = G] = [Q \xleftarrow{p} G \xrightarrow{p} Q]$$

and since every surjective homomorphism $G \xrightarrow{p} Q$ is isomorphic to one of the form $G \to G/N$, the relations (2.6.1) are evidently equivalent to the family of relations

$$[Q \xleftarrow{p} G \xrightarrow{p} Q] \sim \mathrm{id}_Q$$

for all surjective group homomorphisms $p\colon G \to Q$, as stated.

Now we compare the presentation of $\mathrm{Sp}_{\Bbbk}(\mathrm{gr})$ given in Theorem 2.53 and Bouc's presentation of $\mathrm{Bis}_{\Bbbk}(\mathrm{gr})$ recalled in Remark 2.85.

As already mentioned in Remark 2.86, the functor $\mathrm{Re}_{\Bbbk}$ maps generators (the elementary spans) to the homonymous generators (the elementary bisets), and the only difference between the two presentations lies within the relation family 2.(d). Hence it only remains to prove that the relations (2.6.1), together with all the relations of the span category (as in Theorem 2.53), imply the following:

Bouc's 2.(d): For any two normal subgroups $M, N \trianglelefteq G$ we have

$$\mathrm{Def}^G_{G/N} \circ \mathrm{Inf}^G_{G/M} = \mathrm{Inf}^{G/N}_{G/NM} \circ \mathrm{Def}^{G/M}_{G/NM} .$$

To this end, let M, N be any two normal subgroups of a group G, and consider the following diagram of group homomorphisms, where the outer square consists of the quotient maps and the inner square is a pullback:

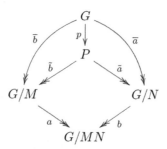

One verifies easily that the comparison map $p\colon G \to P$ is also surjective, hence by (2.6.1) we get the relation $p_* p^* = \mathrm{id}_P$; here and below we use the short-hand notations $(-)_*$ and $(-)^*$ as in Remark 2.43. Now we compute as follows, using also the basic functoriality of $(-)_*$ and $(-)^*$

afforded by the relations 0.(a) and 1.(c):

$$b^* a_* = \tilde{a}_* \tilde{b}^* \qquad \text{by Remark 2.56 and 2.(d) for spans}$$
$$= \tilde{a}_* \operatorname{id}_P \tilde{b}^*$$
$$= \tilde{a}_* p_* p^* \tilde{b}^* \qquad \text{by (2.6.1) for } p$$
$$= (\tilde{a}p)_* (\tilde{b}p)^*$$
$$= (\overline{a})_* (\overline{b})^*$$

This proves Bouc's relation 2.(d), and concludes the proof of the theorem.

$\square$

Remark 2.87 The argument of the above proof shows in fact that, in the presence of his other relation families, Bouc's relations of type 2.(d) for bisets, *i.e.* the relations

$$\operatorname{Def}^G_{G/N} \circ \operatorname{Inf}^G_{G/M} = \operatorname{Inf}^{G/N}_{G/NM} \circ \operatorname{Def}^{G/M}_{G/NM} \qquad \text{for all } N, M \trianglelefteq G,$$

follow already from the special case with $N = M$.

$$* * *$$

In the remainder of this last section we consider a couple of frequently used variants of biset functors and show that the analogue of Theorem 2.82 provides for each of them an *equivalence* with the corresponding notion of Mackey functors.

Terminology 2.88 Recall that (for groups G and H) a G, H-biset $_G U_H$ is *right-free* if H acts freely on U, *left-free* if G acts freely, and *bifree* if both actions are free. Of the five elementary kinds of bisets in Remark 2.85, we see that isomorphisms, restrictions and inductions are bifree; inflations are only right-free; and deflations are only left-free (unless of course, for the latter two kinds, we are in the degenerate case $N = 1$). Both right-free and bifree bisets form classes closed under horizontal composition, so we can consider the 2-full sub-bicategories containing them.

Notation 2.89 In a way which is hopefully self-explanatory, we may write

$$\mathsf{biset}^{\mathsf{rf}}(\mathsf{gr}), \quad \mathsf{biset}^{\mathsf{bif}}(\mathsf{gr}) \qquad \text{and} \qquad \mathsf{biset}^{\mathsf{rf}}(\mathsf{gpd}), \quad \mathsf{biset}^{\mathsf{bif}}(\mathsf{gpd})$$

for the 2-full sub-bicategories of $\mathsf{biset}(\mathsf{gr})$ and $\mathsf{biset}(\mathsf{gpd})$, respectively, where only right-free ("rf"), or bifree ("bif"), bisets are allowed as 1-cells. (These notations are used at the end of the Introduction.) We will be interested in their $\Bbbk$-linearized 1-truncations, for which we use the following notations:

$$\mathrm{Bis}_{\Bbbk}^{\mathrm{rf}}(\mathrm{gr}), \quad \mathrm{Bis}_{\Bbbk}^{\mathrm{bif}}(\mathrm{gr}) \quad \text{and} \quad \mathrm{Bis}_{\Bbbk}^{\mathrm{rf}}(\mathrm{gpd}), \quad \mathrm{Bis}_{\Bbbk}^{\mathrm{bif}}(\mathrm{gpd}) \,.$$

Remark 2.90 Clearly, both $\mathrm{Bis}_{\Bbbk}^{\mathrm{rf}}(\mathrm{gr}) \subset \mathrm{Bis}_{\Bbbk}^{\mathrm{rf}}(\mathrm{gpd})$ and $\mathrm{Bis}_{\Bbbk}^{\mathrm{bif}}(\mathrm{gr}) \subset \mathrm{Bis}_{\Bbbk}^{\mathrm{bif}}(\mathrm{gpd})$ are again the inclusion of a $\Bbbk$-linear category in its additive hull. By construction, moreover, they are generated as $\Bbbk$-linear categories, respectively as additive $\Bbbk$-linear categories, by the following elementary bisets:

- $\mathrm{Bis}_{\Bbbk}^{\mathrm{rf}}(\mathrm{gr})$ and $\mathrm{Bis}_{\Bbbk}^{\mathrm{rf}}(\mathrm{gpd})$: by isomorphisms, restrictions, inductions and inflations (*i.e.* deflations are not allowed).
- $\mathrm{Bis}_{\Bbbk}^{\mathrm{bif}}(\mathrm{gr})$ and $\mathrm{Bis}_{\Bbbk}^{\mathrm{bif}}(\mathrm{gpd})$: by isomorphisms, restrictions and inductions (*i.e.* neither inflations nor deflations are allowed).

Remark 2.91 For even more variation, we may follow [31, § 8] and choose three classes of finite groups $\mathcal{D}, \mathcal{X}, \mathcal{Y}$, with $\mathcal{X}$ and $\mathcal{Y}$ closed under the formation of group extensions and subquotients. Such a triple defines a full subcategory of bisets where the objects are the groups in $\mathcal{D}$, and where the morphisms are those bisets whose right and left isotropy groups belong to $\mathcal{X}$ and $\mathcal{Y}$, respectively. In other words, the morphisms include all inductions, restrictions and isomorphisms between the available groups of $\mathcal{D}$, but only the inflations along quotient homomorphisms with kernel in $\mathcal{X}$ and only deflations for those with kernel in $\mathcal{Y}$. For instance, right-free bisets correspond to choosing $\mathcal{X} = \{1\}$ and $\mathcal{Y} = \{\text{all groups}\}$ and bifree bisets to $\mathcal{X} = \mathcal{Y} = \{1\}$.

For each triple $(\mathcal{D}, \mathcal{X}, \mathcal{Y})$, the $\Bbbk$-linear functors on the associated biset category is equivalent to Webb's category $\mathrm{Mack}_{\Bbbk}^{\mathcal{X}, \mathcal{Y}}(\mathcal{D})$ of *globally defined Mackey functors*. Thus the latter can be identified with a particular kind of biset functors. Hence, by the arguments of this section, they can be seen as a (full subcategory of a) kind of generalized Mackey functors. We leave the details to the interested reader and only treat here the above two chosen special cases.

Theorem 2.92 *The realization pseudo-functor of Theorem 2.80 induces two isomorphisms of $\Bbbk$-linear categories*

$$\mathrm{Sp}_{\Bbbk}(\mathrm{gpd}; \mathrm{gpd}^{\mathrm{f}}) \xrightarrow{\sim} \mathrm{Bis}_{\Bbbk}^{\mathrm{rf}}(\mathrm{gpd}) \quad and \quad \mathrm{Sp}_{\Bbbk}(\mathrm{gpd}^{\mathrm{f}}; \mathrm{gpd}^{\mathrm{f}}) \xrightarrow{\sim} \mathrm{Bis}_{\Bbbk}^{\mathrm{bif}}(\mathrm{gpd})$$

which identify the right-free biset category and the bifree biset category with suitable categories of spans of groupoids (see Examples 2.45 and 2.46).

Therefore, once again, we can subsume the corresponding notion of functors under our generalized Mackey functors:

Corollary 2.93 *Precomposition with the functor of Theorem 2.92 induces equivalences (in fact isomorphisms) of functor categories*

$$\mathrm{Fun}_\Bbbk(\mathrm{Bis}^{\mathrm{rf}}_\Bbbk(\mathrm{gpd}), \Bbbk\text{-Mod}) \xrightarrow{\sim} \mathrm{Fun}_\Bbbk(\mathrm{Sp}_\Bbbk(\mathrm{gpd}; \mathrm{gpd}^{\mathrm{f}}), \Bbbk\text{-Mod})$$

and

$$\mathrm{Fun}_\Bbbk(\mathrm{Bis}^{\mathrm{bif}}_\Bbbk(\mathrm{gpd}), \Bbbk\text{-Mod}) \xrightarrow{\sim} \mathrm{Fun}_\Bbbk(\mathrm{Sp}_\Bbbk(\mathrm{gpd}^{\mathrm{f}}; \mathrm{gpd}^{\mathrm{f}}), \Bbbk\text{-Mod})$$

which identify right-free and bifree biset functors with certain categories of generalized Mackey functors. □

Proof of Theorem 2.92 It is an immediate consequence of Remarks 2.85 and 2.86 that the realization pseudo-functor $\mathcal{R}$ of Theorem 2.80 restricts to pseudo-functors

$$\mathrm{Span}(\mathrm{gpd}; \mathrm{gpd}^{\mathrm{f}}) \xrightarrow{\mathcal{R}} \mathrm{biset}^{\mathrm{rf}}(\mathrm{gpd}) \quad \text{and} \quad \mathrm{Span}(\mathrm{gpd}^{\mathrm{f}}; \mathrm{gpd}^{\mathrm{f}}) \xrightarrow{\mathcal{R}} \mathrm{biset}^{\mathrm{bif}}(\mathrm{gpd}) \, .$$

Indeed, on the side of spans, limiting right (resp. right and left) legs to faithful functors precisely eliminates the elementary spans $G = G \twoheadrightarrow G/N$ (resp. both $G = G \twoheadrightarrow G/N$ and $G/N \twoheadleftarrow G = G$) from the set of generators.

The two 1-truncated functors $\tau_1 \mathcal{R}$ are full, like the one of Theorem 2.80, and so are the induced $\Bbbk$-linear functors

$$\mathrm{Sp}_\Bbbk(\mathrm{gpd}; \mathrm{gpd}^{\mathrm{f}}) \longrightarrow \mathrm{Bis}^{\mathrm{rf}}_\Bbbk(\mathrm{gpd}) \quad \text{and} \quad \mathrm{Sp}_\Bbbk(\mathrm{gpd}^{\mathrm{f}}; \mathrm{gpd}^{\mathrm{f}}) \longrightarrow \mathrm{Bis}^{\mathrm{bif}}_\Bbbk(\mathrm{gpd})$$
$$(2.6.3)$$

both of which will still be denoted by $\mathrm{Re}_\Bbbk$. It only remains to see that these two functors are faithful, and we will do so by comparing two presentations, as in the proof of Theorem 2.82.

To this end, we make the following claim: For each of the four $\Bbbk$-linear categories involved in (2.6.3), we can obtain a presentation simply by 'restricting' the presentation of Theorem 2.53 (for the two categories of spans) or Remark 2.85 (for the two categories of bisets). In other words, it suffices to ignore the irrelevant generators and relations and keep the rest. Thus, explicitly, the categories $\mathrm{Sp}_\Bbbk(\mathrm{gpd}; \mathrm{gpd}^{\mathrm{f}})$ and $\mathrm{Bis}^{\mathrm{rf}}_\Bbbk(\mathrm{gpd})$ are generated by the elementary spans (resp. bisets) other than the deflations, and all relations follow from those of the families 0.-2. other than 1.(c), 2.(b), 2.(d) and some of the ones in 0.(a) and 2.(e-f) (because they involve deflations). Similarly, $\mathrm{Sp}_\Bbbk(\mathrm{gpd}^{\mathrm{f}}; \mathrm{gpd}^{\mathrm{f}})$ and $\mathrm{Bis}^{\mathrm{bif}}_\Bbbk(\mathrm{gpd})$ are generated by the elementary spans/bisets other than inflations and deflations, with relations determined by the relations 0.-2. other than 1.(c), 2.(b), 2.(d-f) and some of those in 0.(a). Since in both cases we are led to ignore the relations of type 2.(d), the corresponding presentations for spans and bisets now look identical, from which it follows that the two functors $\mathrm{Re}_\Bbbk$ of (2.6.3) are indeed $\Bbbk$-linear isomorphisms.

To see why the above claim on presentations is true, first notice that in each of the four categories the retained generators *do* generate, by construction. Moreover, it is immediate to see that in each case the retained relations still allow us to commute or fuse any pair of the remaining generators, hence they still let us reduce an arbitrary string of generators to a linearly independent finite sum of short strings in the length-six canonical form of (2.4.3)

$$\mathrm{Ind}_B^G \circ \mathrm{Iso}(f) \circ \mathrm{Def}_{S/M}^S \circ \mathrm{Inf}_{S/N}^S \circ \mathrm{Iso}(\ell^{-1}) \circ \mathrm{Res}_D^H \qquad \text{(for spans)}$$

respectively in the length-five canonical form of (2.6.2)

$$\mathrm{Ind}_D^G \circ \mathrm{Inf}_{D/C}^D \circ \mathrm{Iso}(f) \circ \mathrm{Def}_{B/A}^B \circ \mathrm{Res}_B^H \qquad \text{(for bisets)}$$

(of course, deflations resp. deflations and inflations are now absent from both forms). For bisets, the length-five canonical form of a string (or of an $\sqcup$-irreducible biset) is unique, hence any set of relations that allows us to bring each string of generators to its canonical form is sufficient to determine all relations that hold within the ambient category $\mathrm{Bis}_\Bbbk(\mathsf{gpd})$.

For spans, the argument is slightly subtler. First notice that we may consider $\mathrm{Sp}_\Bbbk(\mathsf{gpd};\mathsf{gpd}^f)$ and $\mathrm{Sp}_\Bbbk(\mathsf{gpd}^f;\mathsf{gpd}^f)$ to be (non-full) subcategories of $\mathrm{Sp}_\Bbbk(\mathsf{gpd})$, because the morphisms of pairs $(\mathsf{gpd};\mathsf{gpd}^f) \to (\mathsf{gpd};\mathsf{gpd})$ and $(\mathsf{gpd}^f;\mathsf{gpd}^f) \to (\mathsf{gpd};\mathsf{gpd})$ induce faithful functors on span categories. This is because $\mathsf{gpd}^f \subset \mathsf{gpd}$ is a 2-full 2-subcategory containing all equivalences, hence the data of any equivalences between spans with one or two faithful legs is already available in $(\mathsf{gpd};\mathsf{gpd}^f)$, resp. in $(\mathsf{gpd}^f;\mathsf{gpd}^f)$. Now in the ambient category $\mathrm{Sp}_\Bbbk(\mathsf{gpd})$, as we have seen in the proof of Theorem 2.53, the above length-six canonical form of a string (or of an $\sqcup$-irreducible span) is only unique up to changing (D, S, N, M, B, f, ℓ) by a bunch of compatible isomorphisms, as in (2.4.4); but all elementary isomorphisms are still available in the two subcategories, as well as the commutativity relations between isomorphisms and the other retained generators. Thus, once again, we conclude that the retained families of relations allow us to determined all relations between strings as they hold in $\mathrm{Sp}_\Bbbk(\mathsf{gpd})$.

This concludes the proof. $\square$

Remark 2.94   Miller [23] offers a different approach to Theorem 2.92, already at the bicategorical level. More precisely, the main result of *loc. cit.* is an explicit biequivalence $\mathbb{B} \simeq \mathbb{C}$ between the 'Burnside bicategory of groupoids' $\mathbb{B}$ and the 'bicategory of correspondences' (= nice spans) $\mathbb{C}$. Infinite groupoids are allowed in $\mathbb{B}$ and $\mathbb{C}$, but if we restrict attention to finite groupoids we easily find biequivalences

$$(\mathbb{B}|_{\mathsf{fin}})^{\mathrm{op}} \simeq \mathsf{biset}^{\mathsf{rf}}(\mathsf{gpd})_{\mathsf{core}} \quad \text{and} \quad (\mathbb{C}|_{\mathsf{fin}})^{\mathrm{op}} \simeq \mathsf{Span}(\mathsf{gpd};\mathsf{gpd}^{\mathsf{f}})_{\mathsf{core}} .$$

Here $(\dots)|_{\mathsf{fin}}$ denotes the 2-full bicategory of finite groupoids and finite bisets, resp. of spans of finite groupoids; the decoration $(\dots)_{\mathsf{core}}$ indicates that we discard all non-invertible 2-cells in our two bicategories of Notation 2.89 and Remark 2.36. Theorem 2.92 and Corollary 2.93 can then be deduced from Miller's result, by arguing as above. While the first biequivalence above is immediate, the second one involves replacing every span $H \leftarrow P \rightarrow G$ with $P \rightarrow G$ faithful with an equivalent span $H \leftarrow \tilde{P} \rightarrow G$ with the property that $\tilde{P} \rightarrow G$ is a 'weak finite cover' (see [23, Def. 4.1]). To this end, just replace i by $\tilde{i}$ as in the iso-comma square

or, alternatively, use the functorial fibrant replacement in the canonical model structure of groupoids (or categories); see *e.g.* [27]. A final subtle point is that the horizontal composition of $\mathbb{C}$ is defined using strict pullbacks, but since its spans have a leg which is a fibration, this agrees with our composition via iso-commas.

Remark 2.95   (Added 'in proof':) Upgrading Theorem 2.92, and sharpening the bicategorical image drawn at the end of the Introduction, we now know for a fact that the realization pseudo-functor $\mathcal{R}$ actually induces biequivalences

$$\mathsf{Span}(\mathsf{gpd};\mathsf{gpd}^{\mathsf{f}}) \xrightarrow{\sim} \mathsf{biset}^{\mathsf{rf}}(\mathsf{gpd}) \quad \text{and} \quad \mathsf{Span}(\mathsf{gpd}^{\mathsf{f}}) \xrightarrow{\sim} \mathsf{biset}^{\mathsf{bif}}(\mathsf{gpd}) ,$$

without any need to first throw away the non-invertible 2-cells as per Miller's result (see Remark 2.94). This will appear and be put to good use in [2].

References

[1] P. Balmer and I. Dell'Ambrogio. *Mackey 2-functors and Mackey 2-motives*. EMS Monographs in Mathematics. European Mathematical Society (EMS), Zürich, 2020.

[2] P Balmer and I. Dell'Ambrogio. Cohomological Mackey 2-functors. Preprint arXiv:2103.03974, 2021.

[3] F. Borceux. *Handbook of categorical algebra. 1*, volume 50 of *Encyclopedia of Mathematics and its Applications*. Cambridge University Press, Cambridge, 1994. Basic category theory.

[4] S. Bouc. *Biset functors for finite groups*, volume 1990 of *Lecture Notes in Mathematics*. Springer-Verlag, Berlin, 2010.

[5] S. Bouc. Fused Mackey functors. *Geom. Dedicata*, 176:225–240, 2015.

[6] M. Bunge and S. Lack. van Kampen theorems for toposes. *Adv. Math.*, 179(2):291–317, 2003.

[7] A. Carboni, S. Lack, and R. F. C. Walters. Introduction to extensive and distributive categories. *J. Pure Appl. Algebra*, 84(2):145–158, 1993.

[8] G. Carlsson. A survey of equivariant stable homotopy theory. *Topology*, 31(1):1–27, 1992.

[9] I. Dell'Ambrogio and J. Huglo. On the comparison of spans and bisets. *Cahiers Topologie Géom. Différentielle Catég.*, 62(1):63–104, 2021.

[10] A. W. M. Dress. Contributions to the theory of induced representations. In *Algebraic K-theory, II: "Classical" algebraic K-theory and connections with arithmetic (Proc. Conf., Battelle Memorial Inst., Seattle, Wash., 1972)*, pages 183–240. Lecture Notes in Math., Vol. 342, 1973.

[11] N. Ganter. Global Mackey functors with operations and n-special lambda rings. Preprint arXiv:1301.4616v1, 2013.

[12] J. A. Green. Axiomatic representation theory for finite groups. *J. Pure Appl. Algebra*, 1(1):41–77, 1971.

[13] I. Hambleton, L. R. Taylor, and E. B. Williams. Mackey functors and bisets. *Geom. Dedicata*, 148:157–174, 2010.

[14] A. E. Hoffnung. The Hecke bicategory. *Axioms*, 1(3):231–323, 2012.

[15] J. Huglo. Functorial precomposition with applications to Mackey functors and biset functors. PhD thesis, Université de Lille, 2019.

[16] J. Ibarra, A. G. Raggi-Cárdenas, and N. Romero. The additive completion of the biset category. *J. Pure Appl. Algebra*, 222(2):297–315, 2018.

[17] A. Joyal and R. Street. Pullbacks equivalent to pseudopullbacks. *Cahiers Topologie Géom. Différentielle Catég.*, 34(2):153–156, 1993.

[18] L. G. Lewis, Jr. When projective does not imply flat, and other homological anomalies. *Theory and Applications of Categories*, 5:202–250, 1999.

[19] L. G. Lewis, Jr., J. P. May, M. Steinberger, and J. E. McClure. *Equivariant stable homotopy theory*, volume 1213 of *Lecture Notes in Mathematics*. Springer-Verlag, Berlin, 1986. With contributions by J. E. McClure.

[20] H. Lindner. A remark on Mackey-functors. *Manuscripta Math.*, 18(3):273–278, 1976.

[21] J. Lurie. *Higher topos theory*, volume 170 of *Annals of Mathematics Studies*. Princeton University Press, Princeton, NJ, 2009.

[22] S. MacLane. *Categories for the working mathematician.* Springer-Verlag, New York-Berlin, 1971. Graduate Texts in Mathematics, Vol. 5.

[23] H. Miller. The Burnside bicategory of groupoids. *Bol. Soc. Mat. Mex. (3)*, 23(1):173–194, 2017.

[24] H. Nakaoka. Biset functors as module Mackey functors and its relation to derivators. *Comm. Algebra*, 44(12):5105–5148, 2016.

[25] H. Nakaoka. A Mackey-functor theoretic interpretation of biset functors. *Adv. Math.*, 289:603–684, 2016.

[26] E. Panchadcharam and R. Street. Mackey functors on compact closed categories. *J. Homotopy Relat. Struct.*, 2(2):261–293, 2007.

[27] C. Rezk. A model category for categories. Available on the author's homepage https://faculty.math.illinois.edu/~rezk/papers.html, 1999.

[28] S. Schwede. *Global homotopy theory*, volume 34 of *New Mathematical Monographs*. Cambridge University Press, Cambridge, 2018.

[29] P. Symonds. A splitting principle for group representations. *Comment. Math. Helv.*, 66(2):169–184, 1991.

[30] P. Webb. Two classifications of simple Mackey functors with applications to group cohomology and the decomposition of classifying spaces. *J. Pure Appl. Algebra*, 88(1-3):265–304, 1993.

[31] P. Webb. A guide to Mackey functors. In *Handbook of algebra, Vol. 2*, volume 2 of *Handb. Algebr.*, pages 805–836. Elsevier/North-Holland, Amsterdam, 2000.

3
Chromatic Fracture Cubes

Omar Antolín-Camarena[a]

Tobias Barthel[b]

Abstract

In this note, we construct a higher-dimensional version of the chromatic fracture square. We then categorify the resulting chromatic fracture cubes obtaining a decomposition of the category of $E(n)$-local spectra into monochromatic pieces.

3.1 Introduction

The chromatic fracture square displayed in Figure 3.2 is a fundamental tool in the chromatic perspective on stable homotopy theory; originally due to Bousfield, an essentially equivalent statement can be found for example in [11, Thm. 6.19] or [7, Thm. 3.3]. It can be interpreted as the chromatic analogue of the arithmetic pullback square (Figure 3.1) and

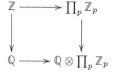

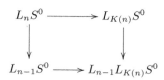

Figure 3.1 Arithmetic pullback square

Figure 3.2 Chromatic fracture square

thus expresses a fundamental local-to-global principle in stable homotopy

[a] Instituto de Matemáticas, National Autonomous University of Mexico
[b] Max Planck Institute for Mathematics

theory. As such, it has been repeatedly used to first study problems in an appropriate local context and to then reassemble the results. Most noticeably, this approach is used in the construction and study of the spectrum of topological modular forms, see Behrens' chapter in [8].

The goal of this note is twofold: First, we construct a higher dimensional version of the chromatic fracture square, well-known to the experts, which allows one to explicitly decompose the $E(n)$-localization of a spectrum into its $K(t)$-local pieces for $0 \le t \le n$. Second, we obtain an ∞-categorical decomposition of the $E(n)$-local category into diagram categories of $K(t)$-local categories for $0 \le t \le n$. The starting point is the homotopy pullback square of ∞-categories

$$
\begin{array}{ccc}
\mathrm{Sp}_{E(n)} & \xrightarrow{\;L_{K(n)}\;} & \mathrm{Sp}_{K(n)} \\
\downarrow & & \downarrow{\scriptstyle L_{n-1}} \\
\mathrm{Sp}_{E(n-1)}^{\Delta^1} & \xrightarrow[\mathrm{proj}_1]{} & \mathrm{Sp}_{E(n-1)},
\end{array}
$$

where $\mathrm{Sp}_{E(n-1)}^{\Delta^1}$ denotes the ∞-category of functors from Δ^1 to $\mathrm{Sp}_{E(n-1)}$, proj_1 is projection on the target, and the left vertical functor sends an $E(n)$-local spectrum X to the map $L_{n-1}X \to L_{n-1}L_{K(n)}X$. Roughly speaking, we then obtain our decomposition of $\mathrm{Sp}_{E(n)}$ iteratively by downward induction on n. We refer to Example 3.8 (objectwise) and Figure 3.4 (categorical) for explicit displays of our chromatic decompositions.

In fact, our methods only require as in input the identity of Bousfield classes

$$
\langle E(n) \rangle = \langle K(0) \oplus K(1) \oplus \cdots \oplus K(n) \rangle
$$

together with the orthogonality relations between Morava K-theories of different heights. Consequently, we work in a slightly more general setting, see Proposition 3.7, Theorem 3.12, and Theorem 3.19 for precise statements of our results.

The proof uses the characterization of the total homotopy fiber of an n-cube as a right adjoint; since we do not know of a published reference for this fact, we include the argument. This then allows one to easily deduce the iterative construction of the total homotopy fiber, as can be found for instance in [10].

3.1.1 Related work

Categorical decompositions of triangulated categories have been studied in a variety of contexts, a prominent example being the work of Beilinson, Bernstein, and Deligne on perverse sheaves [5]. The main contribution of the present chapter is to consider a higher-dimensional reconstruction of a stable ∞-category from its irreducible pieces in the simplest non-trivial case. Subsequently, several significant generalizations appeared, and we include a short and selective overview here.

In unpublished work [9], Glasman constructs ∞-categorical stratifications along finite posets, with applications to equivariant homotopy theory and Goodwillie calculus. Our main result is essentially the special case of the poset $[n] = 1 < 2 < \cdots < n$. Passing from finite posets to Noetherian spectral spaces and in the setting of tensor-triangular geometry [3], Balchin and Greenlees [2] established an adelic reconstruction theorem for a tensor-triangulated category $\mathcal{T}$ over its Balmer spectrum $\mathrm{Spc}(\mathcal{T})$. Weakening the Noetherian hypothesis is part of current work in progress of the second author with Balchin and Greenlees. Finally, Ayala, Mazel-Gee, and Rozenblyum [1] have recently developed a general framework for formulating and studying ∞-categorical decompositions. In particular, rewriting their lax limits in terms of strict homotopy limits as in [1, A.6] recovers the combinatorics of the present chapter.

We refer the interested to these papers for further references.

Acknowledgments

We thank Rune Haugseng for a helpful conversation about coCartesian fibrations, Nicholas Kuhn for comments on an earlier version, and Scott Balchin for useful suggestions and encouraging us to revise our original document. Moreover, we are grateful to the anonymous referee whose comments helped improve an earlier version of this chapter. The second author would also like to thank the Max Planck Institute for Mathematics for its hospitality.

3.2 Cubical homotopy preliminaries

Let $\mathcal{C} = (\mathcal{C}, *)$ be a pointed ∞-category with finite limits, and let $\mathrm{Fun}(\mathcal{P}(T), \mathcal{C})$ be the category of n-cubes in $\mathcal{C}$. Here, T is a fixed set with n elements and $\mathcal{P}(T)$ is its poset of subsets, ordered by inclusion.

We will also use $\mathcal{P}_{\neq\varnothing}(T)$ for the poset of non-empty subsets of T. If $\mathcal{F} \in \mathrm{Fun}(\mathcal{P}(T), \mathcal{C})$, then the total homotopy fiber $\mathrm{tfib}(\mathcal{F})$ of $\mathcal{F}$ is the fiber of the natural map $f \colon \mathcal{F}_\varnothing \to \lim \left(\mathcal{F}|_{\mathcal{P}_{\neq\varnothing}(T)} \right) = \lim_{S \neq \varnothing} \mathcal{F}(S)$ over $*$. Recall that $\mathcal{F}$ is said to be Cartesian if f is an equivalence; this implies that $\mathrm{tfib}(\mathcal{F})$ is contractible. If $\mathcal{C}$ is stable the converse is true: $\mathrm{tfib}(\mathcal{F})$ being contractible implies that f is an equivalence and thus that $\mathcal{F}$ is Cartesian. Indeed, if a morphism $f \colon X \to Y$ in a stable ∞-category has contractible fiber, then f is equivalent to the canonical map from X to the cofiber of $* \to X$, and thus an equivalence.

Proposition 3.1 *The total (homotopy) fiber of a cubical diagram is right adjoint to the functor*

$$\mathcal{C} \xrightarrow{\ \mathrm{in}_\varnothing\ } \mathrm{Fun}(\mathcal{P}(T), \mathcal{C})$$

sending an element $X \in \mathcal{C}$ to the n-cube with initial vertex X and the terminal object $*$ everywhere else.

Proof There are canonical natural transformations

$$\eta \colon \mathrm{id} \xrightarrow{\ \sim\ } \mathrm{tfib} \circ \mathrm{in}_\varnothing \qquad \epsilon \colon \mathrm{in}_\varnothing \circ \mathrm{tfib} \longrightarrow \mathrm{id}.$$

To describe the natural transformation

$$\epsilon \colon \Delta^1 \to \mathrm{Fun}(\mathrm{Fun}(\mathcal{P}(T), \mathcal{C}), \mathrm{Fun}(\mathcal{P}(T), \mathcal{C})),$$

think of it instead as an n-cube of natural transformations between functors on n-cubes, that is, as $\epsilon \colon \mathcal{P}(T) \to \mathrm{Fun}(\mathrm{Fun}(\mathcal{P}(T), \mathcal{C}), \mathcal{C})^{\Delta^1}$. By the universal property of limits of shape $\mathcal{P}_{\neq\varnothing}(T)$, giving such an n-cube ϵ is equivalent to giving $\epsilon(\varnothing)$, the restriction $\epsilon|_{\mathcal{P}_{\neq\varnothing}(T)}$ and a morphism $\epsilon(\varnothing) \to \lim \left(\epsilon|_{\mathcal{P}_{\neq\varnothing}(T)} \right)$.

In our case, this data is given as follows:

- $\epsilon(\varnothing)$ shall be the canonical natural transformation $\mathrm{tfib} \to \mathrm{ev}_\varnothing$;
- the restriction $\epsilon|_{\mathcal{P}_{\neq\varnothing}(T)}$ is simply the canonical natural transformation $* \to \mathrm{res}_{\mathcal{P}_{\neq\varnothing}(T)}$, whose limit is the natural transformation $* \to \lim_{\mathcal{P}_{\neq\varnothing}(T)}$;
- finally, the natural transformation $\epsilon(\varnothing) \to \lim \left(\epsilon|_{\mathcal{P}_{\neq\varnothing}(T)} \right)$ is the pullback square defining the total fiber:

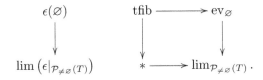

It is not difficult to verify that η and ϵ satisfy the triangle identities, thus providing the unit and counit for the stated adjunction. $\qquad\square$

Remark 3.2　We also outline a slightly more abstract perspective which may be useful in other contexts.

Let $i\colon \{\varnothing\} \hookrightarrow \mathcal{P}(T)$ be the inclusion with "complement"

$$j\colon \mathcal{P}_{\neq\varnothing}(T) \hookrightarrow \mathcal{P}(T).$$

The resulting pullback functor

$$i^*\colon \operatorname{Fun}(\mathcal{P}(T),\mathcal{C}) \longrightarrow \operatorname{Fun}(\{\varnothing\},\mathcal{C}) \simeq \mathcal{C}$$

is evaluation at $\varnothing$ and admits a right adjoint i_* given by Kan extension; in fact, we have $i_* \simeq \operatorname{in}_\varnothing$. Similarly, there is an adjunction (j^*, j_*) associated to j. For every $\mathcal{F} \in \operatorname{Fun}(\mathcal{P}(T),\mathcal{C})$, this "open-closed decomposition" of $\mathcal{P}(T)$ induces a natural pullback square of cubical diagrams

$$
\begin{array}{ccc}
\mathcal{F} & \longrightarrow & j_*j^*\mathcal{F} \\
\downarrow & & \downarrow \\
i_*i^*\mathcal{F} & \longrightarrow & i_*i^*j_*j^*\mathcal{F}.
\end{array}
$$

This can be checked directly using the fact that limits in diagram categories are computed componentwise, or more abstractly by appealing to the notion of recollement, see for example [14, Appendix A.8]. Now given $X \in \mathcal{C}$, apply $\operatorname{Hom}(i_*X, -)$ to this pullback square. Using the adjunction (i^*, i_*), we thus obtain a pullback square

$$
\begin{array}{ccc}
\operatorname{Hom}(i_*X, \mathcal{F}) & \longrightarrow & \operatorname{Hom}(j^*i_*X, j^*\mathcal{F}) \\
\downarrow & & \downarrow \\
\operatorname{Hom}(i^*i_*X, i^*\mathcal{F}) & \longrightarrow & \operatorname{Hom}(i^*i_*X, i^*j_*j^*\mathcal{F}).
\end{array}
$$

The right upper corner is contractible because $j^*i_* \simeq *$. The lower left corner is equivalent to $\operatorname{Hom}(X, \mathcal{F}(\varnothing))$. For the right lower corner, we see that $i^*j_*j^*\mathcal{F} \simeq \lim_{S \neq \varnothing} \mathcal{F}(S)$ by the universal property of the limit of the restriction of $\mathcal{F}$ to the punctured cube $\mathcal{P}_{\neq\varnothing}(T)$. Therefore, we obtain a canonical equivalence

$$\operatorname{Hom}(\operatorname{in}_\varnothing(X), \mathcal{F}) \simeq \operatorname{fib}\Big(\operatorname{Hom}(X, \mathcal{F}(\varnothing)) \to \operatorname{Hom}(X, \lim_{S \neq \varnothing} \mathcal{F}(S))\Big)$$

$$\simeq \operatorname{Hom}(X, \operatorname{tfib}(\mathcal{F})),$$

as desired.

As a consequence of Proposition 3.1 we obtain an easy proof of the fact that the total homotopy fiber can be computed by taking fibers of all the edges of a cube in a fixed direction and then taking total fibers of the resulting cube. More precisely, let $\mathcal{F} \in \mathrm{Fun}(\mathcal{P}(T), \mathcal{C})$ be an n-cube and $T' \subseteq T$. For any $S' \in \mathcal{P}(T')$, we can consider the cube $\mathcal{F}_{T \setminus T', S'} \in \mathrm{Fun}(\mathcal{P}(T \setminus T'), \mathcal{C})$ given by $S \mapsto \mathcal{F}(S \cup S')$. This yields a new cube $\mathrm{tfib}_{T'}(\mathcal{F}) \in \mathrm{Fun}(\mathcal{P}(T'), \mathcal{C})$ whose value on $S' \subseteq T'$ is $\mathrm{tfib}(\mathcal{F}_{T \setminus T', S'})$.

Corollary 3.3 *With notation as above, there is a natural equivalence* $\mathrm{tfib}(\mathcal{F}) = \mathrm{tfib}(\mathrm{tfib}_{T'}(\mathcal{F}))$ *for any* $T' \subseteq T$

Proof Observe that $\mathrm{in}_\varnothing \colon \mathcal{C} \to \mathrm{Fun}(\mathcal{P}(T), \mathcal{C})$ can be decomposed as follows

$$\mathcal{C} \xrightarrow{\mathrm{in}_\varnothing} \mathrm{Fun}(\mathcal{P}(T'), \mathcal{C}) \xrightarrow{\mathrm{in}_\varnothing} \mathrm{Fun}(\mathcal{P}(T \setminus T'), \mathrm{Fun}(\mathcal{P}(T'), \mathcal{C})) \xrightarrow{\sim} \mathrm{Fun}(\mathcal{P}(T), \mathcal{C}),$$

so the same is true for the right adjoint tfib. $\qquad\square$

Corollary 3.4 *If $\mathcal{C}$ is stable and $\mathcal{F} \in \mathrm{Fun}(\mathcal{P}(T), \mathcal{C})$ is a Cartesian n-cube and $T' = \{t\} \subseteq T$ a subset of size 1, then the fiber of $(n-1)$-cubes*

$$\mathrm{fib}(\mathcal{F}_{T \setminus \{t\}, \varnothing} \to \mathcal{F}_{T \setminus \{t\}, \{t\}})$$

is also Cartesian.

Proof This follows immediately from Corollary 3.3 and the fact that a cube in a stable category is Cartesian if and only if its total homotopy fiber is contractible. $\qquad\square$

In the same spirit, but moving away from total fibers, we give a formula for inductively computing limits of partial n-cubes $\mathcal{G} \colon \mathcal{P}_{\neq\varnothing}(T) \to \mathcal{C}$.

Proposition 3.5 *Let $\mathcal{G} \colon \mathcal{P}_{\neq\varnothing}(T) \to \mathcal{C}$ be a partial cube in $\mathcal{C}$. Let $t \in T$ be arbitrary and set $T' = T \setminus \{t\}$. Then there is a pullback square:*

$$\begin{array}{ccc}
\lim_{S \in \mathcal{P}_{\neq\varnothing}(T)} \mathcal{G}(S) & \longrightarrow & \lim_{S \in \mathcal{P}_{\neq\varnothing}(T')} \mathcal{G}(S) \\
\downarrow & & \downarrow \\
\mathcal{G}(\{t\}) & \longrightarrow & \lim_{S \in \mathcal{P}_{\neq\varnothing}(T')} \mathcal{G}(\{t\} \cup S)
\end{array}$$

Proof Let $\mathcal{Q} = \mathcal{P}_{\neq\varnothing}(\{a, b\}) \times \mathcal{P}_{\neq\varnothing}(T')$ and $\psi \colon \mathcal{Q} \to \mathcal{P}_{\neq\varnothing}(T)$ be the map of posets defined by $\psi(\{a\}, I) = \{t\}$, $\psi(\{b\}, I) = I$ and $\psi(\{a, b\}, I) = \{t\} \cup I$. We will show that ψ is (homotopy) initial. It suffices to check that for each $I \in \mathcal{P}_{\neq\varnothing}(T)$, the comma category $\psi \downarrow I$ is contractible. In all cases the comma category is a subposet of $\mathcal{Q}$:

- If $t \notin I$, $\psi \downarrow I$ is the subposet $\{(\{b\}, J) \in \mathcal{Q} : J \subseteq I\}$. This is contractible because it has a largest element, namely $(\{b\}, I)$.
- If $t \in I$, $\psi \downarrow I$ is the subposet $\mathcal{R} = \{(K, J) \in \mathcal{Q} : K = \{a\}$ **or** $J \subseteq I\}$. We can write $\mathcal{R}$ as a pushout of posets:

$$
\begin{array}{ccc}
\{\{a\}\} \times \mathcal{P}_{\neq\varnothing}(I) & \longrightarrow & \{\{a\}\} \times \mathcal{P}_{\neq\varnothing}(T') \\
\downarrow & & \downarrow \\
\mathcal{P}_{\neq\varnothing}(\{a,b\}) \times \mathcal{P}_{\neq\varnothing}(I) & \longrightarrow & \mathcal{R}
\end{array}
$$

The geometric realizations of the three posets besides $\mathcal{R}$ are cubes (of dimensions $|I|$, $n-1$ and $2+|I|$) and the top horizontal and left vertical maps realize to a face inclusion. This shows $\mathcal{R}$ is contractible.

Since ψ is initial, $\lim \mathcal{G}$ can be computed as $\lim(\mathcal{G} \circ \psi)$. This limit we compute as an iterated limit:

$$
\lim_{(K,J)\in\mathcal{Q}} \mathcal{G}(\psi(K,J)) = \lim_{K\in\mathcal{P}_{\neq\varnothing}(\{a,b\})} \left(\lim_{J\in\mathcal{P}_{\neq\varnothing}(T')} \mathcal{G}(\psi(K,J)) \right).
$$

To conclude we identify the three terms in the pullback with the ones in the statement of the proposition:

(1) $\lim_{J\in\mathcal{P}_{\neq\varnothing}(T')} \mathcal{G}(\psi(\{a\}, J)) = \lim_{J\in\mathcal{P}_{\neq\varnothing}(T')} \mathcal{G}(\{1\}) \cong \mathcal{G}(\{1\})$, since the indexing category $\mathcal{P}_{\neq\varnothing}(T')$ is contractible.

(2) $\lim_{J\in\mathcal{P}_{\neq\varnothing}(T')} \mathcal{G}(\psi(\{b\}, J)) = \lim_{J\in\mathcal{P}_{\neq\varnothing}(T')} \mathcal{G}(J)$.

(3) $\lim_{J\in\mathcal{P}_{\neq\varnothing}(T')} \mathcal{G}(\psi(\{a,b\}, J)) = \lim_{J\in\mathcal{P}_{\neq\varnothing}(T')} \mathcal{G}(\{1\} \cup J)$.

$\square$

3.3 The chromatic fracture cube

Let Sp be the stable ∞-category of spectra [14], and denote Bousfield localization [6] at a spectrum E by L_E, with associated category Sp_E of E-local spectra. Two spectra E and F are said to be Bousfield equivalent if $L_E = L_F$; in this case, we write $\langle E \rangle = \langle F \rangle$. An endofunctor on Sp will be called E-local if it takes values in the category of E-local spectra. If we are given a collection $\{F(1), F(2), \ldots, F(n)\}$ of spectra, for any set $S = \{i_1, \ldots, i_k\}$ with $1 \leq i_1 < i_2 < \cdots < i_k \leq n$, we write L_S for the composite $L_{F(i_1)} \ldots L_{F(i_k)}$. The general form of the chromatic fracture cube takes the following form, generalizing the well-known fracture square stated in the introduction.

Construction 3.6 Suppose $\{F(1), \ldots, F(n)\}$ is any collection of spectra. We inductively define an n-cube $\mathcal{F} : \mathcal{P}(\{1, \ldots, n\}) \to \mathrm{End}(\mathrm{Sp})$, whose vertices will turn out to be given by $S \mapsto L_S$, as follows:

- If $n = 1$, the cube is simply the natural localization morphism id $\to L_{F(i)}$.
- For $n > 1$, inductively construct the cube $\mathcal{F}'$ on $\mathcal{P}(\{2, \ldots, n\})$, and get the full n-cube $\mathcal{F}$ as the morphism of $(n-1)$-cubes $\mathcal{F}' \to L_{F(1)}\mathcal{F}'$ given by the naturality of the localization morphisms id $\to L_{F(1)}$.

We refer to Example 3.8 below for examples displaying the combinatorics of this construction.

Proposition 3.7 *Suppose* $\{F(1), \ldots, F(n)\}$ *is a collection of spectra such that* $L_{F(j)}L_{F(i)} = 0$ *for all* $j > i$. *If* $\mathcal{F} \in \mathrm{Fun}(\mathcal{P}(\{1, \ldots, n\}), \mathrm{End}(\mathrm{Sp}))$ *is the* n-*cube given by Construction 3.6 then*

$$\lim_{S \neq \varnothing} \mathcal{F}(S) = L_E,$$

where E *is any spectrum Bousfield equivalent to* $F(1) \oplus \cdots \oplus F(n)$.

Proof Let $P = \lim_{S \neq \varnothing} \mathcal{F}(S)$ be the limit with legs $f_i \colon P \to L_{F(i)}$. The localization maps $\eta_i \colon L_E \to L_{F(i)}L_E = L_{F(i)}$ induce a natural map from L_E to this limit,

$$\eta \colon L_E \longrightarrow P.$$

Since $F(i)$-locals are clearly E-local and locality is preserved under limits, P is E-local. It therefore suffices to show that η is an $F(i)$-equivalence for all $1 \leq i \leq n$. Fix i and consider the commutative triangle

Because η_i is an $F(i)$-equivalence by definition, we only need to show that so is $f_i \colon P \to L_{F(i)}$. To this end, apply L_{F_i} to the limit cube $\tilde{\mathcal{F}}$ with P as initial vertex, $\tilde{\mathcal{F}}(S) = \mathcal{F}(S)$ for $S \neq \varnothing$, and the natural maps. This yields a cube $L_{F_i}\tilde{\mathcal{F}} \in \mathrm{Fun}(\mathcal{P}(\{1, \ldots, n\}), \mathrm{End}(\mathrm{Sp}))$ with three properties:

- It is again Cartesian as L_{F_i} is exact.
- $L_{F(i)}\tilde{\mathcal{F}}(S) = 0$ whenever S contains an element smaller than i, because $L_{F(i)}L_{F(k)} = 0$ for $k < i$.

- For $S \neq \varnothing$ and $i \notin S$, the edges from S to $S \cup \{i\}$ are all equivalences since either both vertices are 0 by the previous point or, if $\min(S) < i$, the edge $\mathcal{F}(S) \to \mathcal{F}(S \cup \{i\})$ of $\mathcal{F}$ is equivalent to the localization map $\mathcal{F}(S) \to L_{F(i)}\mathcal{F}(S)$ by construction, so applying $L_{F(i)}$ will yield an equivalence.

By Corollary 3.4, taking fibers in the $\{i\}$ direction produces a Cartesian $(n-1)$-cube, which by the third item above is $\operatorname{in}_{\varnothing}(\operatorname{fib}(L_{F(i)}P \to L_{F(i)}))$. Therefore, $L_{F(i)}P \cong L_{F(i)}$ and the claim follows. $\qquad\qquad\square$

Example 3.8 The Morava K-theories $K(0), \ldots, K(n)$ satisfy the conditions of Proposition 3.7, as can be seen as follows[1]: Since $K(i)$-local spectra are $E(i)$-local and $L_{E(i)}$ is smashing, the claim that $L_{K(j)}L_{K(i)}X = 0$ for $i < j$ and any $X \in \mathrm{Sp}$ can be reduced to showing that $L_{K(j)}L_{E(i)}S^0 = 0$, which in turn is a consequence of $K(j) \otimes E(i) = 0$ if $i < j$. In particular, if $i < j$, then there is a pullback square

$$
\begin{array}{ccc}
L_{K(i) \oplus K(j)} & \longrightarrow & L_{K(j)} \\
\downarrow & & \downarrow \\
L_{K(i)} & \longrightarrow & L_{K(i)}L_{K(j)}.
\end{array}
$$

Similarly, for $F(1) = E(n-1)$ and $F(2) = K(n)$ we get the usual pullback square

$$
\begin{array}{ccc}
L_{E(n)} & \longrightarrow & L_{K(n)} \\
\downarrow & & \downarrow \\
L_{E(n-1)} & \longrightarrow & L_{E(n-1)}L_{K(n)}
\end{array}
$$

using the identity of Bousfield classes $\langle E(n-1) \oplus K(n) \rangle = \langle E(n) \rangle$.

The three-dimensional chromatic fracture for $L_{E(2)}$ takes the form of the following Cartesian cube:

3.4 A description of the category of local objects

As in the previous section, let $\{F(1), \ldots, F(n)\}$ be a collection of spectra such that $L_{F(j)}L_{F(i)} = 0$ for all $j > i$, fixed for the remainder of the section, and let $T = \{1, \ldots, n\}$. We will inductively construct a category

---

[1] As Nick Kuhn points out, this can also be proven without using consequences of the nilpotence theorem, see for example [8, Ch. 6, Thm 3.6].

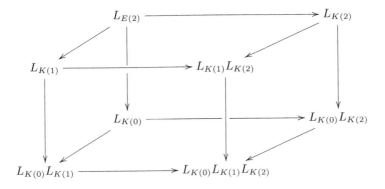

Figure 3.3 A three-dimensional chromatic fracture cube

$\mathcal{C}_S$ of partial cubes $\mathcal{P}_{\neq\varnothing}(S)$ for $S \subset T$, and prove that $\mathcal{C}_T$ is equivalent to Sp_E for any spectrum E with the same Bousfield class as $F(1)\oplus\cdots\oplus F(n)$.

Construction 3.9   For $\varnothing \neq S \subset T$ we let $\mathcal{C}_S$ be the full subcategory of the diagram category $\mathrm{Fun}(\mathcal{P}_{\neq\varnothing}(S), \mathrm{Sp})$ spanned by certain partial cubes $\mathcal{G}$ chosen as follows:

- If $S = \{i\}$ is a singleton, we take all $\mathcal{G}$ such that the unique value of $\mathcal{G}$, namely $\mathcal{G}(S)$, is $F(i)$-local.
- If $S = \{i\} \cup S'$ where $i = \min(S) \notin S'$, we take all $\mathcal{G}$ such that:

  (1) $\mathcal{G}' := \mathcal{G}|_{\mathcal{P}_{\neq\varnothing}(S')}$ belongs to $\mathcal{C}_{S'}$, and
  (2) if we think of $\mathcal{G}|_{\mathcal{P}(S)\setminus\{\varnothing,\{i\}\}}$ as a morphism between diagrams of shape $\mathcal{P}_{\neq\varnothing}(S')$, namely $\mathcal{G}' \to \mathcal{G}'|_{\{U\subseteq S: i\in U\}}$, this morphism is the natural morphism $\mathcal{G}' \to L_{F(i)}\mathcal{G}'$.

Remark 3.10   From the definitions it is clear that for any $\mathcal{G} \in \mathcal{C}_T$ we have that:

- For any $S \subseteq T$, $\mathcal{G}(S)$ is $F(\min(S))$-local.
- For every k, $\mathcal{G}|_{\{S\subseteq T:\max(S)=k\}}$ is just the cube from Construction 3.6 applied to $\mathcal{G}(\{k\})$.

This construction becomes much clearer with an example.

Example 3.11 The ∞-category $\mathcal{C}_{\{1,2,3\}}$ is equivalent to the full subcategory of $\mathrm{Fun}(\mathcal{P}_{\neq\varnothing}(\{1,2,3\}),\mathrm{Sp})$ on the diagrams of the form:

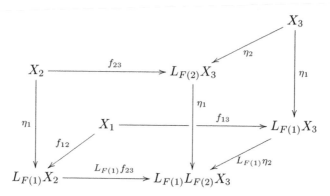

where X_i is $F(i)$-local for $i = 1, 2, 3$ and the morphisms labeled η_i are the natural maps $Y \to L_{F(i)}Y$. Notice that the diagram is determined just by $X_1, X_2, X_3, f_{12}, f_{13}, f_{23}$ and a homotopy showing the bottom square commutes.

As a simpler example, the ∞-category $\mathcal{C}_{\{1,2\}}$ is the category of cospans of the form of the left face of the above partial cube. Those cospans are determined by X_1, X_2 and f_{12}.

Theorem 3.12 *If E is any spectrum Bousfield equivalent to $F(1) \oplus \cdots \oplus F(n)$, the ∞-category $\mathcal{C}_T$ constructed above is equivalent to Sp_E. Moreover, the following functors are mutually inverse equivalences:*

- $\lim: \mathcal{C}_T \to \mathrm{Sp}_E$ *given by* $\mathcal{G} \mapsto \lim_{S \neq \varnothing} \mathcal{G}(S)$, *and*
- $\mathcal{F}_{\neq\varnothing}: \mathrm{Sp}_E \to \mathcal{C}_T$ *given by* $X \mapsto \mathcal{F}[X]|_{\mathcal{P}_{\neq\varnothing}(T)}$ *where $\mathcal{F}$ is the n-cube of functors from Construction 3.6 and $\mathcal{F}[X]$ denotes the cube obtained by applying those functors to X.*

Proof First of all let us show those functors are well defined. This is clear for $\mathcal{F}_{\neq\varnothing}$ by construction. For $\lim$ it is because all spectra in the image of a given partial cube $\mathcal{G}$ are $F(i)$-local for some i and thus also E-local.

Now we will show the functors are mutually inverse. First, for any E-local spectrum X, the canonical map $\lim(\mathcal{F}_{\neq\varnothing}[X]) \to X$ (coming from the diagram $\mathcal{F}[X]$, which is a cone over $\mathcal{F}_{\neq\varnothing}[X]$) is an equivalence by Proposition 3.7.

For the other composite, let $\mathcal{G} \in \mathcal{C}_T$ and let $P = \lim\mathcal{G}$. We must show that $\mathcal{F}_{\neq\varnothing}(P) \cong \mathcal{G}$. Extend $\mathcal{G}$ to a Cartesian cube $\tilde{\mathcal{G}}$ with $\tilde{\mathcal{G}}(\varnothing) = P$. First we show that $L_{F(i)}P \cong \mathcal{G}(\{i\})$, and moreover, that the $\varnothing \to \{i\}$ edge in the n-cube $\tilde{\mathcal{G}}$ is the localization map $P \to L_{F(i)}P$ under this equivalence.

To this end, apply $L_{F(i)}$ to $\tilde{\mathcal{G}}$. For any $i \notin S \subseteq T$ consider the edge of $L_{F(i)}\tilde{\mathcal{G}}$ from S to $S \cup \{i\}$. There are three cases:

- If $S = \varnothing$, the edge has the form $L_{F(i)}P \to L_{F(i)}\mathcal{G}(\{i\})$. This is the morphism we wish to show is an equivalence.

- If $\min(S) < i$, we have $L_{F(i)}\mathcal{G}(S) = L_{F(i)}\mathcal{G}(S \cup \{i\}) = 0$, because both $\mathcal{G}(S)$ and $\mathcal{G}(S \cup \{i\})$ are $F(\min(S))$-local. Therefore these edges are equivalences.

- If $\min(S) > i$, then the edge $\mathcal{G}(S) \to \mathcal{G}(\{i\} \cup S)$ is the natural map $\mathcal{G}(S) \to L_{F(i)}\mathcal{G}(S)$ as this edge is contained in $\mathcal{G}|_{\{U \subset T : \max(U) = \max(S)\}}$ which is a cube obtained from Construction 3.6 applied to $\mathcal{G}(\{\max(S)\})$. This edge, of course, becomes an equivalence after applying $L_{F(i)}$.

Now the cube $L_{F(i)}\tilde{\mathcal{G}}$ is Cartesian because $\tilde{\mathcal{G}}$ was and $L_{F(i)}$ is exact. This means that taking fibers in the direction of $\{i\}$ must lead to a Cartesian $(n-1)$-cube. By the above case analysis, that $(n-1)$-cube is simply in$_\varnothing(\text{fib}(L_{F(i)}P \to L_{F(i)}\mathcal{G}(\{i\})))$, from which we conclude that $\text{fib}(L_{F(i)}P \to L_{F(i)}\mathcal{G}(\{i\}))$ is 0 and thus $L_{F(i)}P \to L_{F(i)}\mathcal{G}(\{i\})$ is an equivalence, as desired.

At this point we are close to showing that $\tilde{\mathcal{G}}$ and $\mathcal{F}(P)$ are equivalent n-cubes: we have shown that they have equivalent objects at all vertices and also many of the maps agree, but we have not shown for example that the map $\mathcal{G}(\{1\}) \to \mathcal{G}(\{1, i\})$ is $L_{F(1)}(P \to \mathcal{G}(\{i\}))$.

To conclude, consider taking fibers of $\tilde{\mathcal{G}}$ in the $\{1\}$ direction to get an $(n-1)$-cube $\tilde{\mathcal{G}}'$. From the argument above and the exactness of $L_{F(1)}$, we know that this $(n-1)$-cube vanishes if we apply $L_{F(1)}$ to it, so that the n-cube $\tilde{\mathcal{G}}$ when thought of as a map of $(n-1)$-cubes $\tilde{\mathcal{G}}|_{\{S : 1 \notin S\}} \to \mathcal{G}|_{\{S : 1 \in S\}}$ is just $L_{F(1)}$-localization.

Now applying $L_{F(2)}$ to $\tilde{\mathcal{G}}$, we get a Cartesian n-cube whose "bottom" face, $L_{F(2)}\tilde{\mathcal{G}}|_{\{S : 1 \in S\}}$, vanishes. This means the top face, $L_{F(2)}\tilde{\mathcal{G}}|_{\{S : 1 \notin S\}}$, is also Cartesian and we can recursively apply the argument of the previous paragraph to conclude that $\tilde{\mathcal{G}} \cong \mathcal{F}(P)$. $\qquad \square$

Remark 3.13 The special case $n = 2$ of the previous result appears as Remark 7 in [13, Lecture 23].

3.5 A decomposition of the category of local objects

As in the previous section, let $\{F(1), \dots, F(n)\}$ be a collection of spectra such that $L_{F(j)} L_{F(i)} = 0$ for all $j > i$, and let $T = \{1, \dots, n\}$. In this section we will describe a partial n-cube of ∞-categories whose limit is Sp_E where E is Bousfield equivalent to $F(1) \oplus \cdots \oplus F(n)$. To do that we will need some combinatorial preliminaries. A two-dimensional instance of the construction has already been displayed in the introduction, while three-dimensional illustrations can be found below in Example 3.18 and Figure 3.4.

Definitions 3.14   For $\varnothing \neq S \subseteq S' \subseteq T$, define $\alpha(S)$, $\beta(S, S')$ and $\theta_{S,S'}$:

- $\alpha(S) = \{U \subseteq T : S \subseteq U, \min(S) = \min(U)\}$
- $\beta(S, S') = \{V \subseteq [\min(S'), \min(S) - 1] : S' \cap [\min(S'), \min(S) - 1] \subseteq V\}$
- $\theta_{S,S'} : \alpha(S') \to \beta(S, S') \times \alpha(S)$
- $U \mapsto (U \cap [\min(S'), \min(S) - 1], U \cap [\min(S), n])$

where we have repurposed traditional interval notation to denote intervals of integers.

We will regard $\alpha(S)$ and $\beta(S, S')$ as posets, ordering them by inclusion, which makes $\theta_{S,S'}$ a map of posets. Notice that $\alpha(S)$ and $\beta(S, S')$ are isomorphic to posets of all subsets of some set, so that diagrams of shape $\alpha(S)$ or $\beta(S, S')$ are cubical diagrams.

Construction 3.15   Let $\iota \colon \mathrm{Sp}_{F(\min(S))} \to \mathrm{Sp}$ denote the natural inclusion functor and write $\mathcal{F}$ for the cube of functors from Construction 3.6 restricted to $\beta(S, S') \subseteq \mathcal{P}(T)$. In more detail, this restriction is a functor $\beta(S, S') \to \mathrm{End}(\mathrm{Sp})$, which we can think of instead as a functor $\mathrm{Sp} \to \mathrm{Fun}(\beta(S, S'), \mathrm{Sp})$ and then replace Sp in the target by $\mathrm{Sp}_{F(\min(S'))}$ since all $V \in \beta(S, S')$ satisfy $\min(V) = \min(S')$.

   We construct a partial n-cube of ∞-categories $\mathcal{G} \colon \mathcal{P}_{\neq \varnothing}(T) \to \mathrm{Cat}_\infty$ as follows:

- The vertices are given by $\mathcal{G}(S) = \mathrm{Sp}_{F(\min(S))}^{\alpha(S)} := \mathrm{Fun}(\alpha(S), \mathrm{Sp}_{F(\min(S))})$.
- For $S \subset S'$, the functor $\mathcal{G}(S \subseteq S') : \mathcal{G}(S) \to \mathcal{G}(S')$ is given by the

composite

$$\mathrm{Fun}(\alpha(S), \mathrm{Sp}_{F(\min(S))}) \xrightarrow{\iota \circ -} \mathrm{Fun}(\alpha(S), \mathrm{Sp})$$

$$\xrightarrow{\mathcal{F}|_{\beta(S,S')} \circ -} \mathrm{Fun}(\alpha(S), \mathrm{Fun}(\beta(S, S'), \mathrm{Sp}_{F(\min(S'))}))$$

$$\xrightarrow{\cong} \mathrm{Fun}(\beta(S, S') \times \alpha(S), \mathrm{Sp}_{F(\min(S'))})$$

$$\xrightarrow{- \circ \theta} \mathrm{Fun}(\alpha(S'), \mathrm{Sp}_{F(\min(S'))}).$$

Remark 3.16 In order to see that Construction 3.15 indeed gives a coherent diagram of ∞-categories, one has to make use of the universal property of localizations. We omit the details here; one way of making it precise can be found in [4]. A more general construction is contained in [1].

Remark 3.17 Unwinding the definitions in Construction 3.15 and Construction 3.6, we see that for, $\varnothing \neq S \subseteq S'$, the functor $\mathcal{G}(S) \to \mathcal{G}(S')$ sends a cube $X \colon \alpha(S) \to \mathrm{Sp}_{F(\min(S))}$ to the cube $X' \colon \alpha(S') \to \mathrm{Sp}_{F(\min(S'))}$ given on vertices by

$$X'(U) = L_{U \cap [\min(S'), \min(S) - 1]} X(U \cap [\min(S), n]).$$

Notice that if $\min(S') < \min(S)$, the formula shows $X'(U)$ is $L_{F(\min(S'))}$-local, as it should be. Also, when $\min(S') = \min(S)$, there is no localization at all and X' is simply the restriction of X to the face $\alpha(S') \subseteq \alpha(S)$ (this inclusion does not hold when $\min(S) \neq \min(S')$).

This definition also becomes much clearer with an example:

Example 3.18 Let $n = 3$. The diagram $\mathcal{G}$ looks like:

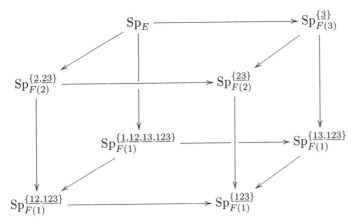

where we have used shorthand for the elements of the various $\alpha(S)$: $\underline{13}$ denotes the set $\{1,3\}$, for example.

Though $\mathcal{G}(\varnothing)$ is not defined, we have put Sp_E in that corner, since Theorem 3.19 will show that this produces a Cartesian cube for any E which is Bousfield equivalent to $F(1) \oplus F(2) \oplus F(3)$.

A square in $\mathrm{Sp}_{F(1)}^{\{1,\underline{12},\underline{13},\underline{123}\}}$ should be thought of as being

$$
\begin{array}{ccc}
L_{F(1)}X & \longrightarrow & L_{F(1)}L_{F(3)}X \\
\downarrow & & \downarrow \\
L_{F(1)}L_{F(2)}X & \longrightarrow & L_{F(1)}L_{F(2)}L_{F(3)}X
\end{array}
$$

for some spectrum X, and the two functors out of $\mathrm{Sp}_{F(1)}^{\{1,\underline{12},\underline{13},\underline{123}\}}$ are projection to the faces

$$L_{F(1)}L_{F(2)}X \longrightarrow L_{F(1)}L_{F(2)}L_{F(3)}X$$

and

$$L_{F(1)}L_{F(3)}X \longrightarrow L_{F(1)}L_{F(2)}L_{F(3)}X.$$

The vertical functor $\mathrm{Sp}_{F(3)}^{\{3\}} \to \mathrm{Sp}_{F(1)}^{\{\underline{13},\underline{123}\}}$ sends a $F(3)$-local spectrum X to the morphism $L_{F(1)}X \to L_{F(1)}L_{F(2)}X$ you get by applying $L_{F(1)}$ to the localization map $X \to L_{F(2)}X$.

Our decomposition of the category of $\bigoplus_{i=1}^{n} F(i)$-local objects can now be stated as follows.

Theorem 3.19 *If E is any spectrum Bousfield equivalent to $F(1) \oplus \cdots \oplus F(n)$, then there is an equivalence of stable presentable ∞-categories*

$$\mathrm{Sp}_E \xrightarrow{\ \sim\ } \lim_{S \neq \varnothing} \mathcal{G}(S)$$

where $\mathcal{G}$ is given by Construction 3.15.

Proof For any $S \in T$, observe that there is a natural functor

$$\Phi_S \colon \mathcal{C}_T \longrightarrow \mathrm{Sp}_{F(\min(S))}^{\alpha(S)},$$

sending a partial T-cube $\mathcal{X}$ to the cube given by restriction,

$$\Phi_S(\mathcal{X}) = \mathcal{X}|_{\alpha(S)}.$$

The collection of these functors induces a natural functor

$$\Phi \colon \mathcal{C}_T \longrightarrow \lim_{S \neq \varnothing} \mathcal{G}(S)$$

and it suffices to show that this is an equivalence by Theorem 3.12.

We will now argue by induction on n, the case $n = 1$ being trivial. Proposition 3.5 applied with $t = 1$ shows that there is a pullback diagram

$$
\begin{array}{ccc}
\lim_{S \neq \varnothing} \mathcal{G}(S) & \longrightarrow & \lim_{S \in \alpha(\underline{1}) \setminus \{\underline{1}\}} \mathcal{G}(S \setminus \underline{1}) \\
\downarrow & & \downarrow \\
\mathrm{Sp}_{F(1)}^{\alpha(\underline{1})} & \longrightarrow & \lim_{S \in \alpha(\underline{1}) \setminus \{\underline{1}\}} \mathcal{G}(S)
\end{array}
$$

(where again $\underline{1} = \{1\}$) and the inductive hypothesis gives

$$
\lim_{S \in \alpha(\underline{1}) \setminus \{\underline{1}\}} \mathcal{G}(S \setminus \underline{1}) = \mathcal{C}_{\{2,\dots,n\}}.
$$

Also, since all the edges in the bottom face of $\mathcal{G}$ are restrictions:

$$
\lim_{S \in \alpha(\underline{1}) \setminus \{\underline{1}\}} \mathcal{G}(S) = \lim_{S \in \alpha(\underline{1}) \setminus \{\underline{1}\}} \mathrm{Sp}_{F(1)}^{\alpha(S)}
$$

$$
= \mathrm{Sp}_{F(1)}^{\mathrm{colim}_{S \in \alpha(\underline{1}) \setminus \{\underline{1}\}} \alpha(S)}
$$

$$
= \mathrm{Sp}_{F(1)}^{\alpha(\underline{1}) \setminus \{\underline{1}\}},
$$

so it suffices to show that the following commutative square is a pullback:

$$
\begin{array}{ccc}
\mathcal{C}_T & \xrightarrow{\ p\ } & \mathcal{C}_{\{2,\dots,n\}} \\
f \downarrow & & \downarrow L_{F(1)} \\
\mathrm{Sp}_{F(1)}^{\alpha(\underline{1})} & \xrightarrow[\ q\ ]{} & \mathrm{Sp}_{F(1)}^{\alpha(\underline{1}) \setminus \{\underline{1}\}}.
\end{array}
$$

This is intuitively clear: objects of the pullback can be described by giving a diagram $\mathcal{Y} \in \mathcal{C}_{\{2,\dots,n\}}$, a diagram $\mathcal{X} \in \mathrm{Sp}_{F(1)}^{\alpha(\underline{1})}$, and an equivalence $\mathcal{X}|_{\alpha(\underline{1}) \setminus \{\underline{1}\}} \to L_{F(1)}\mathcal{Y}$. That data clearly assembles to make a diagram in $\mathcal{Z} \in \mathcal{C}_T$ with top (partial) face $\mathcal{Z}|_{\mathcal{P}(\{2,\dots,n\})} = \mathcal{Y}$ and bottom face $\mathcal{Z}|_{\alpha(\underline{1})} = \mathcal{X}$.

More formally, first note that it is easy to check that the horizontal arrows p and q in the above diagram are coCartesian fibrations and that the left vertical map f preserves coCartesian morphisms. Therefore, we can make use of [12, 2.4.4.4] to reduce the claim to checking that, for every $\mathcal{Y} \in \mathcal{C}_{\{2,\dots,n\}}$, the fiber over $\mathcal{Y}$ in the horizontal direction are equivalent via f, i.e.,

$$
(\mathcal{C}_T)_{\mathcal{Y}} \xrightarrow[f]{\ \sim\ } (\mathrm{Sp}_{F(1)}^{\alpha(\underline{1})})_{L_{F(1)}\mathcal{Y}}.
$$

Let us first consider the case $n = 2$. The fiber over $\mathcal{Y} = Y \in \mathrm{Sp}_{F(2)}$ is given by the full subcategory of $\mathcal{C}_{\{1,2\}}$ on object of shape

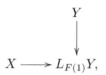

which are thus determined by the bottom morphism $X \to L_{F(1)}Y$. This category is then easily seen to be equivalent to the fiber $(\mathrm{Sp}_{F(1)}^{\alpha(1)})_{L_{F(1)}Y}$, hence the claim holds for $n = 2$. Now we can explain how to deduce the general case from here. We have a commutative diagram of fiber sequences:

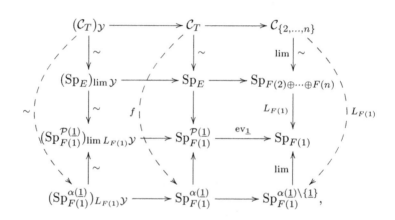

where the top right square is a pullback by Theorem 3.12, the middle right one is by the induction hypothesis applied to the pair $(F(1), F(2) \oplus \cdots \oplus F(n))$, and the the bottom right one is by construction. The claim follows.   □

Following up on Example 3.8 and Example 3.18, we conclude this section by making the decomposition of the category of $E(2)$-local spectra explicit:

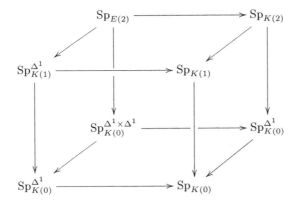

Figure 3.4 The decomposition of the category of $E(2)$-local spectra as a Cartesian cube.

References

[1] D. Ayala, A. Mazel-Gee, and N. Rozenblyum. Stratified noncommutative geometry. *arXiv e-prints*, page arXiv:1910.14602, October 2019.

[2] S. Balchin and J. P. C. Greenlees. Adelic models of tensor-triangulated categories. *Adv. Math.*, 375:107339, 2020.

[3] P. Balmer. Supports and filtrations in algebraic geometry and modular representation theory. *Amer. J. Math.*, 129(5):1227–1250, 2007.

[4] N. Bellumat and N. Strickland. Iterated chromatic localisation. *arXiv e-prints*, page arXiv:1907.07801, July 2019.

[5] A. A. Beĭlinson, J. Bernstein, and P. Deligne. Faisceaux pervers. In *Analysis and topology on singular spaces, I (Luminy, 1981)*, volume 100 of *Astérisque*, pages 5–171. Soc. Math. France, Paris, 1982.

[6] A. K. Bousfield. The localization of spectra with respect to homology. *Topology*, 18(4):257–281, 1979.

[7] A. K. Bousfield. On the telescopic homotopy theory of spaces. *Trans. Amer. Math. Soc.*, 353(6):2391–2426 (electronic), 2001.

[8] C. L. Douglas, J. Francis, A. G. Henriques, and M. A. Hill, editors. *Topological modular forms*, volume 201 of *Mathematical Surveys and Monographs*. American Mathematical Society, Providence, RI, 2014.

[9] S. Glasman. Stratified categories, geometric fixed points and a generalized Arone-Ching theorem. *arXiv e-prints*, page arXiv:1507.01976, July 2015.

[10] T. G. Goodwillie. Calculus. II. Analytic functors. *K-Theory*, 5(4):295–332, 1991/92.

[11] M. Hovey and N. P. Strickland. Morava K-theories and localisation. *Mem. Amer. Math. Soc.*, 139(666):viii+100, 1999.

[12] J. Lurie. *Higher Topos Theory*, volume 170 of *Annals of Mathematics Studies*. Princeton University Press, Princeton, NJ, 2009.

[13] J. Lurie. Chromatic homotopy theory. `https://www.math.ias.edu/~lurie/252x.html`, 2010.

[14] J. Lurie. *Higher Algebra*. 2017. Draft available from author's website as `https://www.math.ias.edu/~lurie/papers/HA.pdf`.

4

An Introduction to Algebraic Models for Rational G-Spectra

David Barnes[a]

Magdalena Kędziorek[b]

Abstract

The project of Greenlees et al. on understanding rational G-spectra in terms of algebraic categories has had many successes, classifying rational G-spectra for finite groups, $SO(2)$, $O(2)$, $SO(3)$, free and cofree G-spectra as well as rational toral G-spectra for arbitrary compact Lie groups.

This chapter provides an introduction to the subject in two parts. The first discusses rational G-Mackey functors, the action of the Burnside ring and change of group functors. It gives a complete proof of the well-known classification of rational Mackey functors for finite G. The second part discusses the methods and tools from equivariant stable homotopy theory needed to obtain algebraic models for rational G-spectra. It gives a summary of the key steps in the classification of rational G-spectra in terms of a symmetric monoidal algebraic category.

Having these two parts in the same place allows one to clearly see the analogy between the algebraic and topological classifications.

4.1 Introduction

Suppose G is a compact Lie group. The project of understanding the homotopy theory of rational G-spectra in terms of algebraic categories was started by Greenlees in 1999 [20]. It has had many successes since, classifying rational G-spectra for finite groups, $SO(2)$, $O(2)$, $SO(3)$, free and cofree G-spectra as well as rational toral G-spectra for an arbitrary compact Lie group G. The project has expanded to consider

[a] Mathematical Sciences Research Centre, Queen's University Belfast
[b] IMAPP, Radboud University Nijmegen

(commutative) ring spectra in terms of these algebraic models. This chapter provides an introduction to this body of work, whose papers often assume a deep familiarity with rational equivariant homotopy theory.

Starting from the definition of rational G-Mackey functors for a finite group G, we explain how the rational Burnside ring acts on this category and how change of groups functors behave. Combining these functors, we give an accessible account of the structure and classification of rational G-Mackey functors in terms of group rings and a comparison of the monoidal structures. We explain how this classification is the template for the classifications of rational G-spectra for varying G.

The second half of the chapter considers rational G-spectra for G a compact Lie group. Here the rational Burnside ring appears as the ring of self maps of the sphere spectrum. We describe the structure of this ring and its idempotents. Following the template, we show how the same approach (Burnside ring actions, restriction to subgroups and fixed points) is used in the various classifications of rational G-spectra. We also discuss the additional complexities (isotropy separation, localisations and cellularisations) that are needed for spectra.

The conjecture by Greenlees states that for any compact Lie group G there is a nice graded abelian category $\mathcal{A}(G)$, such that the category $d\mathcal{A}(G)$ of differential objects in $\mathcal{A}(G)^1$ with a certain model structure is Quillen equivalent to the category of rational G-spectra

$$G\text{-}\mathrm{Sp}_{\mathbb{Q}} \simeq_Q d\mathcal{A}(G).$$

Nice here means that the category $\mathcal{A}(G)$ is of homological dimension (that is, injective dimension) equal to the rank of G and of a form that is easy to use in calculations. If we find such $\mathcal{A}(G)$ and $d\mathcal{A}(G)$ equipped with a model structure Quillen equivalent to $G\text{-}\mathrm{Sp}_{\mathbb{Q}}$, we say that $\mathcal{A}(G)$ is an *abelian model* and $d\mathcal{A}(G)$ is an *algebraic model* for rational G-spectra. The conjecture is known for quite a number of groups in some form. Particularly useful examples are the case of $O(2)$ as given in [4] and [19]; and $SO(3)$ as given in [33] and [21]. We refer to the introduction of [27] for a more complete summary of the known cases.

Since [27] was published, there have been significant developments in the field. This includes extending the existence of algebraic models to profinite groups (see [10] and [48]) as well as taking various complexities with monoidal structure into account (see [5], [7] and [43]). We refer the

---

[1] In other words objects of $\mathcal{A}(G)$ equipped with a differential

reader to [1] for a related result stating that a nice stable, monoidal model category has a model built from categories of modules over completed rings in an adelic fashion. Recently, monoidal algebraic models (when G is a finite abelian group) were used in establishing the uniqueness of naive and genuine commutative ring structures on the rational equivariant complex K-theory spectrum, see Bohmann et al. [12], [11].

The aim of this chapter is to give a new introduction and explanation to some of these existing results while demonstrating the analogy between the algebraic and topological sides. By doing so, we intend to give an overview of the methods and tools used in obtaining algebraic models for rational G-spectra and provide a step-by-step guide, at least in some cases.

Acknowledgements

The second author is grateful for support from the Dutch Research Council (NWO) under Veni grant 639.031.757.

Part 1. The structure of rational Mackey functors

4.2 An introduction to rational Mackey functors

For G a finite group, the category of Mackey functors is an abelian category that is important to group theorists (see Nakaoka [42] for example) and algebraic topologists working equivariantly. Working over the rationals greatly simplifies the category, rationally it splits into a direct product of modules over group rings of the Weyl groups of subgroups of G (counted up to conjugacy). We use the rationals for definiteness, but the splitting holds when working over any commutative ring R where $|G|^{-1} \in R$.

This result is stated formally as Theorem 4.28. It was proven independently by two sources, Greenlees and May [24, Appendix A] and Thévenaz and Webb [49, Theorems 8.3 and 9.1] The former took an approach from equivariant stable homotopy theory, the latter from algebra. We find the former approach simpler, so we follow it, expanding substantially on the proofs. General references for the results on Mackey functors are Greenlees [17], Greenlees and May [23], Thévenaz and Webb [50] and Webb [53]. For a discussion on Mackey functors for compact Lie groups see [34].

From the many equivalent definitions of a Mackey functor, we choose one in terms of induction and restriction maps.

Definition 4.1   A *rational G-Mackey functor M* is:

- a collection of $\mathbb{Q}$-modules $M(G/H)$ for each subgroup $H \leqslant G$,
- for subgroups $K, H \leqslant G$ with $K \leqslant H$ and any $g \in G$ we have a *restriction* map, an *induction* map and a *conjugation* map

$$R_K^H: M(G/H) \to M(G/K),$$
$$I_K^H: M(G/K) \to M(G/H),$$
$$C_g: M(G/H) \to M(G/gHg^{-1}).$$

These maps satisfy the following conditions.

1 For all subgroups H of G and all $h \in H$

$$R_H^H = \mathrm{Id}_{M(G/H)} = I_H^H \quad \text{and} \quad C_h = \mathrm{Id}_{M(G/H)}.$$

2 For $L \leqslant K \leqslant H$ subgroups of G and $g, h \in G$, there are composition rules

$$I_L^H = I_K^H \circ I_L^K, \qquad R_L^H = R_L^K \circ R_K^H, \quad \text{and} \quad C_{gh} = C_g \circ C_h.$$

The first two are *transitivity* of induction and restriction. The last is *associativity* of conjugation.

3 For $g \in G$ and $K \leqslant H$ subgroups of G, there are composition rules

$$R_{gKg^{-1}}^{gHg^{-1}} \circ C_g = C_g \circ R_K^H \quad \text{and} \quad I_{gKg^{-1}}^{gHg^{-1}} \circ C_g = C_g \circ I_K^H.$$

This is the *equivariance* of restriction and induction.

4 For subgroups $K, L \leqslant H$ of G

$$R_K^H \circ I_L^H = \sum_{x \in [K \backslash H / L]} I_{K \cap xLx^{-1}}^K \circ C_x \circ R_{L \cap x^{-1}Kx}^L.$$

This condition is known as the *Mackey axiom*.

We denote the category of rational Mackey functors by Mackey(G).

Many texts shorten the input and write $M(H) := M(G/H)$. This notation aligns better with the terms induction and restriction, but precludes the following remark.

Remark 4.2   Since every finite G-set is (up to non-canonical isomorphism) a disjoint union of orbits G/H, we can (by choosing such an isomorphism) extend any Mackey functor to take input from the category of finite G-sets and G-maps by sending disjoint union to direct sums. We will repeatedly use this extension (without further notice) in the adjunctions on Mackey functors that we define later.

Lindner [36] uses this extension to give an equivalent definition of Mackey functors in terms of a pair of covariant and contravariant functors from finite G-sets to $\mathbb{Q}$-modules. These functors agree on objects, send disjoint unions to direct sums and satisfy a pullback condition (that is equivalent to the Mackey axiom). The equivalence is proven via the decomposition

$$G/K \times G/H = \coprod_{x \in [K \backslash H / L]} G/(H \cap xKx^{-1}).$$

A further definition in terms of spans of G-sets (the Burnside category) is also given in that reference.

We illustrate how the structure works for two small groups.

Example 4.3 Let $G = C_2 = \{1, \sigma\}$. A rational C_2-Mackey functor is a pair of $\mathbb{Q}$-modules $M(C_2/C_2)$ and $M(C_2/\{1\})$. The conjugation maps imply that both $\mathbb{Q}$-modules have an action of C_2, but it is trivial on the first module. There is a restriction map, which commutes with the C_2-actions

$$M(C_2/C_2) \longrightarrow M(C_2/1)^{C_2} \hookrightarrow M(C_2/\{1\}).$$

Similarly there is an induction map, which commutes with the C_2-actions

$$M(C_2/\{1\}) \longrightarrow M(C_2/\{1\})/C_2 \longrightarrow M(C_2/C_2).$$

The Mackey axiom (for $H = C_2$, $K = L = \{1\}$) says that

$$R_{\{1\}}^{C_2} \circ I_{\{1\}}^{C_2} = \sum_{x \in [\{1\} \backslash C_2 / \{1\}]} I_{\{1\}}^{\{1\}} \circ C_x \circ R_{\{1\}}^{\{1\}} = \sum_{x \in C_2} C_x = \mathrm{Id} + C_\sigma$$

Example 4.4 Let $G = C_6$. A rational C_6-Mackey functor consists of four $\mathbb{Q}$-modules with maps between them. We draw this as a Lewis diagram below. The looped arrows indicate the group that acts on each module. Section 4.6 gives several examples of rational C_6-Mackey functors.

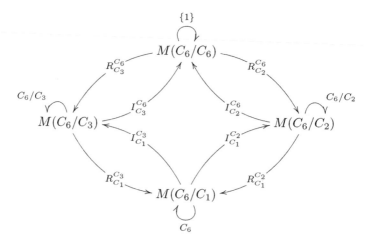

The Mackey axiom also implies that

$$R_{C_3}^{C_6} \circ I_{C_2}^{C_6} = I_{C_1}^{C_3} \circ R_{C_1}^{C_2} \quad \text{and} \quad R_{C_2}^{C_6} \circ I_{C_3}^{C_6} = I_{C_1}^{C_2} \circ R_{C_1}^{C_3}.$$

There are several general constructions that give examples of Mackey functors.

Example 4.5 The *constant Mackey functor* at a $\mathbb{Q}$-module A takes value A at each G/H. The conjugation and restriction maps are the identity map of A, induction from G/K to G/H is multiplication by the index of K inside H. Given that the restriction maps are identities, the Mackey axiom prevents the induction maps from being identity maps.

We may also define the *co-constant Mackey functor* at a $\mathbb{Q}$-module A takes value A at each G/H. The conjugation and induction maps are the identity of A and restriction from G/H to G/K is multiplication by the index of K inside H.

The similarity between the constant and co-constant Mackey functors is an example of duality of Mackey functors. That is, the co-constant Mackey functor is dual to the constant one, in the following sense.

Lemma 4.6 *Given a Mackey functor M, there is a dual Mackey functor DM, that at G/H takes value*

$$DM(G/H) = \operatorname{Hom}(M(G/H), \mathbb{Q}).$$

The conjugation maps for M induce conjugation maps for DM, though the contravariance of $D(-)$ requires us to use $C_{g^{-1}}$ for M to define C_g

*for DM. The induction maps of DM are induced from the restriction
maps of M and the restriction maps are induced from the induction maps
of M.*

Many well-known structures arising from group theory can be assembled into Mackey functors.

Example 4.7   Let $R(G)$ denote the ring of complex representations
of the finite group G. We define a rational Mackey functor M_R by
$M_R(G/H) = R(H) \otimes \mathbb{Q}$, with induction and restriction induced by
induction and restriction of representations.

The ring structure on $R(G)$ gives more structure to this Mackey functor;
it is in fact a *Tambara functor*. See Strickland [47] for a survey of such
functors and related notions like Green functors.

Example 4.8   The zeroth equivariant stable homotopy groups of a
G-spectrum form a Mackey functor. For X an orthogonal G-spectrum
over a complete G-universe, let $[-, X]^G \otimes \mathbb{Q}$ denote the functor that sends
G/H to

$$[\Sigma^\infty G/H_+, X]^G \otimes \mathbb{Q} \cong [\Sigma^\infty S^0, X]^H \otimes \mathbb{Q} \cong \pi_0^H(X) \otimes \mathbb{Q}.$$

We leave the induction, restriction and conjugation maps to the standard
references of May [40, Chapter XIX] and Lewis et al. [35, Section V.9].

We also note that G-equivariant cohomology theories use Mackey
functors as their coefficients, rather than abelian groups.

Example 4.9   Given a $\mathbb{Q}[G]$-module V, we may define a rational Mackey
functor $\mathrm{Mack}_G(V)$ (also called FP_V) as taking value V^H at G/H. The
restriction maps are inclusion of fixed points and the induction maps are
given by coset orbits.

We could also define a Mackey functor FQ_V by taking value V/H at
G/H, with induction maps the quotient maps and restriction given by
summing over a coset. The two functors are related via duality, see [50,
Proposition 4.1].

In the rational case the values V^H and V/H are isomorphic, as we
now explain. Since G is finite, there is a diagram

$$V^H \xrightarrow[\text{inclusion}]{} V \xrightarrow[\text{quotient}]{} V/H$$

with av_H and av'_H the reverse maps.

where

$$\mathrm{av}_H(x) = \frac{1}{|H|} \sum_{h \in H} hx \quad \text{and} \quad \mathrm{av}'_H([x]) = \frac{1}{|H|} \sum_{h \in H} hx.$$

The composite of inclusion and quotient $V^H \cong V/H$ is an isomorphism with inverse given by the composite $\mathrm{av}_H \circ \mathrm{av}'_H$.

When $V = \mathbb{Q}$ with trivial G-action, $\mathrm{Mack}_G(\mathbb{Q})$ is an instance of the constant Mackey functor, see Example 4.5.

Example 4.10 The rational Burnside rings for subgroups of G assemble into a Mackey functor, $\mathsf{A}_\mathbb{Q}(G/H) = \mathsf{A}_\mathbb{Q}(H)$, the rational Grothendieck ring of finite H-sets, see for example [52, Section 1.2] for details. The structure maps of this Mackey functor are the usual restriction and induction of sets with group actions. Moreover, the restriction maps are maps of rings. See Examples 4.36 and 4.37 for worked examples in the case of $G = C_6$.

As is well-known, the rational Burnside ring splits, which is an immediate consequence of the following result.

Lemma 4.11 (tom Dieck's Isomorphism) *For G a finite group, there is an isomorphism of rings*

$$\mathsf{A}_\mathbb{Q}(G) \longrightarrow C(\mathrm{Sub}(G)/G, \mathbb{Q}) = C(\mathrm{Sub}(G), \mathbb{Q})^G = \prod_{(H) \leqslant G} \mathbb{Q}$$

where $\mathrm{Sub}(G)/G$ is the set of conjugacy classes of subgroups of G and $C(\mathrm{Sub}(G)/G, \mathbb{Q})$ is the set of continuous maps between the two spaces (both equipped with the discrete topology). Here (H) denotes the conjugacy class of H in G. We define $C(\mathrm{Sub}(G), \mathbb{Q})$ to have a G-action by conjugation on the domain.

We define $e_H^G \in \mathsf{A}_\mathbb{Q}(G)$ to be the element of the Burnside ring corresponding to the characteristic function of (H) in $C(\mathrm{Sub}(G)/G, \mathbb{Q})$: the function that sends (H) to $1 \in \mathbb{Q}$ and all the other points to 0. We may omit the superscript G from the notation on idempotents when the context is clear.

Proof The isomorphism is defined by sending a G-set T to the map $(H) \mapsto |T^H|$. Since the domain and codomain have the same dimension, the result follows from proving the map is surjective, which follows from the formulas of the following lemma. □

Lemma 4.11 describes idempotents of the rational Burnside ring of G in a simple, but more abstract way. It is often useful to write these idempotents in terms of the additive basis of the Burnside ring. The formula is given by Gluck [15, Section 3].

Lemma 4.12 *For H a subgroup of G, the idempotent $e_H^G \in A_{\mathbb{Q}}(G)$ is given by the formula*

$$e_H^G = \sum_{K \leqslant H} \frac{|K|}{|N_G H|} \mu(K, H) G/K$$

where $\mu(K, H) = \Sigma_i (-1)^i c_i$ for c_i the number of strictly increasing chains of subgroups from K to H of length i. The length of a chain is one less than the number of subgroups involved and $\mu(H, H) = 1$ for all $H \leqslant G$.

For H a subgroup of G, the set G/H can be expressed as a sum of idempotents

$$G/H = \sum_{K \leqslant H} \frac{|N_G K|}{|H|} e_K^G.$$

Example 4.13 Let $G = C_2$, the ring $A_{\mathbb{Q}}(G)$ is additively generated by the one-point space $1 = C_2/C_2$ which is the monoidal unit, and $C_2/\{1\}$. The only non-evident multiplication is

$$C_2/\{1\} \times C_2/\{1\} = 2 C_2/\{1\}.$$

It follows that $e_1 = (1/2)C_2$ is an idempotent, as is $e_{C_2} = 1 - e_1$. Looking at the fixed points of these sets show that the idempotents are correctly named and we recover the isomorphism

$$A_{\mathbb{Q}}(C_2) \cong \mathbb{Q}\langle e_1 \rangle \times \mathbb{Q}\langle e_{C_2} \rangle.$$

See Example 4.36 for the case of $G = C_6$.

Remark 4.14 The restriction map $A_{\mathbb{Q}}(H) \to A_{\mathbb{Q}}(K)$ in terms of

$$C(\mathrm{Sub}(H)/H, \mathbb{Q}) \to C(\mathrm{Sub}(K)/K, \mathbb{Q})$$

corresponds to precomposing with the inclusion $\mathrm{Sub}(K) \to \mathrm{Sub}(H)$ and taking suitable orbits. We can use this description to see how the restriction map interacts with idempotents. For A and H subgroups of G, the restriction of the idempotent e_H^G to A is still an idempotent, but it is not always e_H^A. Instead,

$$R_A^G(e_H^G) = \sum_{\substack{K \leqslant_A A \\ K \in (H)_G}} e_K^A$$

where the sum runs over A-conjugacy classes of subgroups K of A, such that K is G-conjugate to H.

We see that if H is not G-subconjugate to A, this will be zero. Contrastingly, if H is G-conjugate to A, then the only term in the summand will be $K = A$ and $R_A^G(e_H^G) = e_A^A$.

Given a G-Mackey functor M, we can define an action of the Burnside ring $\mathsf{A}_\mathbb{Q}(H)$ on the abelian group $M(G/H)$ by

$$[H/K] := I_K^H \circ R_K^H \colon M(G/H) \longrightarrow M(G/H)$$

and extending linearly from the additive basis for $\mathsf{A}_\mathbb{Q}(H)$ given by H/K for subgroups K of H. The Mackey axiom implies that this action is compatible with the multiplication of $\mathsf{A}_\mathbb{Q}(H)$, so that $M(G/H)$ is a module over $\mathsf{A}_\mathbb{Q}(H)$. Moreover, the following square commutes.

$$
\begin{array}{ccc}
\mathsf{A}_\mathbb{Q}(G/H) \otimes M(G/H) & \longrightarrow & M(G/H) \\
{\scriptstyle R_K^H \otimes R_K^H}\downarrow & & \downarrow{\scriptstyle R_K^H} \\
\mathsf{A}_\mathbb{Q}(G/K) \otimes M(G/K) & \longrightarrow & M(G/K)
\end{array}
$$

The action of Burnside rings is compatible with induction in the sense of the *Frobenius reciprocity* relations. For $\alpha \in \mathsf{A}_\mathbb{Q}(G/H)$, $\beta \in \mathsf{A}_\mathbb{Q}(G/K)$, $m \in M(G/K)$ and $n \in M(G/H)$

$$\alpha \cdot I_K^H(m) = I_K^H(R_K^H(\alpha) \cdot m) \qquad I_K^H(\beta) \cdot n = I_K^H(\beta \cdot R_K^H(n)).$$

See [55, Definition 2.3 and Example 2.11].

Lemma 4.15 *Given an idempotent $e \in \mathsf{A}_\mathbb{Q}(G)$ and a G-Mackey functor M, we can define a new Mackey functor eM by*

$$(eM)(G/H) = R_H^G(e)M(G/H).$$

Proof The conjugation and restriction maps are as for M, since these actions are compatible with restriction.

By Frobenius reciprocity, the induction map for $K \leqslant H$ gives a map

$$R_K^G(e)M(G/K) \xrightarrow{I_K^H} R_H^G(e)M(G/H). \qquad \square$$

4.3 Change of group functors

As one should expect, we have adjunctions coming from inclusions of subgroups and projections onto quotients.

Definition 4.16 Given an inclusion of a subgroup $i\colon H \to G$, there are functors

$$i_\#\colon \mathrm{Mackey}(G) \longrightarrow \mathrm{Mackey}(H) \quad \text{and} \quad i^\#\colon \mathrm{Mackey}(H) \longrightarrow \mathrm{Mackey}(G).$$

Using the extension of Mackey functors to finite G-sets, we may define the functor $i^\#$ as pre-composition with the forgetful functor on sets with group actions. The functor $i_\#$ is defined by pre-composition with extension of groups. Thus for $M \in \mathrm{Mackey}(G)$, $N \in \mathrm{Mackey}(H)$, A a G-set and B a H-set,

$$(i_\# M)(B) = M(G \times_H B) \qquad (i^\# N)(A) = N(i^* A).$$

Similar definitions hold for the induction, restriction and conjugation maps, and for morphisms of Mackey functors.

Lemma 4.17 *Given an inclusion of a subgroup $i\colon H \to G$, there is an adjunction*

$$i_\#\colon \mathrm{Mackey}(G) \rightleftarrows \mathrm{Mackey}(H)\colon i^\#$$

with each functor both left and right adjoint to each other.

Proof To see that this is an adjunction with $i_\#$ as the left adjoint, we take a map $f\colon M \to i^\# N$ and construct a map $\bar{f}\colon i_\# M \to N$. Consider an H-set B, the map $\bar{f}(B)$ is given by the composite

$$M(G \times_H B) \xrightarrow{f(B)} N(i^*(G \times_H B)) \xrightarrow{N(\eta_B)} N(B)$$

where the second map is induced (by using restriction maps) from the canonical map of H-sets $\eta_B\colon B \longrightarrow i^*(G \times_H B)$. Conversely, given $g\colon i_\# M \to N$ we construct $\hat{g}\colon M \to i^\# N$ in a similar way. Given a G-set A, $\hat{g}(A)$ is the composite

$$M(A) \xrightarrow{M(\varepsilon_A)} M(G \times_H i^* A) \xrightarrow{g(i^* A)} N(i^* A)$$

where the first map above is induced (by using restriction maps) from $\varepsilon_A\colon G \times_H i^* A \longrightarrow A$.

Now we show that any map $f\colon M \to i^\# N$ is equal to $\hat{\bar{f}}\colon M \to i^\# N$ (the other case of $\bar{\hat{g}} = g$ is similar). The map $\hat{\bar{f}}$ is defined by taking the lower path in the following diagram.

$$
\begin{array}{ccc}
M(A) & \xrightarrow{\quad f(A) \quad} & N(i^* A) \\
{\scriptstyle M(\varepsilon_A)} \downarrow & & \uparrow {\scriptstyle N(\eta_{i^* A})} \\
M(G \times_H i^* A) & \xrightarrow{\ f(G \times_H i^* A)\ } & N(i^* G \times_H i^* A)
\end{array}
$$

That we have an adjunction follows as

$$N(\eta_{i^*A}) \circ f(G \times_H i^*A) \circ M(\varepsilon_A) = N(\eta_{i^*A}) \circ N(i^*\varepsilon_A) \circ f(A) = f(A)$$

by the triangle identity for sets with group actions.

The proof that $(i^\#, i_\#)$ is an adjunction is very similar to the previous case. The primary difference is that one uses induction maps rather than restriction maps. $\qquad\square$

We want to reproduce this construction for a quotient $\varepsilon \colon G \to G/N$. To make an adjunction, we need to restrict the category of G-Mackey functors somewhat. We take a strong restriction, so that the two functors we produce will be both left and right adjoint to each other.

Definition 4.18 For N a normal subgroup of G, we define the category $\mathrm{Mackey}(G)/N$ as the full subcategory of $\mathrm{Mackey}(G)$ of Mackey functors that are trivial on those G/K where K does not contain N.

Definition 4.19 Given a quotient map $\varepsilon \colon G \to G/N$ for N a normal subgroup of G, there are functors

$$\varepsilon^\# \colon \mathrm{Mackey}(G)/N \longrightarrow \mathrm{Mackey}(G/N),$$
$$\varepsilon_\# \colon \mathrm{Mackey}(G/N) \longrightarrow \mathrm{Mackey}(G)/N.$$

Thus for $M \in \mathrm{Mackey}(G)/N$, $M' \in \mathrm{Mackey}(G/N)$, K a subgroup of G containing N and B a G/N-set, we define

$$\varepsilon^\# M(B) = M(\varepsilon^*B) \quad \text{and} \quad \varepsilon_\# M'(G/K) = M'((G/N)/(K/N)).$$

If K does not contain N we set $\varepsilon_\# M'(G/K) = 0$.

The structure maps of M and M' are defined in terms of these formulas, as are maps of Mackey functors.

Lemma 4.20 *Given N a normal subgroup of G, there is an adjunction*

$$\varepsilon^\# \colon \mathrm{Mackey}(G)/N \rightleftarrows \mathrm{Mackey}(G/N) \colon \varepsilon_\#$$

with each functor both left and right adjoint to each other.

Proof Both cases are similar and use the fact that

$$\varepsilon^*(G/N)/(K/N) = G/K.$$

We give one part of the proof as an illustration.

Take $f \colon M \longrightarrow \varepsilon_\# M'$ a map of G-Mackey functors that are trivial on those G/K where K does not contain N. We want to construct $\bar{f} \colon \varepsilon^{\#} M \longrightarrow M'$. Take a subgroup K/N of G/N, we define

$$\bar{f}((G/N)/(K/N)) = f(G/K) \colon M(G/K) \to M'((G/N)/(K/N)). \quad \square$$

We give one more adjunction, between rational G-Mackey functors and $\mathbb{Q}$-modules with an action of G.

Lemma 4.21 *There is an adjunction*

$$(-)(G/e) \colon \mathrm{Mackey}(G) \rightleftarrows \mathbb{Q}[G]\text{-mod} \colon \mathrm{Mack}_G$$

with each functor both left and right adjoint to each other.

The functor $(-)(G/e)$ sends a G-Mackey functor to the value $M(G/e)$. Its adjoint Mack_G is defined in Example 4.9 and at G/H takes value V^H.

Proof Take a map $f \colon M \to \mathrm{Mack}_G(V)$. Evaluating at G/e gives a map $\bar{f} \colon M(G/e) \to V$. In the other direction, one starts with a map $g \colon M(G/e) \to V$ of $\mathbb{Q}[G]$-modules. The restriction map

$$R_e^H \colon M(G/H) \to M(G/e)$$

takes values in $M(G/e)^H$ as conjugation by elements of H is trivial in $M(G/e)$. We define $\hat{g}$ as $g(G/e)^H \circ R_e^H$.

For the adjunction in the other direction, we use the isomorphic description of $\mathrm{Mack}_G(V)(G/H)$ in terms of V/H and follow a similar pattern, using the induction maps of M to define the adjoint of a map $V \to M(G/e)$. $\quad \square$

4.4 The classification of rational Mackey functors

Let $e_H^G \in \mathsf{A}_{\mathbb{Q}}(G) = \prod_{(H) \leqslant G} \mathbb{Q}$ be the idempotent that is 1 on factor H and zero elsewhere. As described above, we can form a full subcategory of $\mathrm{Mackey}(G)$ consisting of those Mackey functors of the form $e_H^G M$. Applying e_H^G defines a functor $\mathrm{Mackey}(G) \longrightarrow e_H^G \mathrm{Mackey}(G)$. It follows that we have a splitting

$$\mathrm{Mackey}(G) \cong \prod_{H \leqslant_G G} e_H^G \mathrm{Mackey}(G).$$

To classify rational Mackey functors, it therefore suffices to classify the categories $e_H^G \mathrm{Mackey}(G)$. The key step is the following theorem giving a sequence of adjunctions. The proof of the theorem occupies the rest of this section.

Theorem 4.22   *For $H \leqslant G$, there are adjunctions of exact functors*

$$e_H^G \mathrm{Mackey}(G) \xrightarrow[\;i^{\#}\;]{\;i_{\#}\;} R_{N_G H}^G(e_H^G)\mathrm{Mackey}(N_G H)$$

$$\mathrm{Mackey}(N_G H)/H \xrightarrow[\;\varepsilon_{\#}\;]{\;\varepsilon^{\#}\;} \mathrm{Mackey}(N_G H/H)$$

$$\mathrm{Mackey}(N_G H/H) \xrightarrow[\;\mathrm{Mack}_{W_G H}\;]{\;(-)(W_G H/e)\;} \mathbb{Q}[W_G H]\text{-mod}$$

$$R_{N_G H}^G(e_H^G)\mathrm{Mackey}(N_G H) \xrightarrow[\;\varepsilon_* \mathrm{Mack}_{W_G H}\;]{\;(-)(N_G H/H)\;} \mathbb{Q}[W_G H]\text{-mod}$$

with each pair both left and right adjoint to each other.

Lemma 4.23   *The adjunction $(i_{\#}, i^{\#})$ restricts to an adjunction*

$$i_{\#} : e_H^G \mathrm{Mackey}(G) \xrightarrow{\longleftarrow} R_{N_G H}^G(e_H^G)\mathrm{Mackey}(N_G H) : i^{\#} \,.$$

The functors are exact and are both left and right adjoint to each other.

Proof   Take $M \in e_H^G \mathrm{Mackey}(G)$ and $K \leqslant N_G H$. Since $M = e_H^G M$, we have the first equality below

$$
\begin{aligned}
(i_{\#} M)(N_G H/K) &= R_K^G(e_H^G) M(G/K) \\
&= R_K^{N_G H} \circ R_{N_G H}^G(e_H^G) M(G/K) \\
&= \left(R_{N_G H}^G(e_H^G)(i_{\#} M) \right)(N_G H/K).
\end{aligned}
$$

Thus $(i_{\#} M) \in R_{N_G H}^G(e_H^G)\mathrm{Mackey}(N_G H)$.

Conversely, let $M' \in R_{N_G H}^G(e_H^G)\mathrm{Mackey}(N_G H)$ and $K \leqslant G$. Then

$$(i^{\#} M')(G/K) = M'(i^* G/K) = \oplus_{\lambda \in \Lambda} M'(N_G H/L_\lambda)$$

where $i^* G/K$ decomposes as $\coprod_{\lambda \in \Lambda} N_G H/L_\lambda$. Since

$$
\begin{aligned}
M'(N_G H/L_\lambda) &= R_{L_\lambda}^{N_G H} \circ R_{N_G H}^G(e_H^G) M'(N_G H/L_\lambda) \\
&= R_{L_\lambda}^G(e_H^G) M'(N_G H/L_\lambda),
\end{aligned}
$$

it follows that $i^{\#} M' = e_H^G i^{\#} M'$.

The functors are additive and left and right adjoint to each other. Hence they are exact. $\square$

We cannot directly compose the second and third of the above adjunctions as $\mathrm{Mackey}(N_G H)/H$ and $R_{N_G H}^G(e_H^G)\mathrm{Mackey}(N_G H)$ are not easily related. The following lemma remedies this by constructing the fourth adjunction.

Lemma 4.24 *For any $\mathbb{Q}[W_G H]$-module V, there is a canonical isomorphism of $N_G H$-Mackey functors*

$$R^G_{N_G H}(e^G_H)\varepsilon_\# \mathrm{Mack}_{W_G H}(V) \cong \varepsilon_\# \mathrm{Mack}_{W_G H}(V).$$

It follows that we have an adjunction

$$(-)(N_G H/H) : R^G_{N_G H}(e^G_H)\mathrm{Mackey}(N_G H) \rightleftarrows \mathbb{Q}[W_G H]\text{-mod} : \varepsilon_* \mathrm{Mack}_{W_G H}$$

with the functors both left and right adjoint to each other.

Proof The inclusion of an idempotent summand gives the map. To see that this inclusion is an isomorphism, we evaluate both sides at a subgroup $A \leqslant N_G H$ that contains H. Both domain and codomain take value zero on subgroups that do not contain H.

We first decompose $R^G_A(e^G_H)$ into idempotents of the rational Burnside ring of A

$$R^G_A(e^G_H) = \sum_{\substack{K \leqslant_A A \\ K \in (H)_G}} e^A_K.$$

Secondly, by Lemma 4.12 (e^A_K) is a sum of $|W_A K|^{-1}[A/K]$ and rational multiples of basis elements $[A/K']$, for K' a proper subgroup of K. The element $[A/L]$ acts on $\varepsilon_\# \mathrm{Mack}_{W_G H}(V)(N_G H/A)$ through $I^A_L \circ R^A_L$. This is zero unless L contains H. Hence each $[A/K']$ acts as zero, and $[A/K]$ only acts non-trivially when K contains H. Since K is also G-conjugate to H, we see that $K = H$. Hence,

$$\left(\sum_{\substack{K \leqslant_A A \\ K \in (H)_G}} e^A_K \right)\varepsilon_\# \mathrm{Mack}_{W_G H}(V)(N_G H/A) = e^A_H \varepsilon_\# \mathrm{Mack}_{W_G H}(V)(N_G H/A)$$

$$= e^A_H V^{A/H}$$

with e^A_H acting through $|W_A H|^{-1} I^A_H \circ R^A_H$.

Thirdly, $I^A_H \circ R^A_H$ is the composite

$$V^A \longrightarrow V^H \longrightarrow V^A$$

with the first map the inclusion and the second map taking the sum over A/H-coset representatives. Hence this map is multiplication by $|A/H|$. Since H is normal in A, it follows that $|W_A H|^{-1} I^A_H \circ R^A_H$ acts through the identity, giving the first statement.

For the adjunction statement, compose the functor $\varepsilon_{\#}\mathrm{Mack}_{W_G H}$ with the inclusion

$$\mathrm{Mackey}(N_G H)/H \longrightarrow \mathrm{Mackey}(N_G H).$$

By the first statement of the lemma, this defines a functor

$$\varepsilon_*\mathrm{Mack}_{W_G H} \colon \mathbb{Q}[W_G H] \longrightarrow R^G_{N_G H}(e^G_H)\mathrm{Mackey}(N_G H).$$

We can explicitly describe $\varepsilon_*\mathrm{Mack}_{W_G H}(V)$ as the Mackey functor which sends $N_G H/K$ to V^K if K contains H and zero otherwise. The restriction maps are inclusions of fixed points and the induction maps are taking sums over cosets.

Evaluation at $N_G H/H$ is the left and right adjoint to $\varepsilon_*\mathrm{Mack}_{W_G H}$. Take M in $R^G_{N_G H}(e^G_H)\mathrm{Mackey}(N_G H)$, K a subgroup of $N_G H$ which contains H, and maps of $\mathbb{Q}[W_G H]$-modules $f\colon V \to M(N_G H/H)$ and $g\colon M(N_G H/H) \to V$. We claim that the maps

$$M(N_G H/K) \xrightarrow{R^K_H} M(N_G H/H)^K \xrightarrow{g^K} A^K$$

(and zero at those subgroups not containing H) define a map of Mackey functors from M to $\varepsilon_*\mathrm{Mack}_{W_G H}(V)$. Similarly, we claim that the maps

$$V^K \xrightarrow{|K|^{-1}f^K} M(N_G H/H)^K \xrightarrow{I^K_H} M(N_G H/K)$$

(and zero at those subgroups not containing H) define a map of Mackey functors from $\varepsilon_*\mathrm{Mack}_{W_G H}(V)$ to M. Both claims follow from calculations using the Mackey axiom and the fact that the action of the idempotent $R^G_{N_G H}(e^G_H)$ ensures that $M(N_G H/L)$ is zero when L is a proper subconjugate of H.

The functors of the adjunction are additive, and are left and right adjoint to each other. Hence they are exact. $\qquad\qquad\square$

Definition 4.25   For $H \leqslant G$, define functors

$$F_H \colon \mathbb{Q}[W_G H]\text{-mod} \longrightarrow e^G_H\mathrm{Mackey}(G)$$
$$U_H \colon e^G_H\mathrm{Mackey}(G) \longrightarrow \mathbb{Q}[W_G H]\text{-mod}$$

where F_H is the composite of the lower level functors from the diagram in Theorem 4.22 and U_H is the composite of the the upper level functors.

We see immediately that the additive functors F_H and U_H are both left and right adjoint to each other and that $U_H M = M(G/H)$.

Proposition 4.26 *For $K \leqslant G$,*

$$F_H(V)(G/K) = (\mathbb{Q}[(G/K)^H] \otimes V)^{W_G H}.$$

Moreover, the Mackey functor $F_H(V)$ is both projective and injective, and $e_H^G F_H(V) = F_H(V)$.

Proof From the definitions, the composite is given by

$$G/K \mapsto \bigoplus_{\lambda \in \Lambda} V^{L_\lambda/H}$$

where G/K decomposes as $\coprod_{\lambda \in \Lambda} N_G H/L_\lambda$ and K contains H. If K does not contain H, the composite takes value zero. Each factor in this decomposition corresponds to an $N_G H$-orbit in the set of $N_G H$-maps $N_G H/H \to i^* G/K$. The $N_G H$-action is by right multiplication by the inverse on $N_G H/H$. Such a map corresponds to a G-map $G/H \to G/K$, which is simply an element α of the set $(G/K)^H$. By thinking of $(G/K)^H$ as a $W_G H$-set, we can sum over all α to obtain the formula

$$G/K \mapsto \left(\bigoplus_{\alpha \in (G/K)^H} V_\alpha \right)^{W_G H}$$

with $W_G H$ permuting the summands (it acts by right multiplication by the inverse on $(G/K)^H$). Replacing summands by a tensor product gives the formula

$$F_H(V)(G/K) = (\mathbb{Q}[(G/K)^H] \otimes V)^{W_G H}.$$

Every $\mathbb{Q}[W_G H]$-module is both injective and projective. Hence, $F_H(V)$ is both projective and injective as the functors F_H and U_H are exact. Lemmas 4.23 and 4.24 give the statement about idempotents. □

This new formula is very compact, but for calculations (see Examples 4.38 and 4.39) and later theory we will also need simple description of the induction and restriction maps of $F_H(V)$.

Lemma 4.27 *For $L \leqslant K \leqslant G$, the induction map*

$$F_H(V)(G/L) \longrightarrow F_H(V)(G/K)$$

$$\| \qquad\qquad\qquad \|$$

$$(\mathbb{Q}[(G/L)^H] \otimes V)^{W_G H} \qquad (\mathbb{Q}[(G/K)^H] \otimes V)^{W_G H}$$

is induced by the projection $\alpha : G/L \to G/K$.

The restriction map

$$F_H(V)(G/K) \longrightarrow F_H(V)(G/L)$$

is induced by the map $\mathbb{Q}[(G/K)^H] \to \mathbb{Q}[(G/L)^H]$ *that sends an element* gK *to the sum of the elements in its preimage under the projection* $\alpha : G/L \to G/K$.

Proof Write $(\mathbb{Q}[(G/L)^H] \otimes V)^{W_G H}$ as

$$\left(\bigoplus_{\sigma \, : \, G/H \to G/L} V_\sigma \right)^{W_G H},$$

we use the subscript σ on V to keep track of the factors. The action of $W_G H$ is given by both acting on V and permuting the summands. That is, for $w \in W_G H$ and $v \in F_H(V)(G/L)$, we define wv to have component in summand σ given by

$$g(v_{\sigma \circ r(w^{-1})})$$

where $r(w^{-1}) \colon G/H \longrightarrow G/H$ is right multiplication by w^{-1}.

Chasing through the definitions, it follows that the restriction map is given by

$$(R_L^K y)_\sigma = y_{\alpha \circ \sigma}.$$

The induction map is given by

$$(I_L^K y)_\tau = \sum_{\alpha \circ \sigma = \tau} y_\sigma$$

the sum over those summands σ that map to τ by α. $\square$

Theorem 4.28 *For each $H \leqslant G$ there is an equivalence of categories*

$$U_H : e_H^G \mathrm{Mackey}(G) \rightleftarrows \mathbb{Q}[W_G H]\text{-mod} : F_H$$

Hence there is an equivalence of categories

$$\mathrm{Mackey}(G) \cong \prod_{(H) \leqslant G} \mathbb{Q}[W_G H]\text{-mod}$$

where the product runs over G-conjugacy classes of subgroups of G.

Proof We have already seen that U_H and F_H are both left and right adjoint to each other. The unit is an isomorphism:

$$V \longrightarrow U_H F_H V = (\mathbb{Q}[(G/H)^H] \otimes V)^{W_G H} \cong V.$$

It follows that the counit is an isomorphism of Mackey functors of the form $F_H V$.

The rest of the proof shows that any Mackey functor is a finite direct sum of Mackey functors of the form $F_H V$ for varying H and V.

We partition the set of subgroups of G into sets, which we may think of as their height in the subgroup lattice. We start with $S_0 = \{e\}$, then we define S_j as those groups not in S_{j-1} but all of whose subgroups are in S_i for $i < j$. Each S_j is closed under conjugation, with n_j conjugacy classes. Choose a $H_{j,k}$ in each conjugacy class, $1 \leqslant k \leqslant n_j$. We say that a Mackey functor M is of type (j, k) if $M(G/H_{j,k})$ is non-zero, but

$$M(G/H_{j',k'}) = 0 \quad \text{for } j' < j \text{ and for } j' = j \text{ and } k' < k.$$

We argue via descending induction. Starting at the top, if $M(G/H)$ is zero for all proper subgroups H, then $M = F_G M(G/G)$. Fix (j, k) inductively and assume that all Mackey functors of type (j', k') for $j' > j$ and for $j' = j$ and $k' > k$ are finite direct sums of Mackey functors of the form $F_J V_J$ where V_J is a $W_G J$-module and

$$J \in \{H_{j'',k''} \mid j'' > j' \text{ or } j'' = j' \text{ and } k'' \geqslant k'\}.$$

Let M be a Mackey functor of type (j, k). For $H = H_{j,k}$, there is a map of Mackey functors

$$\kappa \colon M \longrightarrow F_H M(G/H) = e_H^G F_H M(G/H)$$

that is the identity on G/H. The kernel and cokernel of κ are, by inductive assumption, finite sums of the form $F_J V_J$. It follows that e_H^G applied to the kernel and cokernel are zero. Thus $e_H^G \kappa$ is an isomorphism and so κ, which is equal to the epimorphism $M \to e_H^G M$ followed by $e_H^G \kappa$, is an epimorphism.

Since $F_H M(G/H)$ is projective, the epimorphism splits and M is a direct sum of $F_H M(G/H)$ and Mackey functors of the form $F_J V_J$.   □

Corollary 4.29   *Every rational Mackey functor is both projective and injective.*

By Remark 4.14, for $H \leqslant G$ we have $R_H^G(e_H^G) = e_H^H$. This gives the following corollary of Theorem 4.28.

Corollary 4.30   *A G-Mackey functor M is uniquely determined by the collection*

$$\{e_H^H M(G/H) \in \mathbb{Q}[W_G H]\text{-mod} \mid (H) \leqslant G\}$$

where we index over G-conjugacy classes of subgroups of G.

The remaining question is how to conveniently find the values $M(G/H)$ of the Mackey functor M from such a collection. The next section gives a formula that provides a satisfying answer.

4.5 The diagonal decomposition

Rational Mackey functors for compact Lie groups are considered in Greenlees [18]. For finite G, we present the decomposition formula for rational G-Mackey functors, following Examples C i) and Corollary 5.3 of [18]. The reference proves the result using equivariant stable homotopy theory; a direct algebraic proof is given by Sugrue [48, Lemma 6.1.9]. We use the structure results to prove it via a calculation on Mackey functors of the form $F_A V$.

Theorem 4.31 *Let M be a G-Mackey functor and let $K \leqslant H$ be subgroups of G. Then*

$$e_K^H M(G/H) \cong \left(e_K^K M(G/K)\right)^{W_H K}.$$

Proof By Theorem 4.28, we may assume M is of the form $F_A V$, for V a $\mathbb{Q}[W_G A]$-module. Lemma 4.32 implies that we only need to consider the case $A = K$.

By Remark 4.14 and Proposition 4.26, we may remove the idempotent e_K^K from the formula. Thus we must prove

$$e_K^H (\mathbb{Q}[(G/H)^K] \otimes V)^{W_G K} = e_K^H F_K(G/H) \cong \left(F_K(G/K)\right)^{W_H K}$$
$$= \left((\mathbb{Q}[(G/K)^K] \otimes V)^{W_G K}\right)^{W_H K}.$$

By Lemma 4.12, (e_K^H) is a sum of $|W_H K|^{-1}[H/K]$ and rational multiples of basis elements $[H/L]$ for L a proper subgroup of K. An element $[H/L]$ acts on $F_K(G/H)$ through $I_L^H \circ R_L^H$. As $F_K(V)(G/L) = 0$ unless L contains K (up to H-conjugacy), (e_K^H) acts as

$$|W_H K|^{-1}[H/K] = |W_H K|^{-1} I_K^H \circ R_K^H.$$

Lemma 4.27 shows how the map $I_K^H \circ R_K^H$ is induced from a composite

$$\mathbb{Q}[(G/H)^K] \longrightarrow \mathbb{Q}[(G/K)^K] \longrightarrow \mathbb{Q}[(G/H)^K]. \qquad (*)$$

We look at this composite in detail. The reader may find it useful to compare the following argument with Example 4.35.

If an element gK is fixed by left multiplication by elements of K (that is, g is an element of $N_G K$), then gH is also K-fixed. Take gK and $g'K$ that are K-fixed with $gH = g'H$. Then $g' = gh$ for some $h \in H$, and

$$K = (g')^{-1} K g' = (gh)^{-1} K gh = h^{-1} K h$$

so $h \in N_H K$. It follows that the second map of $(*)$ passes to an injection

$$\mathbb{Q}[(G/K)^K]/W_H K \longrightarrow \mathbb{Q}[(G/H)^K].$$

Take gH in the image of $(G/K)^K \to (G/H)^K$, the composite $(*)$ sends this to the sum of those aH such that $aK = gK$. By the previous argument, this sum is $|W_H K| gH$. Now take a coset gH that is not in the image of $(G/K)^K \to (G/H)^K$, then the first map of $(*)$ sends this to zero.

It follows that the composite

$$(\mathbb{Q}[(G/H)^K] \otimes V)^{W_G K} \to (\mathbb{Q}[(G/K)^K] \otimes V)^{W_G K} \to (\mathbb{Q}[(G/H)^K] \otimes V)^{W_G K}$$

induced from $(*)$ in turn induces an isomorphism from

$$\left((\mathbb{Q}[(G/K)^K] \otimes V)^{W_G K} \right)^{W_H K}$$

onto the image of $I_K^H \circ R_K^H$ in $(\mathbb{Q}[(G/H)^K] \otimes V)^{W_G K}$. □

Lemma 4.32 *Let A, B and C be subgroups of G, with $C \leqslant B$ and let V be a $\mathbb{Q}[W_G A]$-module. Then*

$$e_C^B \big(F_A(V)(G/B) \big) = 0$$

unless A and C are G-conjugate.

Proof By Proposition 4.26, $e_A^G F_A(V) = F_A(V)$. Hence,

$$e_C^B \big(F_A(V)(G/B) \big) = e_C^B R_B^G (e_A^G) \big(F_A(V)(G/B) \big).$$

The product $e_C^B R_B^G(e_A^G)$ is zero unless A is G-conjugate to a subgroup A' of G and that subgroup A' is B-conjugate to C. We also require that A is G-subconjugate to B, as otherwise $\mathbb{Q}[(G/B)^A] = 0$. This is equivalent to requiring that A is G-conjugate to C. □

We illustrate this decomposition with two examples.

Example 4.33 Let M be a rational Mackey functor for C_{p^3}. Define $\mathbb{Q}$-modules

$$V_0 = M(C_{p^3}/C_1), \quad V_1 = e_{C_{p^1}} M(C_{p^3}/C_{p^1}),$$
$$V_2 = e_{C_{p^2}} M(C_{p^3}/C_{p^2}), \quad V_3 = e_{C_{p^3}} M(C_{p^3}/C_{p^3}).$$

where $e_{C_{p^i}} \in A_{\mathbb{Q}}(C_{p^i})$ is the idempotent with support C_{p^i}, $i \in \{0, 1, 2, 3\}$. Note that for $i = 0$ this is $1 \in A_{\mathbb{Q}}(C_1) = \mathbb{Q}$. The $\mathbb{Q}$-module V_i has an action of $\mathbb{Q}[W_{C_{p^3}} C_{p^i}] = \mathbb{Q}[C_{p^3}/C_{p^i}]$.

The classification theorem implies that

$$M \cong F_e V_0 \oplus F_{C_p} V_1 \oplus F_{C_{p^2}} V_2 \oplus F_{C_{p^3}} V_3.$$

Writing out the values of M at varying subgroups gives the diagram

With vertical maps indicating induction and restriction.

Remark 4.34 Given a G-Mackey functor M and $H \leqslant G$, we can construct $\overline{M}(G/H)$, the quotient of $M(G/H)$ by the images of the induction maps from proper subgroups of H. This example shows how $\overline{M}(G/H) = e_H^H M(G/H)$, so that the classification result is based around stripping out the images of the induction functors.

Example 4.35 We look at a particular instance of Theorem 4.31 in order to illustrate the second half of the proof. Let $G = S_4$ be the symmetric group on four letters, $K = \langle (12) \rangle$, $H = \langle (12), (34) \rangle$. We have

$$N_G K = H$$
$$(G/K)^K = W_G K = W_H K = H/K = \{K, (34)K\},$$
$$(G/H)^K = \{H, (14)(23)H\}.$$

We consider the case of $F_K(\mathbb{Q}[W_G K])$. We write out the two sides of Theorem 4.31 below.

$$e_K^H F_K(\mathbb{Q}[W_G K])(G/H) = e_K^H \mathbb{Q}[(G/H)^K] = I_K^H \circ R_K^H \mathbb{Q}[(G/H)^K]$$

$$\left(e_K^K F_K(\mathbb{Q}[W_G K])(G/K) \right)^{W_H K} \cong (\mathbb{Q}[(G/K)^K]^{W_H K} = \mathbb{Q}[\{K + (34)K\}]$$

To calculate $I_K^H \circ R_K^H \mathbb{Q}[(G/H)^K]$, consider the maps of Lemma 4.27

$$\mathbb{Q}[\{H, (14)(23)H\}] \longrightarrow \mathbb{Q}[\{K, (34)K\}] \longrightarrow \mathbb{Q}[\{H, (14)(23)H\}].$$

The first map (restriction) sends H to $K + (34)K$ and $(14)(23)H$ to zero. The second map (induction) sends K and $(34)K$ to H. The image of the composite is therefore the submodule of $\mathbb{Q}[\{H, (14)(23)H\}]$ generated by H. Hence, $I_K^H \circ R_K^H \mathbb{Q}[(G/H)^K] = \mathbb{Q}[\{H\}]$ and the map induced by restriction

$$
\begin{array}{ccc}
\mathbb{Q}[\{H\}] & \longrightarrow & \mathbb{Q}[\{K + (34)K\}] \\
\| & & \|{\scriptstyle \wr} \\
e_K^H F_K(\mathbb{Q}[W_G K])(G/H) & & \left(e_K^K F_K(\mathbb{Q}[W_G K])(G/K)\right)^{W_H K}
\end{array}
$$

is an isomorphism.

4.6 Worked examples for $G = C_6$

Fixing the group to be C_6, the cyclic group of order 6, we give a detailed investigation of rational C_6-Mackey functors and their corresponding decomposition as modules over group rings. We choose a small group so that group-theoretic complexities do not obstruct the overall picture. We organise this section as a series of examples.

For clarity, write $C_6 = \{0, 1, 2, 3, 4, 5\}$, $C_3 = \{0, 2, 4\}$, $C_2 = \{0, 3\}$, $C_1 = \{0\}$ and quotients via underlining, so $C_6/C_3 = \{\underline{0}, \underline{1}\}$. We start by describing the rational Burnside ring of C_6 and its idempotents.

Example 4.36 The rational Burnside ring of C_6 is additively generated by the sets C_6/C_1, C_6/C_3, C_6/C_2 and C_6/C_6, which is the multiplicative unit. The multiplicative properties of these sets are:

$$
\begin{array}{ll}
C_6/C_1 \times C_6/C_1 = 6C_6/C_1 & C_6/C_1 \times C_6/C_3 = 2C_6/C_1 \\
C_6/C_1 \times C_6/C_2 = 3C_6/C_1 & C_6/C_3 \times C_6/C_3 = 2C_6/C_3 \\
C_6/C_2 \times C_6/C_3 = C_6/C_1 & C_6/C_2 \times C_6/C_2 = 3C_6/C_2.
\end{array}
$$

Evidently, we have non-zero idempotents C_6/C_6, $(1/6)C_6/C_1$, $(1/2)C_6/C_3$, $(1/3)C_6/C_2$, but these are not orthogonal to each other. Some algebraic manipulation gives an orthogonal decomposition of the unit as

$$
\begin{aligned}
e_{C_1}^{C_6} &= (1/6)C_6/C_1 \\
e_{C_2}^{C_6} &= (1/3)C_6/C_2 - (1/6)C_6/C_1 \\
e_{C_3}^{C_6} &= (1/2)C_6/C_3 - (1/6)C_6/C_1 \\
e_{C_6}^{C_6} &= C_6/C_6 - (1/3)C_6/C_2 - (1/2)C_6/C_3 + (1/6)C_6.
\end{aligned}
$$

One can check that these formulas match those of Lemma 4.12 and verify the reverse formulas describing the homogeneous sets in terms of sums of idempotents. Furthermore, one can confirm that the map of Lemma 4.11 is indeed an isomorphism in this case.

Now we have an understanding of the rational Burnside ring of C_6, we can look at the Burnside ring Mackey functor.

Example 4.37   The rational Burnside ring Mackey functor of Example 4.10 is defined by

$$
A_{\mathbb{Q}}(C_6/H) = A_{\mathbb{Q}}(H),
$$

with structure maps defined in terms of restriction and induction of sets with group actions. For example, $C_6/C_3 \in A_{\mathbb{Q}}(C_6)$ restricts to $2C_3/C_3 \in A_{\mathbb{Q}}(C_3)$ and $C_2/C_1 \in A_{\mathbb{Q}}(C_2)$ inducts to $C_6/C_1 \in A_{\mathbb{Q}}(C_6)$.

We can calculate the idempotents of the rational Burnside rings of C_3 and C_2 as we did for C_6.

$$
\begin{aligned}
A_{\mathbb{Q}}(C_3) \quad & e_{C_1}^{C_3} = (1/3)C_3/C_1 \quad e_{C_3}^{C_3} = C_3/C_3 - (1/3)C_3/C_1 \\
A_{\mathbb{Q}}(C_2) \quad & e_{C_1}^{C_2} = (1/2)C_2/C_1 \quad e_{C_2}^{C_2} = C_2/C_2 - (1/2)C_2/C_1
\end{aligned}
$$

We present the rational Burnside ring Mackey functor as the following Lewis diagram, phrased in terms of idempotents. The downward arrows indicate that an idempotent restricts to another idempotent, where an idempotent restricts to zero, we omit an arrow. The upward arrows indicate that an idempotent inducts to a multiple of another idempotent.

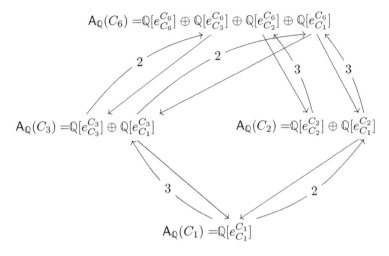

As described at the start of Section 4.4, we may use the idempotents of $A_{\mathbb{Q}}(C_6)$ to split the category of rational C_6-Mackey functors:

$$\mathrm{Mackey}(C_6) \cong e_{C_6}^{C_6}\mathrm{Mackey}(C_6)$$
$$\oplus \, e_{C_3}^{C_6}\mathrm{Mackey}(C_6)$$
$$\oplus \, e_{C_2}^{C_6}\mathrm{Mackey}(C_6)$$
$$\oplus \, e_{C_1}^{C_6}\mathrm{Mackey}(C_6).$$

Hence, any rational C_6-Mackey functor M can be written as

$$M \cong e_{C_6}^{C_6}M \oplus e_{C_3}^{C_6}M \oplus e_{C_2}^{C_6}M \oplus e_{C_1}^{C_6}M.$$

Identifying the action of idempotents on a given Mackey functor M from knowledge of the induction and restriction maps can be time-consuming. It is often easier to produce examples of Mackey functors in the subcategories of the splitting. We do so using the equivalences of categories of Theorem 4.28:

$$U_6 : e_{C_6}^{C_6}\mathrm{Mackey}(C_6) \rightleftarrows \mathbb{Q}[C_6/C_6]\text{-mod} : F_6$$

$$U_3 : e_{C_3}^{C_6}\mathrm{Mackey}(C_6) \rightleftarrows \mathbb{Q}[C_6/C_3]\text{-mod} : F_3$$

$$U_2 : e_{C_2}^{C_6}\mathrm{Mackey}(C_6) \rightleftarrows \mathbb{Q}[C_6/C_2]\text{-mod} : F_2$$

$$U_1 : e_{C_1}^{C_6}\mathrm{Mackey}(C_6) \rightleftarrows \mathbb{Q}[C_6/C_1]\text{-mod} : F_1$$

where U_H is evaluation at G/H and

$$F_H(V)(G/K) = (\mathbb{Q}[(G/K)^H] \otimes V)^{W_G H}.$$

Example 4.38   Take $\mathbb{Q} \in \mathbb{Q}[C_6/C_2]$-mod (with trivial group action). We have

$$
\begin{aligned}
F_2(\mathbb{Q})(C_6/C_6) &= (\mathbb{Q}[(C_6/C_6)^{C_2}] \otimes \mathbb{Q})^{C_6/C_2} = \mathbb{Q} \\
F_2(\mathbb{Q})(C_6/C_3) &= (\mathbb{Q}[(C_6/C_3)^{C_2}] \otimes \mathbb{Q})^{C_6/C_2} = 0 \\
F_2(\mathbb{Q})(C_6/C_2) &= (\mathbb{Q}[(C_6/C_2)^{C_2}] \otimes \mathbb{Q})^{C_6/C_2} = \mathbb{Q}[C_6/C_2]^{C_6/C_2} = \mathbb{Q} \\
F_2(\mathbb{Q})(C_6/C_1) &= (\mathbb{Q}[(C_6/C_1)^{C_2}] \otimes \mathbb{Q})^{C_6/C_2} = 0.
\end{aligned}
$$

The only non-trivial restriction map is $F_2(\mathbb{Q})(C_6/C_6) \to F_2(\mathbb{Q})(C_6/C_2)$. By Lemma 4.27 this map is induced from the map

$$
\mathbb{Q}[(C_6/C_6)^{C_2}] \longrightarrow \mathbb{Q}[(C_6/C_2)^{C_2}]
$$

sending an element $q\underline{0}$ to $q(\underline{0} + \underline{1} + \underline{2})$. Hence, the restriction map is the identity. It follows (either by using the Mackey axiom, or another application of Lemma 4.27) that the non-trivial induction map is multiplication by 3 (the subgroup index). We summarise this calculation, along with the analogous calculations for $F_1(\mathbb{Q})$, $F_3(\mathbb{Q})$ and $F_6(\mathbb{Q})$, in the following Lewis diagrams.

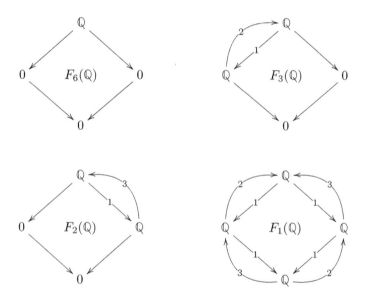

We can relate this to Example 4.37, where our idempotent calculations make the splitting evident. Indeed,

$$
\mathsf{A}_{\mathbb{Q}}(C_6) \cong F_1(\mathbb{Q}) \oplus F_2(\mathbb{Q}) \oplus F_3(\mathbb{Q}) \oplus F_6(\mathbb{Q}).
$$

Rather than starting with $\mathbb{Q}$, the other obvious example is to take $\mathbb{Q}[C_6/C_3]$ in $\mathbb{Q}[C_6/C_3]$-mod.

Example 4.39 We can calculate the following:

$$F_3(\mathbb{Q}[C_6/C_3])(C_6/C_6) = (\mathbb{Q}[(C_6/C_6)^{C_3}] \otimes \mathbb{Q}[C_6/C_3])^{C_6/C_3} = \mathbb{Q}$$
$$F_3(\mathbb{Q}[C_6/C_3])(C_6/C_3) = (\mathbb{Q}[(C_6/C_3)^{C_3}] \otimes \mathbb{Q}[C_6/C_3])^{C_6/C_3} = \mathbb{Q}[C_6/C_3]$$
$$F_3(\mathbb{Q}[C_6/C_3])(C_6/C_2) = (\mathbb{Q}[(C_6/C_2)^{C_3}] \otimes \mathbb{Q}[C_6/C_3])^{C_6/C_3} = 0$$
$$F_3(\mathbb{Q}[C_6/C_3])(C_6/C_1) = (\mathbb{Q}[(C_6/C_1)^{C_3}] \otimes \mathbb{Q}[C_6/C_3])^{C_6/C_3} = 0.$$

The non-trivial restriction map is the inclusion of the fixed points. The induction map is the augmentation map (which sends both $\underline{0}$ and $\underline{1}$ to $1 \in \mathbb{Q}$).

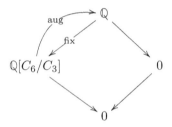

Example 4.40 Similarly, we have the Lewis diagram of $F_1(\mathbb{Q}[C_6])$

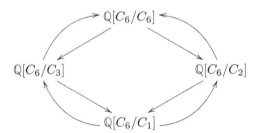

As with previous examples, the induction maps are given by the projections $C_6/K \to C_6/H$. The restriction maps are given by summing over the preimages of that projection.

One may check this is isomorphic to the Mackey functor $\mathrm{Mack}_{C_6}(\mathbb{Q}[C_6])$ of Example 4.9, which is defined by $H \mapsto \mathbb{Q}[C_6]^H$. The claimed isomorphism of Mackey functors is induced from the isomorphism

$$\mathbb{Q}[C_6/H] \longrightarrow \mathbb{Q}[C_6]^H, \quad gH \mapsto \sum_{h \in H} hg.$$

4.7 Comparison to equivariant spectra

For G a finite group, we have a classification of rational G-spectra in terms of an algebraic model, see Theorem 4.74. The algebraic model is built from chain complexes of $\mathbb{Q}[W_G H]$-modules for H running over conjugacy classes of subgroups of G.

The most modern approach to the classification takes several steps

- idempotent splitting
- restriction to normalisers
- passing to Weyl groups (by taking fixed points)
- algebraicisation

which we see are analogous to our classification of rational Mackey functors. At the level of homotopy categories, one takes a spectrum X, and then splits it into $e_H^G X$ for varying H. Then one forgets to $N_G H$-spectra and takes H-fixed points. Taking homology of the algebraicisation of a spectrum $e_H^G X$ gives the homotopy groups of the spectrum. This results in a graded $\mathbb{Q}[W_G H]$-module

$$\pi_*\left(\left(i^*(e_H^G X)\right)^H\right) = i^*(e_H^G)\pi_*^H(X) = \left(e_H^G([-, X]_*^G \otimes \mathbb{Q})\right)(G/H).$$

This is exactly the functor U_H applied to the e_H^G-part of the Mackey functor $[-, X]_*^G \otimes \mathbb{Q}$.

A major difference between the method we use for Mackey functors and the approach for rational G-spectra is that in the latter one proves that the various model categories are Quillen equivalent at each stage, rather than arguing via the composite functor. This is partly due to adjunctions in the topological setting not being both left and right adjoint and partly due to the difficulty of working with complex composite functors in model categories. For Mackey functors, we see that the adjunctions

$$e_H^G \mathrm{Mackey}(G) \xrightleftharpoons[i^\#]{i_\#} R_{N_G H}^G(e_H^G)\mathrm{Mackey}(N_G H) \xrightleftharpoons[\varepsilon_* \mathrm{Mack}_{W_G H}]{(-)(N_G H/H)} \mathbb{Q}[W_G H]\text{-mod}$$

are not equivalences. To resolve this, we restrict $R_{N_G H}^G(e_H^G)\mathrm{Mackey}(N_G H)$ to the full subcategory of the image of the functor from $\mathbb{Q}[W_G H]$-mod. We write $R_{N_G H}^G(e_H^G)\overline{\mathrm{Mackey}}(N_G H)$ for this categories. The counit of $((-)(N_G H/H), \varepsilon_* \mathrm{Mack}_{W_G H})$ is an isomorphism, hence $\varepsilon_* \mathrm{Mack}_{W_G H}$ is full and faithful. It follows that we have equivalences of categories.

$$e_H^G \mathrm{Mackey}(G) \xrightleftharpoons[i^\#]{i_\#} R_{N_G H}^G(e_H^G)\overline{\mathrm{Mackey}}(N_G H) \xrightleftharpoons[\varepsilon_* \mathrm{Mack}_{W_G H}]{(-)(N_G H/H)} \mathbb{Q}[W_G H]\text{-mod}$$

Passing to a full subcategory is the algebraic equivalent of localisation at an idempotent as used in the classification of rational G-spectra for finite G.

We will discuss the topological analogues of these results in more detail in Part 2.

4.8 Monoidal properties

We end this part with a discussion of the monoidal structure on Mackey functors. Details can be found in Green [16] and Luca [37]. Given G-Mackey functors M and N, we define

$$T(H) = \bigoplus_{K \leqslant H} M(G/K) \otimes_{\mathbb{Q}} N(G/K) \qquad (M \square N)(G/H) = T(H)/I(H)$$

where $I(H)$ is the $\mathbb{Q}$-submodule of $T(H)$ generated by

$$R_L^K(x) \otimes y' - x \otimes I_L^K(y') \qquad \text{for } x \in M(G/K),\ y' \in N(G/L),$$
$$x' \otimes R_L^K(y) - I_L^K(x') \otimes y \qquad \text{for } x' \in M(G/L),\ y \in N(G/K),$$
$$C_h(x) \otimes y - x \otimes C_h^{-1}(y) \qquad \text{for } x \in M(G/K),\ y \in N(G/hKh^{-1}).$$

where $L \leqslant K \leqslant H$.

Theorem 4.41 *For M and N G-Mackey functors, the construction*

$$(M \square N)(G/H) = T(H)/I(H)$$

defines a Mackey functor when equipped with the conjugation, restriction and induction maps described below. We call this Mackey functor the box product *of M and N.*

Conjugation is given by the diagonal action

$$C_h(x \otimes y) = C_h(x) \otimes C_h(y).$$

Induction from H to H' is given by the inclusion

$$\left(\bigoplus_{K \leqslant H} M(G/K) \otimes_{\mathbb{Q}} N(G/K) \right) \longrightarrow \left(\bigoplus_{K \leqslant H'} M(G/K) \otimes_{\mathbb{Q}} N(G/K) \right)$$

followed by taking quotients with respect to $I(H)$ and $I(H')$. Restriction from H' to H is induced by the map $T(H') \to T(H)/I(H)$ given by

$$x \otimes y \longmapsto \sum_{l \in [H \backslash H' / K]} R_{H \cap lKl^{-1}}^{lKl^{-1}} C_l(x) \otimes R_{H \cap lKl^{-1}}^{lKl^{-1}} C_l(y)$$

for $x \in M(K)$, $y \in N(K)$ and $K \leqslant H$.

One can also define the box product via a convolution product (a left Kan extension over the product of G-sets), using the definition of Mackey functors in terms of spans of G-sets (the Burnside category), see [37, Chapter 3]. The unit for the box product is the Burnside ring Mackey functor.

The equivalence of [37, Chapter 3] also implies that a (commutative) monoid for the box product is a rational Mackey functor M, such that each $M(G/H)$ is a (commutative) $\mathbb{Q}$-algebra, the conjugation and restriction maps are maps of algebras and for $K \leqslant H$, the *Frobenius relations* hold:

$$x \cdot I_K^H(y) = I_K^H(R_K^H(x) \cdot y) \qquad I_K^H(y) \cdot x = I_K^H(y \cdot R_K^H(x))$$

for $x \in M(G/H)$ and $y \in M(G/K)$. We call such a (commutative) monoid Mackey functor a (commutative) Green functor.

The category of $\mathbb{Q}[W_G H]$-modules has a monoidal product, given by tensoring two modules over $\mathbb{Q}$ and equipping the result with the diagonal $\mathbb{Q}[W_G H]$-action. We then see that U_H sends (commutative) Green functors to (commutative) monoids in $\mathbb{Q}[W_G H]$-modules. In fact, we show that U_H is a symmetric monoidal functor.

Lemma 4.42 *Let M and N be Mackey functors. Then*

$$e_H^G(M \square N) \cong (e_H^G M \square N) \cong (M \square e_H^G N) \cong (e_H^G M \square e_H^G N).$$

Hence

$$\left(e_H^G M \square e_H^G N\right)(G/H) = e_H^G M(G/H) \otimes_{\mathbb{Q}} e_H^G N(G/H).$$

Proof The first statement is a calculation of the action of the Burnside ring on the box product.

For the second, the Mackey functor $e_H^G M$ is trivial on proper subgroups of H, from which it follows that

$$T(H) = e_H^G M(G/H) \otimes_{\mathbb{Q}} e_H^G N(G/H) \quad \text{and} \quad I(H) = 0. \qquad \square$$

The first statement of Lemma 4.42 implies that $e_H^G \mathrm{Mackey}(G)$ is monoidal with respect to $\square$ with the unit $e_H^G A_{\mathbb{Q}}$.

Corollary 4.43 *For each $H \leqslant G$ the equivalence of categories*

$$U_H : e_H^G \mathrm{Mackey}(G) \overset{\longrightarrow}{\longleftarrow} \mathbb{Q}[W_G H]\text{-mod} : F_H$$

is strong symmetric monoidal.

Moreover, the splitting result

$$\text{Mackey}(G) \cong \prod_{H \leqslant_G G} e_H^G \text{Mackey}(G)$$

is strong symmetric monoidal.

The topological equivalent of this result is Barnes, Greenlees and Kędziorek [5]. This paper gives a description of E_∞-algebras in rational G-spectra in terms of differential graded algebras in

$$\prod_{(H) \leqslant G} \mathbb{Q}[W_G H]\text{-mod.}$$

The more complicated case of commutative ring G-spectra (or algebras over E_∞^G, the operad governing the highest level of equivariant commutativity) is considered in work of Wimmer [54]. The extra data here comes from multiplicative norm maps, which are related to Tambara functors (commutative Green functors with additional structure), see Strickland [47], Mazur [41] and Hill and Mazur [28]. The idempotent splitting result we use destroys the additional structure of a Tambara functor, leaving only a commutative Green functor. Hence, there is no immediate extension of the above results to Tamabara functors. The question of which idempotents and splittings persevere norms in the Burnside ring is answered fully in work of Böhme [13].

Part 2. The structure of rational G-spectra

For G a compact Lie group, it is natural to study the homotopy theory of G-spectra as Brown representability holds equivariantly, see [40, Section XIII.3]. That is, G-equivariant cohomology theories are represented by G-spectra, so the category of G-equivariant cohomology theories and stable natural transformations between them, is equivalent to the homotopy category of G-spectra. Due to the complexity of the non-equivariant case, one cannot expect a complete analysis of either G-equivariant cohomology theories or G-spectra integrally. However, if we restrict ourselves to G-equivariant cohomology theories with values in rational vector spaces, the situation is greatly simplified, whilst valuable geometric and group theoretic structures remain. For this reason, the programme of understanding G-equivariant cohomology theories begun by Greenlees restricts attention to rational G-equivariant cohomology theories and rational G-spectra.

In this part, we discuss the methods and tools used to obtain algebraic models for rational equivariant spectra. Recall from the introduction that an algebraic model for rational G-spectra is a model category $d\mathcal{A}(G)$, that is Quillen equivalent to G-spectra. This category must consist of differential objects (and morphisms) in a graded abelian category $\mathcal{A}$. To start our journey we begin by recalling some useful facts about G-spectra.

4.9 Preliminaries on G-spectra

Let G be a compact Lie group. We work with orthogonal G-spectra, see Mandell and May [38] for more details. Unless otherwise stated, our categories of G-spectra will be indexed on a complete G-universe $\mathcal{U}$.

For H a closed subgroup of G, one can define homotopy groups of an orthogonal G-spectrum X with structure map σ as

$$\pi_0^H(X) = \operatorname{colim}_V [S^V, X(V)]^H$$

where the maps in the colimit send a map $\alpha : S^V \longrightarrow X(V)$ to the composite

$$S^W \cong S^V \wedge S^{V^\perp} \xrightarrow{\alpha \wedge \mathrm{Id}} X(V) \wedge S^{V^\perp} \xrightarrow{\sigma} X(V \oplus V^\perp) \cong X(W).$$

Here V runs through the G representations in the universe $\mathcal{U}$. More generally, the integer graded homotopy groups of a G-spectrum X are defined using shift and loop functors on spectra and the formula above. A map f of G-spectra is a weak equivalence, also called a *stable equivalence*, in orthogonal G-spectra if and only if $\pi_p^H(f)$ is an isomorphism for all closed subgroups H of G and all integers p. The class of stable equivalences is part of a stable model structure on G-spectra, G-Sp$^\mathcal{O}$.

Orthogonal G-spectra with the stable model structure is a convenient model category for G-equivariant homotopy theory. In particular, the homotopy category is a symmetric monoidal triangulated category (see [39, Appendix 2] for introduction to triangulated categories) with unit the sphere spectrum $\mathbb{S}$, see Hovey, [30, Section 7]. Furthermore, the stable equivalences can be detected by objects in the category in the following sense. For a closed subgroup H in G, an orthogonal spectrum X and integers $p \geq 0$ and $q > 0$

$$[\Sigma^p S^0 \wedge G/H_+, X]^G \cong \pi_p^H(X) \qquad [F_q S^0 \wedge G/H_+, X]^G \cong \pi_{-q}^H(X)$$
$$\tag{4.9.1}$$

where $[-, -]^G$ denotes morphisms in the homotopy category of G-Sp$^\mathcal{O}$

and $F_q(-)$ is the left adjoint to the evaluation functor at $\mathbb{R}^q$, namely: $Ev_{\mathbb{R}^q}(X) = X(\mathbb{R}^q)$. In particular, $F_q(S^0)$ models $\mathbb{S}^{-q}$, the q-fold desuspension of the sphere spectrum. We can put this relation between the shifts of G/H_+ and the weak equivalences into the formalism of [45, Section 2].

Definition 4.44 Let $\mathcal{C}$ be a triangulated category with infinite coproducts. A full triangulated subcategory of $\mathcal{C}$ (with shift and triangles induced from $\mathcal{C}$) is called *localising* if it is closed under coproducts in $\mathcal{C}$. A set $\mathcal{P}$ of objects of $\mathcal{C}$ is called a *set of generators* if the only localising subcategory of $\mathcal{C}$ containing objects of $\mathcal{P}$ is the whole of $\mathcal{C}$. An object of a stable model category is called a *generator* if it is so when considered as an object of the homotopy category.

An object X in $\mathcal{C}$ is *homotopically compact*[2] if for any family of objects $\{A_i\}_{i \in I}$ the canonical map

$$\bigoplus_{i \in I} [X, A_i]^{\mathcal{C}} \longrightarrow [X, \coprod_{i \in I} A_i]^{\mathcal{C}}$$

is an isomorphism in the homotopy category of $\mathcal{C}$.

The set of suspensions and desuspensions of G/H_+, where H varies through all closed subgroups of G, is a set of homotopically compact generators in the stable model category $G\text{-Sp}^{\mathcal{O}}$. Those objects are compact since homotopy groups commute with coproducts and it is clear from [45, Lemma 2.2.1] and Equation (4.9.1) that this is a set of generators for $G\text{-Sp}^{\mathcal{O}}$.

There is an easy-to-check condition for a Quillen adjunction between stable model categories with sets of homotopically compact generators to be a Quillen equivalence. It is used often in the setting of algebraic models. Also notice that the derived functors of Quillen equivalences preserve homotopically compact objects.

Lemma 4.45 *Suppose $F : \mathcal{C} \rightleftarrows \mathcal{D} : U$ is a Quillen pair between stable model categories with sets of homotopically compact generators, such that the right derived functor RU preserves coproducts (or equivalently, such that the left derived functor sends homotopically compact generators to homotopically compact objects).*

If the derived unit and counit are weak equivalences for the respective sets of generators, then (F, U) is a Quillen equivalence.

---

[2] There are different names used in the literature - compact, small. We chose to use the name *homotopically compact* here.

Proof The result depends upon the fact that the homotopy category of a stable model category is a triangulated category. First notice that since the derived functor RU preserves coproducts, the derived unit and counit are triangulated natural transformations. If the derived unit condition is an isomorphism for a set of objects $\mathcal{K}$ then it is also an isomorphism for every object in the localising subcategory for $\mathcal{K}$. Since we assume that $\mathcal{K}$ consists of homotopically compact generators, the localising subcategory for $\mathcal{K}$ is the whole category and the derived unit is an isomorphism. The same argument applies to the counit and the result follows.   □

To construct a model category of *rational* G-spectra we will need to introduce the language of Bousfield localisations, see Section 4.11. Since we will often localise the model category of rational G-spectra at idempotents of the rational Burnside ring, we first look at this ring.

4.10 Idempotents of the rational Burnside ring

For G a compact Lie group, the Burnside ring $\mathsf{A}(G)$ was defined by tom Dieck in [51] in terms of G-manifolds. For a survey on the subject see, for example, Fausk [14]. When working rationally, several descriptions of this ring exist. We give these descriptions and use them to understand the idempotents of the rational Burnside ring. These idempotents are fundamental to the construction of the algebraic model and the calculations therein.

4.10.1 Two ways of understanding rational Burnside ring

Recall that for H a subgroup of G, $N_G H = \{g \in G \mid gH = Hg\}$ is the normaliser of H in G. We write $W = W_G H = N_G H / H$ for the Weyl group of H in G.

Let $\mathcal{F}(G)$ be the set of closed subgroups of G with finite index in their normalizer. That is, all closed $H \leqslant G$ such that $N_G H / H$ is finite. We give this set the topology induced by the Hausdorff metric, see [35, Section V.2].

The isomorphism of tom Dieck (see Lemma 4.11) can be extended to compact Lie groups by [52, Propositions 5.6.4 and 5.9.13] giving an isomorphism of rings

$$\mathsf{A}(G) \otimes \mathbb{Q} \cong C(\mathcal{F}(G)/G, \mathbb{Q}), \qquad (4.10.1)$$

where $C(\mathcal{F}(G)/G, \mathbb{Q})$ denotes the ring of continuous functions on the orbit space $\mathcal{F}(G)/G$ with values in discrete space $\mathbb{Q}$. From now on, we will use notation $\mathsf{A}_{\mathbb{Q}}(G)$ for $\mathsf{A}(G) \otimes \mathbb{Q}$. This isomorphism generalises that of Lemma 4.11. Notice that if G is a finite group, then $\mathrm{Sub}(G) = \mathcal{F}(G)$, where $\mathrm{Sub}(G)$ is the set of all subgroups of G.

From the ring isomorphism above, it is clear that idempotents of the rational Burnside ring of G correspond to the characteristic functions of open and closed subspaces of the orbit space $\mathcal{F}(G)/G$. By the characteristic function of a set V, we mean the function that takes the value 1 for every point in V and 0 otherwise. Equivalently, an idempotent corresponds to an open and closed G-invariant subspace of $\mathcal{F}(G)$. We write e_V, for the idempotent corresponding to an open and closed subset V of $\mathcal{F}(G)/G$.

Every inclusion $i : H \longrightarrow G$ induces a ring homomorphism of rational Burnside rings $i^* : \mathsf{A}_{\mathbb{Q}}(G) \longrightarrow \mathsf{A}_{\mathbb{Q}}(H)$. In general, it is difficult to explicitly describe the image of a given idempotent in terms of open and closed sets under

$$i^* : C(\mathcal{F}(G)/G, \mathbb{Q}) \longrightarrow C(\mathcal{F}(H)/H, \mathbb{Q}).$$

Even before taking conjugacy classes into account, notice that a subgroup $K \leqslant H$ with finite index in the normaliser $N_H K$ does not have to have a finite index in the normaliser $N_G K$. Thus the map

$$i^* : C(\mathcal{F}(G)/G, \mathbb{Q}) \longrightarrow C(\mathcal{F}(H)/H, \mathbb{Q})$$

is not always induced by a map from $\mathcal{F}(H)$ to $\mathcal{F}(G)$. The exception is of course, when G, H are finite groups, as we discussed in Remark 4.14.

A better approach to investigate the action of i^* on idempotents is to view idempotents as corresponding to certain subspaces of the space of *all closed subgroups* of G as follows. We put a topology on the set of all closed subgroups of G, $\mathrm{Sub}(G)$. This topology is called the f-topology in Greenlees [18, Section 8].

For a closed subgroup $H \leqslant G$ and $\varepsilon > 0$ we define a ball

$$O(H, \varepsilon) = \{K \in \mathcal{F}(H) \mid d(H, K) < \varepsilon\}$$

in $\mathrm{Sub}(G)$, where the distance above is measured with respect to the Hausdorff metric. Thus, subgroups close to H that have infinite Weyl groups are ignored, for example if $H = SO(2)$ is a torus then $O(SO(2), \varepsilon)$ is a singleton. Given also a neighbourhood A of the identity in G consider

$$O(H, \varepsilon, A) = \cup_{a \in A} O(H, \varepsilon)^a,$$

where $O(H, \varepsilon)^a$ is the set of a-conjugates of elements of $O(H, \varepsilon)$.

Definition 4.46 For G a compact Lie group, the f-topology on $\mathrm{Sub}(G)$ is generated by the sets $O(H, \varepsilon, A)$ as H, ε, A vary. We write $\mathrm{Sub}_{\mathrm{f}}(G)$ for this topological space.

We say that subgroups $K \leqslant H$ of G are *cotoral* if H/K is a torus. We write $K \sim H$ for the equivalence relation generated by the cotoral pairs. An idempotent in a rational Burnside ring $\mathsf{A}_{\mathbb{Q}}(G)$ corresponds to an open and closed, G-invariant subspace of $\mathrm{Sub}_{\mathrm{f}}(G)$ that is a union of $\sim$-equivalence classes, see Section 4.10.3 for some examples.

Let V be an open and closed G-invariant set in $\mathrm{Sub}_{\mathrm{f}}(G)$ that is a union of $\sim$-equivalence classes. Let i^*V be the preimage of V under $i^* \colon \mathrm{Sub}_{\mathrm{f}}(H) \longrightarrow \mathrm{Sub}_{\mathrm{f}}(G)$. We then let $\overline{i^*V}$ be the smallest G-invariant open and closed set of $\mathrm{Sub}_{\mathrm{f}}(H)$, that is the union of $\sim$-equivalence classes containing i^*V. Using the techniques of Greenlees [18, Section 8] one can show the following.

Lemma 4.47 *Let $i \colon H \to G$ be an inclusion of a closed subgroup. Let e_V be an idempotent of $\mathsf{A}_{\mathbb{Q}}(G)$ corresponding to V, an open and closed G-invariant set in $\mathrm{Sub}_{\mathrm{f}}(G)$ that is a union of $\sim$-equivalence classes. Then $i^*(e_V) = e_{\overline{i^*V}}$.*

Remark 4.48 As it is more common in the literature, an idempotent e_V will come from a open and closed subset V of $\mathcal{F}(G)/G$ unless otherwise stated.

4.10.2 Special idempotents

There are two situations of particular interest to us. In these cases we have an idempotent in the rational Burnside ring and we can provide an algebraic model for the piece of homotopy theory of rational G-spectra that this idempotent governs.

The first situation is where there is an idempotent that remembers only one, special subgroup. The second author called such a subgroup *exceptional* in [32]. The second situation is where the idempotent corresponds to the maximal torus $\mathbb{T}$ in G and all its subgroups. This is called the *toral part* of rational G-spectra in [6].

When we look at idempotents defined by subsets of $\mathcal{F}(G)/G$ the above two cases look identical at first glance: both idempotents are indexed by one subgroup. However in the case of a torus, there are subgroups of the torus that are "hidden" in the torus idempotent. This is visible when one uses the space $\text{Sub}_f(G)$ to describe the idempotent. The subgroups that are cotoral in $\mathbb{T}$ are responsible for making the algebraic model for that part substantially more difficult than in the case of an exceptional subgroup.

We will start our analysis with the case of an exceptional subgroup of G.

Definition 4.49 Suppose G is a compact Lie group. We say that a closed subgroup $H \leqslant G$ is *exceptional*[3] in G if $W_G H$ is finite, there exists an idempotent e_H^G in the rational Burnside ring of G corresponding to the conjugacy class of H in G (via tom Dieck's isomorphism, Equation 4.10.1) and H has no cotoral subgroups.

If H is an exceptional subgroup of G, then $\{K \mid K \in (H)_G\}$ is an open and closed G-invariant subspace of $\text{Sub}_f(G)$, which already is a union of $\sim$-equivalence classes, since H does not contain any cotoral subgroup and $W_G H$ is finite. The other implication also holds; if there is an idempotent corresponding to $\{K | K \in (H)_G\}$ in $\text{Sub}_f(G)$, then H is an exceptional subgroup of G. Thus we could rephrase the definition in terms of the space $\text{Sub}_f(G)$, but we decided to use the more familiar $\mathcal{F}(G)/G$ with the topology given by the Hausdorff metric.

Any subgroup of a finite group G is exceptional. In $O(2)$ only finite dihedral subgroups are exceptional; in particular none of the finite cyclic subgroups are exceptional (since finite cyclic subgroups do not have idempotents in the rational Burnside ring of $O(2)$). The maximal torus $SO(2)$ in $O(2)$ has an idempotent in the rational Burnside ring of $O(2)$, however it is not an exceptional subgroup, since it contains cotoral subgroups, for example the trivial one. In $SO(3)$ all finite dihedral subgroups are exceptional except for D_2, which is conjugate to C_2 and therefore is a cotoral subgroup of a torus. There are four more conjugacy classes of exceptional subgroups: A_4, Σ_4, A_5 and $SO(3)$, where A_4 denotes rotations of a tetrahedron, Σ_4 denotes rotations of a cube and A_5 denotes rotations of a dodecahedron, see [33, Section 2].

If a trivial subgroup is exceptional in G, then G has to be finite. This holds as the normaliser of a trivial subgroup is the whole G, $W_G\{1\} = G$ and the condition that the Weyl group is finite implies that G is a finite group.

Given an exceptional subgroup H, we may use the corresponding idempotent in the rational Burnside ring to split (see Section 4.11) the

---

[3] the name was motivated by the *exceptional* behaviour of the algebraic model over such a subgroup.

category of rational G-spectra into the part over an exceptional subgroup H and its complement. [32] presents the model for rational G-spectra over an exceptional subgroup H.

The exceptional subgroups of a group G can be divided into two sets, according to how their idempotent behaves once restricted to the normaliser of the exceptional subgroup. We closely follow [32] in analysis of these different behaviours.

Definition 4.50   Suppose $H \leqslant A$ are closed subgroups of G such that H is exceptional in G. Suppose further that $i : A \longrightarrow G$ is an inclusion. We say that H is A-*good* in G if $i^*(e_H^G) = e_H^A$ and A-*bad* in G if it is not A-good, i.e. $i^*(e_H^G) \neq e_H^A$.

Notice that the above definition is all about subgroups conjugate to H in A and in G and their relation to each other. If $L \leqslant A$ is such that L is conjugate to H in A, then it is also true that L is conjugate to H in G. Thus if H is A-bad in G it just means that there exists $L' \leqslant A$ such that $(L')_G = (H)_G$ and $(L')_A \neq (L)_A$. An exceptional subgroup H in a compact Lie group G is always H-good in G.

Lemma 4.51   *[32, Lemma 2.3] For the exceptional subgroups in $SO(3)$, we have the following relation between H and its normaliser $N_G H$:*

1 A_5 is A_5-good in $SO(3)$.
2 Σ_4 is Σ_4-good in $SO(3)$.
3 A_4 is Σ_4-good in $SO(3)$.
4 D_4 is Σ_4-bad in $SO(3)$.

Proof   We only need to prove Part (3) and (4), since any exceptional subgroup H in a compact Lie group G is H-good in G. Part (3) follows from the fact that there is one conjugacy class of A_4 in Σ_4, as there is just one subgroup of index 2 in Σ_4. Part (4) follows from the observation that there are two subgroups of order 4 in D_8 (so also in Σ_4) and they are conjugate by an element $g \in D_{16}$, which is the generating rotation by 45 degrees (thus $g \notin D_8$ and thus $g \notin \Sigma_4$).   □

Remark 4.52   Notice that we can generalise Definition 4.50 to non-exceptional subgroups using the equivalent description in terms of conjugacy classes of H in A and in G. In that case, if $G = SO(3)$, $A = O(2)$ and $H = C_2 \leqslant A$, then H is A-bad in G, which follows from the fact that $D_2 \leqslant A$ is G-conjugate to H, but not A-conjugate. This *bad* behaviour of C_2 in $SO(3)$ is visible in the adjunctions used to obtain the algebraic model for toral part of rational $SO(3)$-spectra in [33], which we recall in Proposition 4.66.

Finishing the discussion about idempotents of rational Burnside ring, we note that there is always an idempotent corresponding to the maximal torus $\mathbb{T}$ in G and all its subgroups. This fact was used in [6] to obtain an algebraic model for rational toral G-spectra, thus the ones that have geometric isotropy contained in the set of subgroups of the maximal torus.

4.10.3 Examples

Closed subgroups of $SO(2)$

Recall that $SO(2)$ is the group of rotations of $\mathbb{R}^2$. The closed subgroups of $SO(2)$ are the finite cyclic groups C_n. Each C_n is cotoral in $SO(2)$, that is, it is normal in $SO(2)$ and $SO(2)/C_n \cong SO(2)$. The only subgroup of $SO(2)$ with finite index in its normaliser is $SO(2)$ itself. Hence, the space $\mathcal{F}(SO(2))/SO(2)$ is a single point and the rational Burnside ring of $SO(2)$ is $\mathbb{Q}$. Similar arguments show that $\mathsf{A}_\mathbb{Q}(\mathbb{T}) = \mathbb{Q}$ for $\mathbb{T}$ a torus of any rank.

Closed subgroups of $O(2)$

Recall that $O(2)$ is the group of rotations and reflections of $\mathbb{R}^2$. The closed subgroups are the finite cyclic groups, $T = SO(2)$, $O(2)$ and finite dihedral groups. For fixed n, the finite dihedral groups of order $2n$ are all conjugate. We write D_{2n} for this conjugacy class. The space $\mathcal{F}(O(2))/O(2)$ consists of two parts, which we call the toral part and the dihedral part. The toral part $\widetilde{\mathcal{T}}$, is just one point T corresponding to the maximal torus and all its subgroups. The dihedral part $\widetilde{\mathcal{D}}$, is the set of all dihedral subgroups together with their limit point $O(2)$. Thus, we have idempotents $e_{\widetilde{\mathcal{T}}}$ and $e_{\widetilde{\mathcal{D}}}$ in the rational Burnside ring of $O(2)$ that sum to the identity.

The toral idempotents for $O(2)$ and $SO(3)$ will behave very differently when we discuss the interactions between localisations and change of group functors in Section 4.12. To help the notation for this comparison, we use a tilde to denote the dihedral and toral parts of $\mathcal{F}(O(2))/O(2)$ and no tilde for $SO(3)$.

Part	Space $\mathcal{F}(O(2))/O(2)$

$\widetilde{\mathcal{T}}$: T
                                                •

$\widetilde{\mathcal{D}}$: D_2 D_4 D_6 $D_8 D_{10}$ ... $O(2)$
                            •   •   •   • •   •

Closed subgroups of $SO(3)$

Recall that $SO(3)$ is a group of rotations of $\mathbb{R}^3$. We choose a maximal torus T in $SO(3)$ with rotation axis the z-axis. We divide the closed subgroups of G into three types: *toral* $\mathcal{T}$, *dihedral* $\mathcal{D}$ and *exceptional* $\mathcal{E}$. This division is motivated by our preferred splitting of the category of rational $SO(3)$-spectra. The toral part consist of all tori in $SO(3)$ and all cyclic subgroups of these tori. Note that for any natural number n there is one conjugacy class of subgroups from the toral part of order n in $SO(3)$.

The dihedral part consists of all dihedral subgroups D_{2n} (dihedral subgroups of order $2n$) of $SO(3)$ where n is greater than 2, together with all subgroups isomorphic to $O(2)$. Note that $O(2)$ is the normaliser for itself in $SO(3)$. Moreover, there is only one conjugacy class of a dihedral subgroup D_{2n} for each n greater than 2. The normaliser of D_{2n} in $SO(3)$ is D_{4n} for $n > 2$.

We deliberately exclude the conjugacy classes of D_2 and D_4 from the dihedral part. Conjugates of D_2 are excluded from the dihedral part, as D_2 is conjugate to C_2 in $SO(3)$ and that subgroup is already taken into account in the toral part. Conjugates of D_4 are excluded from the dihedral part since its normaliser in $SO(3)$ is Σ_4 (symmetries of a cube), thus its Weyl group Σ_4/D_4 is of order 6, whereas all other finite dihedral subgroups $D_{2n}, n > 2$ have Weyl groups of order 2. For simplicity we decided to treat D_4 separately.

There are five conjugacy classes of subgroups which we call exceptional, namely $SO(3)$ itself, the rotation group of a cube Σ_4, the rotation group of a tetrahedron A_4, the rotation group of a dodecahedron A_5 and D_4, the dihedral group of order 4. Normalisers of these exceptional subgroups are as follows: Σ_4 is equal to its normaliser, A_5 is equal to its normaliser and the normaliser of A_4 is Σ_4, as is the normaliser of D_4.

Consider the space $\mathcal{F}(SO(3))/SO(3)$ of conjugacy classes of subgroups of $SO(3)$ with finite index in their normalisers. Recall that the topology on this space is induced by the Hausdorff metric. The division into these parts is an indication of idempotents of the rational Burnside ring for $SO(3)$ that are chosen to obtain an algebraic model for rational $SO(3)$-spectra.

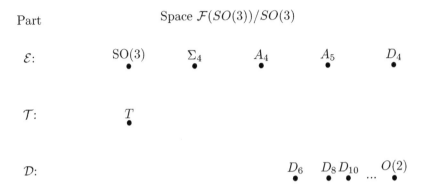

| Part | Space $\mathcal{F}(SO(3))/SO(3)$ | | | | |

The topology on $\mathcal{E}$ is discrete, $\mathcal{T}$ consists of one point T and $\mathcal{D}$ forms a sequence of points converging to $O(2)$.

Note the difference between the dihedral parts for $O(2)$ and $SO(3)$: the conjugacy class of D_2 and D_4. At a first glance, the toral part for $SO(3)$ looks the same as the toral part for $O(2)$. However, for $SO(3)$ it contains information about $D_2 \leqslant O(2)$ (since D_2 is conjugate to C_2 in $SO(3)$), whereas for $O(2)$ it does not. These differences will become significant when we look at the interactions between localisations at idempotents and change of groups functors in Section 4.12.

We use the following idempotents in the rational Burnside ring of $SO(3)$: $e_{\mathcal{T}}$ corresponding to the characteristic function of the toral part $\mathcal{T}$, $e_{\mathcal{D}}$ corresponding to the characteristic function of the dihedral part $\mathcal{D}$ and $e_{\mathcal{E}}$ corresponding to the characteristic function of the exceptional part $\mathcal{E}$. Since $\mathcal{E}$ is a disjoint union of five points, it is in fact a sum of five idempotents, one for every (conjugacy class of a) subgroup in the exceptional part: $e_{SO(3)}$, e_{Σ_4}, e_{A_4}, e_{A_5} and e_{D_4}. We use a simplified notation e_H to mean $e_H^{SO(3)}$ here.

Remark 4.53 All finite dihedral subgroups in $SO(3)$ are exceptional, hence each has an idempotent corresponding to it. However, as there are countably many conjugacy classes of dihedral subgroups, we cannot write $e_{\mathcal{D}}$ as the sum of all these idempotents. Similarly, the characteristic function of the point $O(2)$ is not a continuous map to $\mathbb{Q}$, hence it does not correspond to an idempotent.

4.11 Left and right Bousfield localisations and splittings

There are two well-understood ways of making a homotopy category of a given model category *smaller*. Both ways boil down to adding weak equivalences in a tractable way. The first one keeps the cofibrations the same and is called a left Bousfield localisation (the particular version we use is also called a *homological localisation*). The second one keeps the fibrations the same and is called the right Bousfield localisation (or cellularisation).

4.11.1 Left Bousfield localisation

The general theory of left Bousfield localisations is given in Hirschhorn [29]. For homological localisation we use the following result, which is [38, Chapter IV, Theorem 6.3].

Theorem 4.54 *Suppose E is a cofibrant object in G-$\mathrm{Sp}^{\mathcal{O}}$ or a cofibrant based G-space. There is a new model structure called the E-local model structure on G-$\mathrm{Sp}^{\mathcal{O}}$, denoted $L_E(G\text{-}\mathrm{Sp}^{\mathcal{O}})$, defined as follows. A map $f\colon X \longrightarrow Y$ is*

- *a weak equivalence if it is an E-equivalence, that is,*

$$\mathrm{Id}_E \wedge f\colon E \wedge X \longrightarrow E \wedge Y$$

is a stable equivalence,
- *a cofibration if it is a cofibration with respect to the stable model structure,*
- *a fibration if it has the right lifting property with respect to all trivial cofibrations.*

The E-fibrant objects Z are the fibrant G-spectra that are E-local, that is, the map

$$[f, Z]^G\colon [Y, Z]^G \longrightarrow [X, Z]^G$$

is an isomorphism for all E-equivalences f. For X a G-spectrum, an E-fibrant approximation gives Bousfield localisation $\lambda\colon X \longrightarrow L_E X$ of X at E.

We will refer to the above model structure as the left Bousfield locali-sation of the category of G-spectra at E. This model category is proper, stable, symmetric monoidal and cofibrantly generated. An E-equivalence between E-local objects is a weak equivalence by [29, Theorems 3.2.13 and 3.2.14].

As previously mentioned, the first simplification of the category of G-spectra is rationalisation. This means localisation at the Moore spectrum for $\mathbb{Q}$, $\mathbb{S}_\mathbb{Q}$. For details see [3, Definition 5.1]. This spectrum has the property that $\pi_*(X \wedge \mathbb{S}_\mathbb{Q}) = \pi_*(X) \otimes \mathbb{Q}$. We refer to this model category as the model category of rational G-spectra.

The self-maps of the rational sphere spectrum in the homotopy category of G-spectra are given by the rational Burnside ring

$$\mathsf{A}_\mathbb{Q}(G) \cong [\mathbb{S}, \mathbb{S}]^{G\text{-}\mathrm{Sp}^{\mathcal{O}}} \otimes \mathbb{Q} \cong [\mathbb{S}, \mathbb{S}]^{L_{\mathbb{S}_\mathbb{Q}} G\text{-}\mathrm{Sp}^{\mathcal{O}}} \cong [\mathbb{S}_\mathbb{Q}, \mathbb{S}_\mathbb{Q}]^{G\text{-}\mathrm{Sp}^{\mathcal{O}}}.$$

It follows that $e \in \mathsf{A}_\mathbb{Q}(G)$ can be represented by a map $e\colon \mathbb{S}_\mathbb{Q} \longrightarrow \mathbb{S}_\mathbb{Q}$. We define $e\mathbb{S}_\mathbb{Q}$ to be the homotopy colimit (a mapping telescope) of the diagram

$$\mathbb{S}_\mathbb{Q} \xrightarrow{\ e\ } \mathbb{S}_\mathbb{Q} \xrightarrow{\ e\ } \mathbb{S}_\mathbb{Q} \xrightarrow{\ e\ } \dots \ .$$

We ask for this spectrum to be cofibrant either by choosing a good construction of homotopy colimit, or by cofibrantly replacing the result in the stable model structure for G-spectra. We thus have model structures $L_{e\mathbb{S}_\mathbb{Q}}(G\text{-}\mathrm{Sp}^{\mathcal{O}})$ and $L_{(1-e)\mathbb{S}_\mathbb{Q}}(G\text{-}\mathrm{Sp}^{\mathcal{O}})$. Fibrant replacement in $L_{e\mathbb{S}_\mathbb{Q}}(G\text{-}\mathrm{Sp}^{\mathcal{O}})$ is given by taking the fibrant replacement of $X \wedge e\mathbb{S}_\mathbb{Q}$. Since this commutes with taking infinite coproducts, the localisation is smashing in the sense of Ravenel [44] and Hovey et al. [31]). In particular, this localisation preserves homotopically compact generators.

We know from Section 4.10 that e corresponds to an open and closed, G-invariant subspace of $\mathrm{Sub}_\mathrm{f}(G)$ that is a union of $\sim$-equivalence classes, call it V_e. By considering the geometric fixed point functors Φ^H, for all $H \leqslant G$ (see [38, Section V.4]), we can see that the homotopy category of $L_{e\mathbb{S}_\mathbb{Q}}(G\text{-}\mathrm{Sp}^{\mathcal{O}})$ is the homotopy category of rational G-spectra X with geometric isotropy

$$\mathcal{GI}(X) = \{H \leqslant G \mid \Phi^H(X) \not\simeq *\},$$

concentrated over the subgroups H that are in V_e.

4.11.2 Splitting

A common step in the classification of rational G-spectra is to split the category using idempotents of the rational Burnside ring. Work of the first author [3] allows us to perform a compatible splitting at the level of model categories.

Theorem 4.55 *[3, Theorem 4.4] Let e be an idempotent in the rational Burnside ring $A_{\mathbb{Q}}(G)$. There is a strong symmetric monoidal Quillen equivalence:*

$$\triangle \ : \ L_{\mathbb{S}_{\mathbb{Q}}}(G\text{-Sp}^{\mathcal{O}}) \ \underset{\longleftarrow}{\overset{\longrightarrow}{\rightleftharpoons}} \ L_{e\mathbb{S}_{\mathbb{Q}}}(G\text{-Sp}^{\mathcal{O}}) \times L_{(1-e)\mathbb{S}_{\mathbb{Q}}}(G\text{-Sp}^{\mathcal{O}}) \ : \Pi$$

The left adjoint is a diagonal functor, the right adjoint is a product and the product category on the right is considered with the objectwise model structure (a map (f_1, f_2) is a weak equivalence, a fibration or a cofibration if both factors f_i are so).

One can also look at splittings non-rationally, as in Böhme [13].

4.11.3 Cellularisation

A cellularisation of a model category is a right Bousfield localisation at a set of objects. Such a localisation exists by [29, Theorem 5.1.1] whenever the model category is right proper and cellular. When we are in a stable context, the results of [9, Section 5] can be used, which allows us to relax the cellularity condition.

The most common use of cellularisation in the context of algebraic models is the Cellularisation Principle, which we recall in Theorem 4.58.

Definition 4.56 Let $\mathcal{C}$ be a stable model category and K a stable set of objects of $\mathcal{C}$, i.e. a set such that a class of K-cellular objects of $\mathcal{C}$ is closed under desuspension (Note that this class is always closed under suspension.) We call K a set of *cells*. We say that a map $f \colon A \longrightarrow B$ of $\mathcal{C}$ is a K-*cellular equivalence* if the induced map

$$[k, f]_*^{\mathcal{C}} \colon [k, A]_*^{\mathcal{C}} \longrightarrow [k, B]_*^{\mathcal{C}}$$

is an isomorphism of graded abelian groups for each $k \in K$. An object $Z \in \mathcal{C}$ is said to be K-*cellular* if

$$[Z, f]_*^{\mathcal{C}} \colon [Z, A]_*^{\mathcal{C}} \longrightarrow [Z, B]_*^{\mathcal{C}}$$

is an isomorphism of graded abelian groups for any K-cellular equivalence f.

The following is Hirschhorn [29, Theorem 5.1.1].

Theorem 4.57 *For K a set of objects in a right proper, cellular model category $\mathcal{C}$, the* right Bousfield localisation *or* cellularisation *of $\mathcal{C}$ with respect to K is the (right proper) model structure K-cell-$\mathcal{C}$ on $\mathcal{C}$ defined as follows.*

- *The weak equivalences are K-cellular equivalences,*
- *the fibrations of K-cell-$\mathcal{C}$ are the fibrations of $\mathcal{C}$,*
- *the cofibrations of K-cell-$\mathcal{C}$ are defined via left lifting property.*

The cofibrant objects of K-cell-$\mathcal{C}$ are called K-cofibrant and are precisely the K-cellular and cofibrant objects of $\mathcal{C}$.

When $\mathcal{C}$ is stable and K is a stable set of cofibrant objects, then the cellularisation of a proper, cellular stable model category is proper, cellular and stable by Barnes and Roitzheim [9, Theorem 5.9].

We can further ask the cells K to be homotopically compact objects. By [9, Section 9] the homotopy category K-cell-$\mathcal{C}$ is the full triangulated subcategory of the homotopy category of $\mathcal{C}$ generated by K. In particular, K is a set of homotopically compact generators for K-cell-$\mathcal{C}$. These ideas lead to the following theorem. For examples of its use, see Section 4.11.4, Theorem 4.68 or Theorem 4.75.

Theorem 4.58 (The Cellularisation Principle) *Let M and N be right proper, stable, cellular model categories with (F, U) a Quillen adjunction between M and N. Let $\mathcal{Q}$ be a cofibrant replacement functor in M and $\mathcal{R}$ a fibrant replacement functor in N.*

- *Let $\mathcal{K}$ be a set of objects in M with $F\mathcal{QK}$ its image in N. Then F and U induce a Quillen adjunction*

$$F : \mathcal{K}\text{-cell-}M \rightleftarrows F\mathcal{QK}\text{-cell-}N : U$$

between the $\mathcal{K}$-cellularisation of M and the $F\mathcal{QK}$-cellularisation of N.
- *If $\mathcal{K}$ is a stable set of homotopically compact objects in M such that for each A in $\mathcal{K}$ the object $F\mathcal{Q}A$ is homotopically compact in N and the derived unit $\mathcal{Q}A \to U\mathcal{R}F\mathcal{Q}A$ is a weak equivalence in M, then F and U induce a Quillen equivalence between the cellularisations:*

$$\mathcal{K}\text{-cell-}M \simeq F\mathcal{QK}\text{-cell-}N.$$

- *If L is a stable set of homotopically compact objects in N such that for each B in L the object $U\mathcal{R}B$ is homotopically compact in M and* *the derived counit $FQU\mathcal{R}B \to \mathcal{R}B$ is a weak equivalence in N, then F and U induce a Quillen equivalence between the cellularisations:*

$$U\mathcal{R}\mathcal{L}\text{-cell-}M \simeq L\text{-cell-}N.$$

4.11.4 Alternatives to splitting

In the case of $SO(2)$, the rational Burnside ring is $\mathbb{Q}$, so there are no idempotents to give a splitting. Instead, one must look for replacements for the idempotents or other methods of simplifying the category of rational $SO(2)$-spectra. One approach comes from inducing idempotents from the smaller subgroups. Suppose H is a subgroup of $SO(2)$ such that $A_{\mathbb{Q}}(H)$ has an idempotent e. Then $SO(2)_+ \wedge_H e\mathbb{S}$ is a retract of $SO(2)/H_+$ that does not come from an idempotent of $A_{\mathbb{Q}}(SO(2))$. The set of these spectra as H and e vary give a better behaved set of homotopically compact generators for rational $SO(2)$-spectra. We can think of this construction as applying an *induced idempotent* to $SO(2)/H_+$. While they are not used directly in constructing the algebraic model for rational $SO(2)$-spectra, they are highly useful in understanding it.

Generalising the situation above, the rational Burnside ring of any torus $\mathbb{T}$ has no idempotents. Greenlees and Shipley [27] provided a new method of obtaining an algebraic model in this case. Suppose $\mathcal{F}$ is the family of all proper subgroups of $\mathbb{T}$, we define the universal space $E\mathcal{F}_+$ as a $\mathbb{T}$-CW-complex with the following universal property

$$(E\mathcal{F}_+)^H \simeq \begin{cases} S^0 & \text{iff } H \in \mathcal{F} \\ * & \text{otherwise.} \end{cases}$$

The universal space $E\mathcal{F}_+$ is part of a cofiber sequence called the *isotropy separation sequence*

$$E\mathcal{F}_+ \longrightarrow S^0 \longrightarrow \widetilde{E}\mathcal{F}$$

which can be turned into a homotopy pullback diagram in $\mathbb{T}$-spectra[4]. In the case of $\mathbb{T} = SO(2)$, this is also called the *Hasse square*:

$$\begin{array}{ccc} \mathbb{S} & \longrightarrow & \widetilde{E}\mathcal{F} \\ \downarrow & & \downarrow \\ DE\mathcal{F}_+ & \longrightarrow & DE\mathcal{F}_+ \wedge \widetilde{E}\mathcal{F}. \end{array}$$

---

[4] We slightly abuse the notation and whenever we write a $\mathbb{T}$-space we actually mean its suspension spectrum.

The diagram with $\mathbb{S}$ removed is called the punctured cube and is denoted by $\mathbb{S}^{\lrcorner}$. Using [26], we may construct a model category of modules over $\mathbb{S}^{\lrcorner}$ in rational $SO(2)$-spectra, which we call $\mathbb{S}^{\lrcorner}$-mod (slightly abusing notation and not mentioning the ambient category). Any $SO(2)$-spectrum X defines a module over the diagram by smashing with the ring spectra $\widetilde{E\mathcal{F}}$, $DE\mathcal{F}_+$ and $DE\mathcal{F}_+ \wedge \widetilde{E\mathcal{F}}$. This functor has a right adjoint that is a type of pullback, giving an adjunction between $\mathbb{S}^{\lrcorner}$-mod and rational $SO(2)$-spectra. The Cellularisation Principle, Theorem 4.58, can be used to construct a Quillen equivalence from this adjunction, see either [27, Sections 4-6] or [8, Section 3.2] for details.

In case of a torus of rank r, repeatedly using the isotropy separation sequence one can obtain a $r + 1$-dimensional cube diagram. The terms of this cube are all genuine-commutative equivariant ring $\mathbb{T}$-spectra by Greenlees [22]. We again use the notation $\mathbb{S}^{\lrcorner}$ for the punctured cube of these ring $\mathbb{T}$-spectra and obtain the following theorem.

Theorem 4.59 *[27, Proposition 4.1] There is a strong symmetric monoidal Quillen equivalence*

$$L_{\mathbb{S}_{\mathbb{Q}}}(\mathbb{T}\text{-Sp}^{\mathcal{O}}) \simeq_{QE} \mathcal{K}\text{-cell-}\mathbb{S}^{\lrcorner}\text{-mod}$$

where $\mathcal{K}$ is the image in $\mathbb{S}^{\lrcorner}$-mod of the set of compact generators for $L_{\mathbb{S}_{\mathbb{Q}}}(\mathbb{T})\text{-Sp}^{\mathcal{O}}$.

When G is a finite group, we let $\mathcal{F}$ be the family of all proper subgroups of G. The homotopy pullback diagram obtained by using the isotropy separation sequence gives exactly the idempotent splitting, since

$$E\mathcal{F}_+ \simeq \prod_{(H),\ H \in \mathcal{F}} e_H^G \mathbb{S} \simeq DE\mathcal{F}_+,$$

$\widetilde{E\mathcal{F}} \simeq e_G^G \mathbb{S}$ and $DE\mathcal{F}_+ \wedge \widetilde{E\mathcal{F}} \simeq *$.

However, the spectra $e_H^G \mathbb{S}$ are not genuine-commutative equivariant ring spectra (they are only naïve-commutative). Hence, it is easier to use the splitting approach for finite G. See Böhme [13] for a complete explanation of the relation between genuine-commutative equivariant ring spectra and localisation at idempotents.

An interesting case when there are some, but not enough idempotents, is the case of the dihedral part of $O(2)$-spectra, see [4]. In that case, there is no idempotent whose support is exactly $O(2)$. The abelian (resp. algebraic) model for the dihedral part of rational $O(2)$-spectra is given in terms of sheaves of $\mathbb{Q}[W]$-modules (resp. differential $\mathbb{Q}[W]$-modules) over the space $\widetilde{\mathcal{D}}$, where the stalk over the point $O(2)$ has a trivial W-action.

The stalk over $O(2)$ can be described in terms of a *virtual idempotent* – a colimit of idempotents, see [4, Section 5].

A similar approach occurs for profinite groups in work of Barnes and Sugrue [10] and Sugrue [48].

4.12 Change of groups and localisations

Once we split the category of rational G-spectra using idempotents, our main aim is to get rid of the remaining equivariance in each piece separately by applying certain fixed points functors. Assume we are working with the category $L_{e_{\mathbb{S}_{\mathbb{Q}}}}(G\text{-}\mathrm{Sp}^{\mathcal{O}})$ and we want to take H fixed points. First we must move to the category $N\text{-}\mathrm{Sp}^{\mathcal{O}}$ where N is the normaliser of H in G, appropriately localised. We need N, since we want to have a residual Weyl group $(W = N_G H/H)$ action. At the same time we need to localise $N\text{-}\mathrm{Sp}^{\mathcal{O}}$ at some idempotent of the rational Burnside ring of N corresponding to e, since we want to obtain a Quillen equivalence with $L_{e_{\mathbb{S}_{\mathbb{Q}}}}(G\text{-}\mathrm{Sp}^{\mathcal{O}})$.

In work of the second author [32] and [33], there was a precise analysis of two adjunctions: the induction–restriction and restriction–coinduction adjunctions in relation to localisations of categories of equivariant spectra at idempotents. Below we summarise how these results allow us to make the restriction–coinduction adjunction into a Quillen equivalence in suitable situations. Our examples are based on finite groups, $O(2)$ and $SO(3)$.

4.12.1 Restriction–coinduction adjunction and localisations

Suppose we have an inclusion $i\colon N \hookrightarrow G$ of a subgroup N in a group G. This gives a pair of adjoint functors at the level of orthogonal spectra (see for example [38, Section V.2]), namely induction, restriction and coinduction as below (the left adjoint is above the corresponding right adjoint). We note here, that for the induction functor to be a left Quillen functor we must take care over the universes involved.

$$
G\text{-}\mathrm{Sp}^{\mathcal{O}} \quad \overset{\overset{\textstyle G_+ \wedge_N -}{\xleftarrow{\hspace{2cm}}}}{\underset{\underset{\textstyle F_N(G_+,-)}{\xleftarrow{\hspace{2cm}}}}{\xrightarrow[\hspace{2cm}]{\quad i^* \quad}}} \quad N\text{-}\mathrm{Sp}^{\mathcal{O}}
$$

We assume that G-spectra are indexed over a complete G-universe $\mathcal{U}$ and N-spectra are indexed over one of two universes. In the case where we want to use the restriction functor as a right adjoint, we use the restriction of $\mathcal{U}$ to an N-universe. If we consider restriction as a left Quillen functor we use a complete N-universe. With these conventions, the two pairs of adjoint functors are Quillen pairs with respect to stable model structures by [38, Chapter V, Proposition 2.3 and 2.4]. Given this, we slightly abuse the notation by not mentioning universes or the change of universe functors of [38, Section V.2].

The restriction functor as a right adjoint is often used when we want to take (both categorical and geometric) H-fixed points of G-spectra, where H is not a normal subgroup of G. The procedure is to restrict to $N_G H$-spectra and then to take H-fixed points to land in $W_G H$-spectra. This is usually done in one go, since the restriction functor and the H-fixed points functor are both right Quillen functors.

It is natural to ask when the pair of adjunctions above passes to the localised categories, in our case localised at $e_H^G \mathbb{S}_{\mathbb{Q}}$ and $e_H^N \mathbb{S}_{\mathbb{Q}}$ respectively. The answer is related to H being a good or bad subgroup in G. The induction–restriction adjunction does not always induce a Quillen adjunction on the localised categories, unless H is N-good in G. However, the restriction–coinduction adjunction induces a Quillen adjunction on these localised categories, for all exceptional subgroups H. Before we discuss this particular adjunction we state a general result.

Lemma 4.60 *Suppose that $F : \mathcal{C} \rightleftarrows \mathcal{D} : R$ is a Quillen adjunction of model categories where the left adjoint is strong (symmetric) monoidal. Suppose further that E is a cofibrant object in $\mathcal{C}$ and that both $L_E \mathcal{C}$ and $L_{F(E)} \mathcal{D}$ exist. Then*

$$F : L_E \mathcal{C} \rightleftarrows L_{F(E)} \mathcal{D} : U$$

is a strong (symmetric) monoidal Quillen adjunction. Furthermore, if the original adjunction was a Quillen equivalence, then the induced adjunction on localised categories is as well.

Proof Since the localisation did not change the cofibrations, the left adjoint F still preserves them. To show that it also preserves acyclic cofibrations, take an acyclic cofibration $f : X \longrightarrow Y$ in $L_E \mathcal{C}$. By definition, $f \wedge \mathrm{Id}_E$ is an acyclic cofibration in $\mathcal{C}$. Since F was a left Quillen functor before localisation, $F(f \wedge \mathrm{Id}_E)$ is an acyclic cofibration in $\mathcal{D}$. As F was strong monoidal, we have $F(f \wedge \mathrm{Id}_E) \cong F(f) \wedge \mathrm{Id}_{F(E)}$, so $F(f)$ is an acyclic cofibration in $L_{F(E)} \mathcal{D}$ which finishes the proof of the first part.

To prove the second part of the statement we use Part (2) from [30, Corollary 1.3.16]. Since F is strong monoidal, and the original adjunction was a Quillen equivalence, F reflects $F(E)$-equivalences between cofibrant objects. It remains to check that the derived counit of this adjunction is an $F(E)$-equivalence. An $F(E)$-fibrant object is fibrant in $\mathcal{D}$ and the cofibrant replacement functor remains unchanged by localisation. Thus the claim follows from the fact that (F, U) was a Quillen equivalence before localisations. □

We will use this result in several cases. We start with the restriction–coinduction adjunction.

Corollary 4.61   *Let* $i\colon N \longrightarrow G$ *denote the inclusion of a subgroup and let* E *be a cofibrant object in* G-$\mathrm{Sp}^{\mathcal{O}}$. *Then*

$$i^* \;:\; L_E(G\text{-}\mathrm{Sp}^{\mathcal{O}}) \; \overset{\longrightarrow}{\longleftarrow} \; L_{i^*(E)}(N\text{-}\mathrm{Sp}^{\mathcal{O}}) \;:\; F_N(G_+, -)$$

is a strong symmetric monoidal Quillen pair.

Notice that if $E = e\mathbb{S}_{\mathbb{Q}}$ for some idempotent $e \in \mathsf{A}_{\mathbb{Q}}(G)$ then we get the following

Corollary 4.62   *Suppose* G *is any compact Lie group,* $i\colon N \longrightarrow G$ *is an inclusion of a subgroup and* e *is an idempotent in* $\mathsf{A}_{\mathbb{Q}}(G)$. *Then the adjunction*

$$i^* \;:\; L_{e\mathbb{S}_{\mathbb{Q}}}(G\text{-}\mathrm{Sp}^{\mathcal{O}}) \; \overset{\longrightarrow}{\longleftarrow} \; L_{i^*(e)\mathbb{S}_{\mathbb{Q}}}(N\text{-}\mathrm{Sp}^{\mathcal{O}}) \;:\; F_N(G_+, -)$$

is a Quillen pair.

4.12.2 Exceptional part of rational G-spectra

We will repeatedly use the above result, mainly in situations where after further localisation of the right hand side we will get a Quillen equivalence.

Corollary 4.63   *[32, Corollary 4.7] Suppose* G *is a compact Lie group and* H *is an exceptional subgroup of* G. *Then*

$$i^* \;:\; L_{e_H^G \mathbb{S}_{\mathbb{Q}}}(G\text{-}\mathrm{Sp}^{\mathcal{O}}) \; \overset{\longrightarrow}{\longleftarrow} \; L_{e_H^N \mathbb{S}_{\mathbb{Q}}}(N\text{-}\mathrm{Sp}^{\mathcal{O}}) \;:\; F_N(G_+, -)$$

is a Quillen pair.

Proof For N-good H, the result follows from the fact that the idempotent on the right hand side satisfies $e_H^N = i^*(e_H^G)$. For N-bad H, it is true since the left hand side is a further localisation of $L_{i^*(e_H^G)S_\mathbb{Q}}(N - \mathrm{Sp}^\mathcal{O})$ at the idempotent e_H^N:

$$L_{e_H^G S_\mathbb{Q}}(G\text{-}\mathrm{Sp}^\mathcal{O}) \xrightleftharpoons[F_N(G_+,-)]{i^*} L_{i^*(e_H^G)S_\mathbb{Q}}(N\text{-}\mathrm{Sp}^\mathcal{O}) \xrightleftharpoons[\mathrm{Id}]{\mathrm{Id}} L_{e_H^N S_\mathbb{Q}}(N\text{-}\mathrm{Sp}^\mathcal{O}).$$

Note that since H is bad, $e_H^N \neq i^*(e_H^G)$ and $e_H^N i^*(e_H^G) = e_H^N$. ☐

Theorem 4.64 *[32, Theorem 4.8] Suppose H is an exceptional subgroup of G. Then the adjunction*

$$i_N^* : L_{e_H^G S_\mathbb{Q}}(G\text{-}\mathrm{Sp}^\mathcal{O}) \xrightleftharpoons{\hspace{1.5cm}} L_{e_H^N S_\mathbb{Q}}(N\text{-}\mathrm{Sp}^\mathcal{O}) : F_N(G_+, -)$$

is a strong symmetric monoidal Quillen equivalence.

Part of the difficulty in providing an algebraic model for a *piece* of homotopy category of rational G-spectra governed by an idempotent e comes from balancing two things. On the one hand, one wants to simplify the ambient category as much as one can. On the other one must preserve all the relevant homotopical information. This balancing act requires deep understanding of the homotopy category of $L_{eS_\mathbb{Q}}G$-Sp. In the case of an exceptional subgroup H of G, this is achieved by passing to $L_{e_H^N S_\mathbb{Q}}N$-Sp using restriction as a left Quillen functor, as we described above.

There was a reason why we considered restriction to be a left Quillen functor and it is related to the good and bad exceptional subgroups in G.

Proposition 4.65 *[32, Proposition 4.5] Suppose H is an exceptional subgroup of G that is N-bad in G. Then*

$$i^* : L_{e_H^G S_\mathbb{Q}}(G\text{-}\mathrm{Sp}^\mathcal{O}) \xrightleftharpoons{\hspace{1.5cm}} L_{e_H^N S_\mathbb{Q}}(N\text{-}\mathrm{Sp}^\mathcal{O}) : G_+ \wedge_N -$$

is not a Quillen adjunction. If H is an N-good subgroup in G, then the above adjunction is a Quillen pair.

4.12.3 Toral part of rational $SO(3)$-spectra

The functor i^* is not always a right Quillen functor when considered between categories localised at the toral idempotents. One can argue that this is because the toral idempotents do not always correspond with each other. One example is when $G = SO(3)$, $\mathbb{T} = SO(2)$ and $N = O(2)$. In that case the proof is based on the fact that D_2 is conjugate to C_2 in $SO(3)$ and thus $i^*(e_\mathcal{T}) \neq e_{\tilde{\mathcal{T}}}$.

Proposition 4.66 *[33, Proposition 2.7] Suppose $e_{\mathcal{T}}$ is the toral idempotent of $SO(3)$ and $e_{\widetilde{\mathcal{T}}}$ is the toral idempotent of $O(2)$. That is, $e_{\mathcal{T}}$ is the idempotent in $A_{\mathbb{Q}}(SO(3))$ corresponding to the characteristic function of the toral part $\mathcal{T}$ (i.e. all subconjugates of the maximal torus of $SO(3)$) and $e_{\widetilde{\mathcal{T}}}$ is the idempotent in $A_{\mathbb{Q}}(O(2))$ corresponding to the characteristic function of the toral part $\widetilde{\mathcal{T}}$, i.e. all subconjugates of the maximal torus of $O(2)$ (see Sections 4.10.3 and 4.10.3). Then*

$$i^* \; : \; L_{e_{\mathcal{T}}S_{\mathbb{Q}}}(SO(3)\text{-Sp}) \; \underset{\longrightarrow}{\longleftarrow} \; L_{e_{\widetilde{\mathcal{T}}}S_{\mathbb{Q}}}(O(2)\text{-Sp}) \; : \; SO(3)_+ \wedge_{O(2)} -$$

is not a Quillen adjunction.

The restriction–coinduction adjunction is often better behaved with respect to localisation at idempotents.

Proposition 4.67 *Let $i\colon O(2) \longrightarrow SO(3)$ be the inclusion. Then the following adjunction*

$$i^* \; : \; L_{e_{\mathcal{T}}S_{\mathbb{Q}}}(SO(3)\text{-Sp}) \; \underset{\longleftarrow}{\longrightarrow} \; L_{e_{\widetilde{\mathcal{T}}}S_{\mathbb{Q}}}(O(2)\text{-Sp}) \; : \; F_{O(2)}(SO(3)_+, -)$$

is a strong symmetric monoidal Quillen adjunction.

The proof follows the same argument as Corollary 4.63 above, in the sense that the adjunction is a composite of the restriction–coinduction adjunction localised at an idempotent $e_{\mathcal{T}}$ (and its restriction $i^*(e_{\mathcal{T}})$) followed by a further localisation of $O(2)$-spectra (which excludes the subgroup D_2).

This adjunction of restriction and coinduction is not quite a Quillen equivalence. Denote by $\mathcal{K}$ any set of homotopically compact generators for rational toral $SO(3)$-spectra. Cellularising the right hand side at $\mathcal{K}$ and using the Cellularisation Principle (see Theorem 4.58) gives a Quillen equivalence.

Theorem 4.68 *[33, Theorem 3.28] The following adjunction*

$$L_{e_{\mathcal{T}}S_{\mathbb{Q}}}(SO(3)\text{-Sp}) \; \underset{F_{O(2)}(SO(3)_+, -)}{\overset{i^*}{\rightleftarrows}} \; i^*(\mathcal{K})\text{-cell-}L_{e_{\widetilde{\mathcal{T}}}S_{\mathbb{Q}}}(O(2)\text{-Sp})$$

is a Quillen equivalence, where $\mathcal{K}$ denotes the set of homotopically compact generators for the model category $L_{e_{\mathcal{T}}S_{\mathbb{Q}}}(SO(3)\text{-Sp})$.

4.12.4 Dihedral part of rational $SO(3)$-spectra

In other cases of idempotents it is not always clear to which category one should restrict. For the dihedral idempotent in rational $SO(3)$-spectra, restricting to certain part of the rational dihedral $O(2)$-spectra is the correct choice, but in general there is no good recipe for obtaining an algebraic model.

In the dihedral part of $SO(3)$ we can use restriction as a right or left Quillen functor, we chose the following one, which also follows from Lemma 4.60.

Corollary 4.69 *Let $\mathcal{D}$ denote the dihedral part of $SO(3)$ and $e_\mathcal{D}$ the corresponding idempotent. Let $i\colon O(2) \longrightarrow SO(3)$ be the inclusion. Then*

$$i^* \ : \ L_{e_\mathcal{D}S_\mathbb{Q}}(SO(3)\text{-Sp}) \ \underset{\longleftarrow}{\overset{\longrightarrow}{\rightleftharpoons}} \ L_{i^*(e_\mathcal{D})S_\mathbb{Q}}(O(2)\text{-Sp}) \ : \ F_{O(2)}(SO(3)_+, -)$$

is a Quillen adjunction.

Remark 4.70 The idempotent on the right hand side $i^*(e_\mathcal{D})$ corresponds to the dihedral part of $O(2)$ *excluding* all subgroups D_2 and D_4. Thus, $i^*(e_\mathcal{D}) = i^*(e_\mathcal{D})e_{\widetilde{\mathcal{D}}}$.

4.12.5 Inflation and fixed point adjunction

Suppose H is a normal subgroup of N and consider the natural projection $\varepsilon\colon N \longrightarrow N/H = W$. Then there is a pair of adjoint functors

$$\varepsilon^* \ : \ W\text{-Sp}^\mathcal{O} \ \underset{\longleftarrow}{\overset{\longrightarrow}{\rightleftharpoons}} \ N\text{-Sp}^\mathcal{O} \ : (-)^H$$

where the right adjoint is the H fixed points functor and the left adjoint is called inflation. For details see [38, Section V.3].

We would like to understand the interaction between the localisation at idempotents and the above adjunction. Notice that since inflation is strong symmetric monoidal, the result below follows from Lemma 4.60.

Corollary 4.71 *Let $\varepsilon\colon N \longrightarrow W$ denote the projection of groups, where H is normal in N and $W = N/H$. Let E be a cofibrant object in $W\text{-Sp}^\mathcal{O}$. Then*

$$\varepsilon^* \ : \ L_E(W\text{-Sp}^\mathcal{O}) \ \underset{\longleftarrow}{\overset{\longrightarrow}{\rightleftharpoons}} \ L_{\varepsilon^*(E)}(N\text{-Sp}^\mathcal{O}) \ : (-)^H$$

is a strong symmetric monoidal Quillen pair.

Lemma 4.72 *[32, Theorem 5.2] For H an exceptional subgroup of N, the adjunction*

$$\varepsilon^* \ : \ L_{e_1^W S_\mathbb{Q}}(W\text{-Sp}^\mathcal{O}) \ \underset{\longleftarrow}{\overset{\longrightarrow}{\rightleftharpoons}} \ L_{e_H^N S_\mathbb{Q}}(N\text{-Sp}^\mathcal{O}) \ : (-)^H$$

is a Quillen equivalence. Here e_1^W denotes an idempotent for the trivial subgroup $\{1\} \leqslant W$.

In case of a torus $\mathbb{T}$, we define $(\mathbb{S}^{\lrcorner})^{\mathbb{T}}$ to be the diagram of commutative ring spectra obtained by taking objectwise $\mathbb{T}$-fixed points of $\mathbb{S}^{\lrcorner}$ (from Section 4.11.4). We illustrate this in the case $\mathbb{T} = SO(2)$.

$$\mathbb{S}^{\lrcorner} = \left(\begin{array}{ccc} & & \widetilde{E}\mathcal{F} \\ & & \downarrow \\ DE\mathcal{F}_+ & \longrightarrow & DE\mathcal{F}_+ \wedge \widetilde{E}\mathcal{F} \end{array} \right)$$

$$(\mathbb{S}^{\lrcorner})^{\mathbb{T}} = \left(\begin{array}{ccc} & & \widetilde{E}\mathcal{F}^{\mathbb{T}} \\ & & \downarrow \\ DE\mathcal{F}_+^{\mathbb{T}} & \longrightarrow & (DE\mathcal{F}_+ \wedge \widetilde{E}\mathcal{F})^{\mathbb{T}} \end{array} \right)$$

The inflation–fixed point adjunction lifts to the level of module categories over the diagrams of rings $\mathbb{S}^{\lrcorner}$ and $(\mathbb{S}^{\lrcorner})^{\mathbb{T}}$ by [25]. This adjunction is a Quillen equivalence and by the Cellularisation Principle, Theorem 4.58, it induces a Quillen equivalence on the cellularised categories as follows. We refer the reader to [27, Section 7] for more details.

Theorem 4.73 *Let $\mathbb{T}$ be a torus. The fixed point functor induces strong symmetric monoidal Quillen equivalences*

$$\mathbb{S}^{\lrcorner}\text{-}mod \simeq_{QE} (\mathbb{S}^{\lrcorner})^{\mathbb{T}}\text{-}mod$$
$$\mathcal{K}\text{-}cell\text{-}\mathbb{S}^{\lrcorner}\text{-}mod \simeq_{QE} \mathcal{K}^{\mathbb{T}}\text{-}cell\text{-}(\mathbb{S}^{\lrcorner})^{\mathbb{T}}\text{-}mod$$

where $\mathcal{K}$ is the image in $\mathbb{S}^{\lrcorner}$-mod of the set of compact generators for $L_{\mathbb{S}_{\mathbb{Q}}}(\mathbb{T}\text{-}\mathrm{Sp}^{\mathcal{O}})$ and $\mathcal{K}^{\mathbb{T}}$ its image in $(\mathbb{S}^{\lrcorner})^{\mathbb{T}}$-mod.

The advantage of this last theorem is that it gives a model for rational $\mathbb{T}$-spectra in terms of non-equivariant spectra.

The base idea for the toral part of rational N-spectra (where $\mathbb{T}$ is normal in N) is to use the same steps, but in a context where after taking $\mathbb{T}$-fixed points we land in a category of spectra with an action of $W = N/\mathbb{T}$. This requires some very detailed constructions to make precise, which we leave to [6].

4.13 An algebraic model for rational G-spectra - overview of some cases

In this section we provide a summary of the necessary steps to obtain an algebraic model for a (part of) rational G-spectra in two cases. The first case is when G is a finite group and we follow the steps presented in the algebraic case in Part 1. The second case is when we are interested in the toral part of rational G-spectra, for any compact Lie group G. We discuss briefly the series of simplifications required for the classification result in this case.

4.13.1 An algebraic model for rational G-spectra for finite G

Building on the results of Sections 4.12.2 and 4.12.5 we can sketch the passage to the algebraic model for rational G-spectra when G is a finite group.

Theorem 4.55 allows us to split the category of rational G-spectra into a finite product

$$\prod_{(H) \leqslant G} L_{e_H^G S_{\mathbb{Q}}}(G\text{-Sp}^{\mathcal{O}}).$$

The next step uses the restriction–coinduction Quillen equivalence

$$L_{e_H^G S_{\mathbb{Q}}}(G\text{-Sp}^{\mathcal{O}}) \simeq_{QE} L_{e_H^N S_{\mathbb{Q}}}(N\text{-Sp}^{\mathcal{O}})$$

for each factor of the product seperately. We then follow with the inflation-fixed point Quillen equivalence

$$L_{e_H^N S_{\mathbb{Q}}}(N\text{-Sp}^{\mathcal{O}}) \simeq_{QE} L_{e_1^W S_{\mathbb{Q}}}(W\text{-Sp}^{\mathcal{O}})$$

of the previous section.

The model category $L_{e_1^W S_{\mathbb{Q}}}(W\text{-Sp}^{\mathcal{O}})$ obtained after taking H-fixed points of $L_{e_H^N S_{\mathbb{Q}}}(N\text{-Sp}^{\mathcal{O}})$ can be described in a much easier way. It is Quillen equivalent to the model category $\text{Sp}^{\mathcal{O}}[W]$ of orthogonal spectra with the W action, where the model structure is created from the one on $\text{Sp}^{\mathcal{O}}$ by the forgetful functor $U \colon \text{Sp}^{\mathcal{O}}[W] \longrightarrow \text{Sp}^{\mathcal{O}}$. This allows us to remove the equivariance from *inside* of the complicated category $W\text{-Sp}^{\mathcal{O}}$ (where it appeared in the indexing spaces for the spectrum) to the *outside* of much simpler $\text{Sp}^{\mathcal{O}}[W]$.

Shipley [46] gives a (zig-zag of weak) symmetric monoidal Quillen equivalences between rational spectra and chain complexes of $\mathbb{Q}$-modules (with the projective model structure). This is often referred to in the literature as a *algebraicisation*. This result readily extends to a Quillen equivalence between rational spectra with a finite group action and rational chain complexes with a finite group action. Hence, we obtain an algebraic model for $L_{e_H^N S_{\mathbb{Q}}}(N\text{-Sp}^{\mathcal{O}})$ in terms of chain complexes of $\mathbb{Q}[W_G H]$-modules.

Combining all the steps mentioned in this section we obtain the following result.

Theorem 4.74 *For G a finite group, there is a zig-zag of symmetric monoidal Quillen equivalences between $L_{e_H^G S_{\mathbb{Q}}}(G\text{-Sp}^{\mathcal{O}})$ and $\mathrm{Ch}(\mathbb{Q}[W_G H])$.*
The algebraic model for rational G-spectra is therefore

$$\prod_{(K) \leqslant G} \mathrm{Ch}(\mathbb{Q}[W_G K]).$$

If X is a rational G-spectrum with corresponding object $(A_K)_{(K) \leqslant G}$ in the algebraic model, then

$$\pi_*\left(\left(i^*(e_K^G X)\right)^K\right) \cong \pi_*(\Phi^K X) \cong H_*(A_K).$$

Here i^* and $(-)^K$ denote derived functors of restriction and fixed points discussed in Sections 4.12.2 and 4.12.5, respectively and H_* denotes homology.

4.13.2 Morita equivalences

A different approach to obtaining an algebraic model for rational G-spectra for a finite group G is presented in [2] and uses Morita equivalences developed in the spectral setting by Schwede and Shipley [45].

The idea is to present $L_{e_H^G S_{\mathbb{Q}}}(G\text{-Sp}^{\mathcal{O}})$ as a category of modules over the endomorphism ring spectrum of the compact generator $e_H^G G/H_+$.

Let $\underline{\mathrm{Hom}}(-, =)$ denote the enrichment of G-spectra in non-equivariant spectra. Then

$$E_H = \underline{\mathrm{Hom}}(e_H^G G/H_+, e_H^G G/H_+)$$

(with fibrant replacements omitted from the notation) is a ring spectrum under composition. Furthermore, the model category of modules over E_H (in non-equivariant spectra) is Quillen equivalent to $L_{e_H^G S_{\mathbb{Q}}}(G\text{-Sp}^{\mathcal{O}})$. One can then use algebraicisation (the results of Shipley [46]) to obtain an algebraic model for this part of rational G-spectra.

However, E_H it is not (in general) a *commutative* ring spectrum in orthogonal spectra. The problem is fundamental and can be seen by looking at homotopy groups. The homotopy groups of E_H are non-trivial only in degree 0, where they take value

$$\pi_0(\underline{\mathrm{Hom}}(e_H^G G/H_+, e_H^G G/H_+)) = \mathbb{Q}[W_G H].$$

While this has a cocommutative Hopf algebra structure, it does not have a commutative ring structure in general.

This makes it much harder to obtain a comparison that takes into account the monoidal structures. In particular, we would need to check that the algebraicisation of the ring spectrum E_H also has a cocommutative Hopf algebra structure. As we only have control over the homology of the algebraicised object, we would also need a formality argument that preserves the cocommutative Hopf algebra structure.

4.13.3 An algebraic model for the toral part of rational G-spectra

We give a brief overview of the remaining steps needed to classify rational toral G-spectra. Details are left to the references. While reading the summary, the reader may like to keep in mind the case $\mathbb{T} = SO(2)$, $G = SO(3)$ and $N = O(2)$. These are the easiest cases of interest and have been discussed in previous sections.

Greenlees and Shipley [27] construct an algebraic model for rational $\mathbb{T}$-spectra, where $\mathbb{T}$ is a torus. See also [20] for a full explanation of the algebraic model and [8] for the classification of rational $SO(2)$-spectra. The first two steps of the classification are to apply Theorems 4.59 and 4.73. The next step is to algebraicise using work of Shipley [46]. This gives an algebraic model for rational $\mathbb{T}$-spectra in terms of (a cellularisation of) a category of modules over a diagram of commutative differential graded algebras. Formality of these commutative dgas allows us to simplify the rings in the diagram. An additional simplification of the algebra removes the cellularisation and gives the algebraic model for rational $\mathbb{T}$-spectra.

Work of the authors and Greenlees gives an algebraic model for the toral part of rational G-spectra for any compact Lie group G, see [6]. Given G, we let $\mathbb{T}$ be a maximal torus and N its normaliser in G. We can lift the classification for rational $\mathbb{T}$-spectra to a classification of rational toral N-spectra. We then use the following result to reduce problem of classifying rational toral G-spectra to understanding a cellularisation of rational toral N-spectra.

Theorem 4.75 *[6, Theorem 2.2] The following adjunction*

$$i^* \; : \; L_{e_{\mathbb{T}}^G S_{\mathbb{Q}}}(G\text{-Sp}^{\mathcal{O}}) \; \underset{\longleftarrow}{\overset{\longrightarrow}{\rule{0pt}{0pt}}} \; i^*(\mathcal{L})\text{-cell-}L_{e_{\mathbb{T}}^N S_{\mathbb{Q}}}(N\text{-Sp}) \; : \; F_N(G_+, -)$$

is a Quillen equivalence, where the idempotent on both sides corresponds to the families of all subgroups of maximal torus $\mathbb{T} \leq N \leq G$ *and* $\mathcal{L}$ *denotes the set of homotopically compact generators for* $L_{e_{\mathbb{T}}^G S_{\mathbb{Q}}}(G\text{-Sp}^{\mathcal{O}})$.

By the Cellularisation Principle, Theorem 4.58, we can cellularise each term of the classification of rational toral N-spectra at the derived images of the cells $\mathcal{L}$. This gives a classification of rational toral G-spectra in terms of a cellularisation of the algebraic model for rational toral N-spectra. The final simplification is to remove this cellularisation, which is based on another formality argument.

References

[1] S. Balchin and J.P.C. Greenlees. Adelic models of tensor-triangulated categories. *Adv. Math.*, 375:107339, 45, 2020.

[2] D. Barnes. Classifying rational G-spectra for finite G. *Homology, Homotopy Appl.*, 11(1):141–170, 2009.

[3] D. Barnes. Splitting monoidal stable model categories. *J. Pure Appl. Algebra*, 213(5):846–856, 2009.

[4] D. Barnes. Rational $O(2)$-equivariant spectra. *Homology Homotopy Appl.*, 19(1):225–252, 2017.

[5] D. Barnes, J. P. C. Greenlees, and M. Kędziorek. An algebraic model for rational naïve-commutative G–equivariant ring spectra for finite G. *Homology, Homotopy and Applications*, (21(1)):73–93, 2018.

[6] D. Barnes, J. P. C. Greenlees, and M. Kędziorek. An algebraic model for rational toral G–spectra. *Algebr. Geom. Topol.*, 19(7):3541–3599, 2019.

[7] D. Barnes, J. P. C. Greenlees, and M. Kędziorek. An algebraic model for rational naïve-commutative ring SO(2)-spectra and equivariant elliptic cohomology. *Math. Z.*, 297(3-4):1205–1235, 2020.

[8] D. Barnes, J. P. C. Greenlees, M. Kędziorek, and B. Shipley. Rational SO(2)-equivariant spectra. *Algebr. Geom. Topol.*, 17(2):983–1020, 2017.

[9] D. Barnes and C. Roitzheim. Stable left and right Bousfield localisations. *Glasg. Math. J.*, 56(1):13–42, 2014.

[10] D. Barnes and D. Sugrue. The equivalence between rational G-sheaves and rational G-Mackey functors for profinite G. arXiv: 2002.11745, 2020.

[11] A.M. Bohmann, C. Hazel, J. Ishak, M. Kędziorek, and C. May. Genuine-commutative structure on rational equivariant K-theory for abelian groups. arXiv:2104.01079, 2021.

[12] A.M. Bohmann, C. Hazel, J. Ishak, M. Kędziorek, and C. May. Naive-commutative structure on rational equivariant K-theory for abelian groups. Accepted for a special volume of *Topology and its Applications*, arXiv:2002.01556, 2020.

[13] B. Böhme. Multiplicativity of the idempotent splittings of the Burnside ring and the G-sphere spectrum. *Adv. Math.*, 347:904–939, 2019.

[14] H. Fausk. Survey on the Burnside ring of compact Lie groups. *J. Lie Theory*, 18(2):351–368, 2008.

[15] D. Gluck. Idempotent formula for the Burnside algebra with applications to the p-subgroup simplicial complex. *Illinois J. Math.*, 25(1):63–67, 1981.

[16] J. A. Green. Axiomatic representation theory for finite groups. *J. Pure Appl. Algebra*, 1(1):41–77, 1971.

[17] J. P. C. Greenlees. Some remarks on projective Mackey functors. *J. Pure Appl. Algebra*, 81(1):17–38, 1992.

[18] J. P. C. Greenlees. Rational Mackey functors for compact Lie groups. I. *Proc. London Math. Soc. (3)*, 76(3):549–578, 1998.

[19] J. P. C. Greenlees. Rational O(2)-equivariant cohomology theories. In *Stable and unstable homotopy (Toronto, ON, 1996)*, volume 19 of *Fields Inst. Commun.*, pages 103–110. Amer. Math. Soc., Providence, RI, 1998.

[20] J. P. C. Greenlees. Rational S^1-equivariant stable homotopy theory. *Mem. Amer. Math. Soc.*, 138(661):xii+289, 1999.

[21] J. P. C. Greenlees. Rational SO(3)-equivariant cohomology theories. In *Homotopy methods in algebraic topology (Boulder, CO, 1999)*, volume 271 of *Contemp. Math.*, pages 99–125. Amer. Math. Soc., Providence, RI, 2001.

[22] J. P. C. Greenlees. Couniversal spaces which are equivariantly commutative ring spectra. *Homology Homotopy Appl.*, 22(1):69–75, 2020.

[23] J. P. C. Greenlees and J. P. May. Some remarks on the structure of Mackey functors. *Proc. Amer. Math. Soc.*, 115(1):237–243, 1992.

[24] J. P. C. Greenlees and J. P. May. Generalized Tate cohomology. *Mem. Amer. Math. Soc.*, 113(543):viii+178, 1995.

[25] J. P. C. Greenlees and B. Shipley. Fixed point adjunctions for equivariant module spectra. *Algebr. Geom. Topol.*, 14(3):1779–1799, 2014.

[26] J. P. C. Greenlees and B. Shipley. Homotopy theory of modules over diagrams of rings. *Proc. Amer. Math. Soc. Ser. B*, 1:89–104, 2014.

[27] J.P.C. Greenlees and B. Shipley. An algebraic model for rational torus-equivariant spectra. *J. Topol.*, 11(3):666–719, 2018.

[28] M. A. Hill and K. Mazur. An equivariant tensor product on Mackey functors. *J. Pure Appl. Algebra*, 223(12):5310–5345, 2019.

[29] P. S. Hirschhorn. *Model categories and their localizations*, volume 99 of *Mathematical Surveys and Monographs*. American Mathematical Society, Providence, RI, 2003.

[30] M. Hovey. *Model categories*, volume 63 of *Mathematical Surveys and Monographs*. American Mathematical Society, Providence, RI, 1999.

[31] M. Hovey, J. H. Palmieri, and N. P. Strickland. Axiomatic stable homotopy theory. *Mem. Amer. Math. Soc.*, 128(610):x+114, 1997.

[32] M. Kędziorek. An algebraic model for rational G-spectra over an exceptional subgroup. *Homology Homotopy Appl.*, 19(2):289–312, 2017.

[33] M. Kędziorek. An algebraic model for rational SO(3)-spectra. *Algebr. Geom. Topol.*, 17(5):3095–3136, 2017.

[34] L. G. Lewis, Jr. The category of Mackey functors for a compact Lie group. In *Group representations: cohomology, group actions and topology (Seattle, WA, 1996)*, volume 63 of *Proc. Sympos. Pure Math.*, pages 301–354. Amer. Math. Soc., Providence, RI, 1998.

[35] L. G. Lewis, Jr., J. P. May, M. Steinberger, and J. E. McClure. *Equivariant stable homotopy theory*, volume 1213 of *Lecture Notes in Mathematics*. Springer-Verlag, Berlin, 1986. With contributions by J. E. McClure.

[36] H. Lindner. A remark on Mackey-functors. *Manuscripta Math.*, 18(3):273–278, 1976.

[37] F. Luca. *The algebra of Green and Mackey functors*. ProQuest LLC, Ann Arbor, MI, 1996. Thesis (Ph.D.)–University of Alaska Fairbanks.

[38] M. A. Mandell and J. P. May. Equivariant orthogonal spectra and S-modules. *Mem. Amer. Math. Soc.*, 159(755):x+108, 2002.

[39] H. R. Margolis. *Spectra and the Steenrod algebra*, volume 29 of *North-Holland Mathematical Library*. North-Holland Publishing Co., Amsterdam, 1983. Modules over the Steenrod algebra and the stable homotopy category.

[40] J. P. May. *Equivariant homotopy and cohomology theory*, volume 91 of *CBMS Regional Conference Series in Mathematics*. Published for the Conference Board of the Mathematical Sciences, Washington, DC, 1996. With contributions by M. Cole, G. Comezaña, S. Costenoble, A. D. Elmendorf, J. P. C. Greenlees, L. G. Lewis, Jr., R. J. Piacenza, G. Triantafillou, and S. Waner.

[41] K. Mazur. *On the Structure of Mackey Functors and Tambara Functors*. ProQuest LLC, Ann Arbor, MI, 2013. Thesis (Ph.D.)–University of Virginia.

[42] H. Nakaoka. A Mackey-functor theoretic interpretation of biset functors. *Adv. Math.*, 289:603–684, 2016.

[43] L. Pol and J. Williamson. The left localization principle, completions, and cofree G-spectra. *J. Pure Appl. Algebra*, 224(11):106408, 33, 2020.

[44] D. C. Ravenel. Localization with respect to certain periodic homology theories. *Amer. J. Math.*, 106(2):351–414, 1984.

[45] S. Schwede and B. Shipley. Stable model categories are categories of modules. *Topology*, 42(1):103–153, 2003.

[46] B. Shipley. $H\mathbb{Z}$-algebra spectra are differential graded algebras. *Amer. J. Math.*, 129(2):351–379, 2007.

[47] N. Strickland. Tambara functors. arXiv: 1205.25161, 2012.

[48] D. Sugrue. Rational G-spectra for profinite G. arXiv: 1910.12951, 2019.

[49] J. Thévenaz and P. Webb. Simple Mackey functors. In *Proceedings of the Second International Group Theory Conference (Bressanone, 1989)*, number 23, pages 299–319, 1990.

[50] J. Thévenaz and P. Webb. The structure of Mackey functors. *Trans. Amer. Math. Soc.*, 347(6):1865–1961, 1995.

[51] T. tom Dieck. The Burnside ring of a compact Lie group. I. *Math. Ann.*, 215:235–250, 1975.

[52] T. tom Dieck. *Transformation groups and representation theory*, volume 766 of *Lecture Notes in Mathematics*. Springer, Berlin, 1979.

[53] P. Webb. A guide to Mackey functors. In *Handbook of algebra, Vol. 2*, volume 2 of *Handb. Algebr.*, pages 805–836. Elsevier/North-Holland, Amsterdam, 2000.

[54] C. Wimmer. A model for genuine equivariant commutative ring spectra away from the group order. 2019. arXiv.org:1905.12420 [math.AT].

[55] T. Yoshida. On *G*-functors. I. Transfer theorems for cohomological *G*-functors. *Hokkaido Math. J.*, 9(2):222–257, 1980.

5

Monoidal Bousfield Localizations and Algebras over Operads

David White[a]

Abstract

We give conditions on a monoidal model category $\mathcal{M}$ and on a set of maps $\mathcal{C}$ so that the Bousfield localization of $\mathcal{M}$ with respect to $\mathcal{C}$ preserves the structure of algebras over various operads. This problem was motivated by an example that demonstrates that, for the model category of equivariant spectra, preservation does not come for free, even for cofibrant operads. We discuss this example in detail and provide a general theorem regarding when localization preserves P-algebra structure for an arbitrary operad P.

We characterize the localizations that respect monoidal structure and prove that all such localizations preserve algebras over cofibrant operads. As a special case we recover numerous classical theorems about preservation of algebraic structure under localization, in the context of spaces, spectra, chain complexes, and equivariant spectra. We also provide several new results in these settings, and we sharpen a recent result of Hill and Hopkins regarding preservation for equivariant spectra. To demonstrate our preservation result for non-cofibrant operads, we work out when localization preserves commutative monoids and the commutative monoid axiom, and again numerous examples are provided. Finally, we provide conditions so that localization preserves the monoid axiom.

[a] Department of Mathematics and Computer Science, Denison University

5.1 Introduction

Bousfield localization is a powerful tool in homotopical algebra, with classical applications to homology localization for spaces and spectra [11], to cellularization and nullification [21], and to p-localization and completion [12]. Hirschhorn generalized the machinery of Bousfield localization to the setting of model categories [32], inverting any class of morphisms generated by a set. This general framework has seen tremendous applications: it allows for the passage from levelwise model structures to stable model structures [38], it is used to set up motivic homotopy theory [35], and it allows for the study of combinatorial model categories via simplicial presheaves [16]. The interplay between left Bousfield localization and monoidal structure has often proven fruitful, e.g. to put an E_∞-algebra structure on connective K-theory [19], for homotopy theoretic computations involving generalized Eilenberg-Maclane spaces [13, 15, 21], and, recently, to create an equivariant spectrum with a certain periodicity that is used to resolve the Kervaire invariant one problem [31]. In this chapter, we further the study of the interplay between left Bousfield localization and monoidal model categories, we provide conditions so that left Bousfield localization preserves algebras over operads (and several important model categorical axioms), and we apply our results to numerous classical and new examples of interest.

Structured ring spectra have had numerous applications in stable homotopy theory [19, 38, 43]. Nowadays, structured ring spectra are often thought of as algebras over operads acting in any of the monoidal model categories for spectra. It is therefore natural to ask the extent to which Bousfield localization preserves such algebraic structure. For Bousfield localizations at homology isomorphisms this question is answered in [19] and [43]. The case for spaces is subtle and is addressed in [13], [15], and [21]. More general Bousfield localizations are considered in [14].

The preservation question may also be asked in the context of equivariant and motivic spectra, and it turns out the answer is far more subtle. In Example 5.36, we discuss an example of a naturally occurring Bousfield localization of equivariant spectra that preserves the type of algebraic structure considered in [19] but fails to preserve the equivariant commutativity needed for the landmark results in [31]. We generalize this example in 5.37.

In order to understand this and related examples, we find conditions on a model category $\mathcal{M}$ and on a class of maps $\mathcal{C}$ so that the left Bousfield localization $L_\mathcal{C}$ with respect to $\mathcal{C}$ preserves the structure of algebras over

various operads. After a review of the pertinent terminology in Section
5.2 we give our general preservation result in Section 5.3, which we state
here for the reader's convenience.

Theorem 1 Let $\mathcal{M}$ be a monoidal model category and $\mathcal{C}$ a class of
morphisms such that the Bousfield localization $L_{\mathcal{C}}(\mathcal{M})$ exists and is
a monoidal model category. Let P be an operad valued in $\mathcal{M}$. If the
categories of P-algebras in $\mathcal{M}$ and in $L_{\mathcal{C}}(\mathcal{M})$ inherit transferred semi-
model structures from $\mathcal{M}$ and $L_{\mathcal{C}}(\mathcal{M})$ (with weak equivalences and
fibrations defined via the forgetful functor) then $L_{\mathcal{C}}$ preserves P-algebras.

In general, it is difficult to check that P-algebras in $L_{\mathcal{C}}(\mathcal{M})$ inherit
a transferred semi-model structure. To make it easier to check this
hypothesis, in Section 5.4 we characterize when $L_{\mathcal{C}}(\mathcal{M})$ is a monoidal
model category, proving the following theorem.

Theorem 2 Suppose $\mathcal{M}$ is a cofibrantly generated monoidal model
category in which cofibrant objects are flat (i.e., for all weak equivalences
f and all cofibrant K, $f \otimes id_K$ is a weak equivalence). Then $L_{\mathcal{C}}(\mathcal{M})$ is a
monoidal model category with cofibrant objects flat if and only if every
map of the form $f \otimes id_K$, where f is in $\mathcal{C}$ and K is cofibrant, is a $\mathcal{C}$-local
equivalence. If the domains of the generating cofibrations I are cofibrant,
it suffices to consider K in the set of domains and codomains of the
morphisms in I.

Most monoidal model categories encountered in nature satisfy the
property that cofibrant objects are flat, as examples in Section 5.5
demonstrate. Furthermore, given a set of morphisms $\mathcal{C}$, the condition that
$f \otimes id_K$ be a $\mathcal{C}$-local equivalence is easy to check in practice (for example,
it is true for every localization of spaces and for every stable localization
of spectra). In Section 5.5 we apply these theorems to numerous model
categories and localizations of interest, obtaining preservation results for
Σ-cofibrant operads such as A_{∞} and E_{∞} in model categories of spaces,
spectra, chain complexes, and equivariant spectra. We recover several
classical preservation results, and prove several new preservation results.
We also provide counterexamples, such as Example 5.7 and Example
5.26, to show that the hypotheses of these theorems are really necessary.

In Section 5.5 we present a lattice of equivariant operads that interpolate between non-equivariant E_{∞}-algebra structure and equivariant
E_{∞}-algebra structure. We apply our results to determine which localizations preserve the type of algebraic structure encoded by these operads.
This new collection of operads is different from the N_{∞}-operads of [10]

(that interpolate between equivariant E_∞-algebra structure and genuine commutative structure), and a common generalization of both types of operads is discussed in [27]. Example 5.36 demonstrates that it is possible to preserve equivariant E_∞-algebra structure, but fail to preserve genuine commutative structure. This motivates the latter half of the chapter.

In Section 5.6, we turn to preservation of structure over non-cofibrant operads, specifically, preservation of commutative monoids. For categories of spectra the phenomenon known as rectification means that preservation of strict commutativity is equivalent to preservation of E_∞-structure, but for general model categories (including equivariant spectra) there can be Bousfield localizations that preserve the latter type of structure and not the former. In the companion paper [58] we introduced a condition on a monoidal model category called the *commutative monoid axiom*, that guarantees that the category of commutative monoids inherits a model structure. We build on this work in Section 5.6 by providing conditions on the maps in $\mathcal{C}$ so that Bousfield localization preserves the commutative monoid axiom, proving the following theorem.

Theorem 3 Assume $\mathcal{M}$ is a cofibrantly generated monoidal model category satisfying the strong commutative monoid axiom and with domains of the generating cofibrations cofibrant. Suppose that $L_\mathcal{C}(\mathcal{M})$ is a monoidal Bousfield localization with generating trivial cofibrations $J_\mathcal{C}$. Then $L_\mathcal{C}(\mathcal{M})$ satisfies the strong commutative monoid axiom if and only if $\mathrm{Sym}^n(f)$ is a $\mathcal{C}$-local equivalence for all $n \in \mathbb{N}$ and for all $f \in J_\mathcal{C}$. This occurs if and only if $\mathrm{Sym}(-)$ preserves $\mathcal{C}$-local equivalences between cofibrant objects.

The hypotheses of this theorem are difficult to check, requiring complex arguments unraveling the symmetric group actions. However, in Section 5.7, we apply Theorems 1 and 3 to obtain preservation results for commutative monoids in spaces, spectra, chain complexes, and equivariant spectra. We recover classical preservation results, and several new preservation results, including Theorem 5.56, which sharpens and generalizes the main theorem of [30]. This is the main application of the chapter, and gives a concrete explanation of what goes wrong in Example 5.36 when a specific localization fails to preserve equivariant commutative monoid structure.

Finally, in Section 5.8 we provide conditions so that $L_\mathcal{C}(\mathcal{M})$ satisfies the monoid axiom when $\mathcal{M}$ does, proving the following theorem (see Section 5.8 for definitions of the unfamiliar terms, from [5]).

Theorem 4 Suppose $\mathcal{M}$ is a cofibrantly generated, h-monoidal, left proper model category such that the (co)domains of I are cofibrant and are finite relative to the h-cofibrations and cofibrant objects are flat. Then for any monoidal Bousfield localization $L_\mathcal{C}$, the model category $L_\mathcal{C}(\mathcal{M})$ satisfies the monoid axiom.

In general, it is difficult to check the hypotheses of this theorem. Fortunately, because Theorem 1 only requires transferred semi-model structures, Theorem 4 is not required for preservation, but is required in order to have a comprehensive study of the relationship between left Bousfield localization and monoidal structure, as the monoid axiom is often required for purposes other than transferring model structures. As always, we provide applications of Theorem 4 to the examples of interest in this chapter: spectra, spaces, chain complexes, and equivariant spectra. Roughly half of this chapter consists of applications to these examples. In the setting of Theorem 4, this requires some new results, including a verification that the commonly studied model structures on symmetric spectra are h-monoidal, and the introduction of new model structures on equivariant spectra that are combinatorial and h-monoidal.

Acknowledgments

The author would like to gratefully acknowledge the support and guidance of his advisor Mark Hovey as this work was completed. The author is also indebted to Mike Hill, Carles Casacuberta, Justin Noel, and Clemens Berger for numerous helpful conversations. The author thanks Clark Barwick for catching an error in an early version of this work, Martin Franklin for suggesting applications of this work to simplicial sets, and Boris Chorny for suggesting a simplification in the proof of Theorem 5.65. This draft was improved by comments from Javier Gutiérrez, Brooke Shipley, Cary Malkiewich, and an anonymous referee. A User's Guide is available for this chapter [57], where the interested reader can learn more, created as part of the User's Guide Project presented in [41].

5.2 Preliminaries

We assume the reader is familiar with basic facts about model categories. Excellent introductions to the subject can be found in [18], [32], and [34]. Throughout the chapter we will assume $\mathcal{M}$ is a cofibrantly generated model category [34, Section 2.1], with generating cofibrations I and generating trivial cofibrations J.

Let I-cell denote the class of transfinite compositions of pushouts of maps in I, and let I-cof denote retracts of such. In order to run the small object argument, we will assume the domains K of the maps in I (and J) are κ-small relative to I-cell (resp. J-cell), i.e. given a regular cardinal $\lambda \geq \kappa$ and any λ-sequence $X_0 \to X_1 \to \dots$ formed of maps $X_\beta \to X_{\beta+1}$ in I-cell, then the map of sets $\varinjlim_{\beta<\lambda} \mathcal{M}(K, X_\beta) \to \mathcal{M}(K, \varinjlim_{\beta<\lambda} X_\beta)$ is a bijection. An object is small if there is some κ for which it is κ-small. See Chapter 10 of [32] for a more thorough treatment of this material. For any object X we have a cofibrant replacement $QX \to X$ and a fibrant replacement $X \to RX$.

We will at times also need the hypothesis that $\mathcal{M}$ possesses sets of generating (trivial) cofibrations I and J with domains (hence codomains) cofibrant. This hypothesis is satisfied by all model categories of interest in this chapter, but does not come for free, even for combinatorial model categories $\mathcal{M}$. An example, due to Carlos Simpson, is discussed in Remark 5.22. A method for finding sets I and J with cofibrant domains is given in Lemma 5.19.

Our model category $\mathcal{M}$ will be a closed symmetric monoidal category with product $\otimes$ and unit $S \in \mathcal{M}$. Additionally, we assume the following two axioms, which make $\mathcal{M}$ a *monoidal model category* [34, Chapter 4].

1 Unit Axiom: For any cofibrant X, the map $QS \otimes X \to S \otimes X \cong X$ is a weak equivalence.

2 Pushout Product Axiom: Given any $f : X_0 \to X_1$ and $g : Y_0 \to Y_1$ cofibrations, $f \square g : X_0 \otimes Y_1 \coprod_{X_0 \otimes Y_0} X_1 \otimes Y_0 \to X_1 \otimes Y_1$ is a cofibration. Furthermore, if, in addition, f or g is a trivial cofibration, then $f \square g$ is a trivial cofibration.

Note that the pushout product axiom is equivalent to the statement that $- \otimes -$ is a Quillen bifunctor. Furthermore, it is sufficient to check the pushout product axiom on the generating maps I and J, by Proposition 4.2.5 of [34]. When we need $\mathcal{M}$ to be a simplicial model category, we require the SM7 axiom, which is analogous to the pushout product axiom. We refer the reader to Definition 4.2.18 in [34] for details.

We will at times also need to assume that *cofibrant objects are flat* in $\mathcal{M}$, i.e. that whenever X is cofibrant and f is a weak equivalence then $f \otimes X$ is a weak equivalence. When a monoidal model category satisfies this condition, it is called a *tensor model category* in [22] (Section 12).

Finally, we remind the reader of the monoid axiom of Definition 3.3 in [50].

Given a class of maps $\mathcal{C}$ in $\mathcal{M}$, let $\mathcal{C} \otimes \mathcal{M}$ denote the class of maps $f \otimes id_X$ where $f \in \mathcal{C}$ and $X \in \mathcal{M}$. A model category is said to satisfy the *monoid axiom* if every map in (Trivial-Cofibrations $\otimes \mathcal{M}$)-cell is a weak equivalence.

We will be discussing preservation of algebraic structure as encoded by an operad. Let P be an operad valued in $\mathcal{M}$ (for a general discussion of the interplay between operads and homotopy theory see [9]). Let P-alg$(\mathcal{M})$ denote the category whose objects are P-algebras in $\mathcal{M}$ (i.e. admit an action of P) and whose morphisms are P-algebra homomorphisms (i.e. respect the P-action). The free P-algebra functor from $\mathcal{M}$ to P-alg$(\mathcal{M})$ is left adjoint to the forgetful functor. We will say that P-alg$(\mathcal{M})$ *inherits* a model structure from $\mathcal{M}$ if the model structure is transferred across this adjunction, i.e. if a P-algebra homomorphism is a weak equivalence (resp. fibration) if and only if it is so in $\mathcal{M}$. In Section 4 of [9], an operad P is said to be *admissible* if P-alg$(\mathcal{M})$ inherits a model structure in this way.

Finally, we remind the reader about the process of Bousfield localization as discussed in [32]. This is a general machine that starts with a (nice) model category $\mathcal{M}$ and a set of morphisms $\mathcal{C}$ and produces a new model structure $L_{\mathcal{C}}(\mathcal{M})$ on the same category in which maps in $\mathcal{C}$ are now weak equivalences. Furthermore, this is done in a universal way, introducing the smallest number of new weak equivalences as possible. When we say Bousfield localization we will always mean left Bousfield localization. So the cofibrations in $L_{\mathcal{C}}(\mathcal{M})$ will be the same as the cofibrations in $\mathcal{M}$.

Bousfield localization proceeds by first constructing the fibrant objects of $L_{\mathcal{C}}(\mathcal{M})$ and then constructing the weak equivalences. In both cases this is done via homotopy function complexes map$(-, -)$. If $\mathcal{M}$ is a simplicial or topological model category then one can use the hom-object in $sSet$ or Top. Otherwise a framing is required to construct the homotopy function complex. We refer the reader to [34] or [32] for details on this process.

An object N is said to be *$\mathcal{C}$-local* if it is fibrant in $\mathcal{M}$ and if for all $g : X \to Y$ in $\mathcal{C}$, map$(g, N) : $ map$(Y, N) \to $ map(X, N) is a weak equivalence in $sSet$. These objects are precisely the fibrant objects in $L_{\mathcal{C}}(\mathcal{M})$. A map $f : A \to B$ is a *$\mathcal{C}$-local equivalence* if for all N as above, map$(f, N) : $ map$(B, N) \to $ map(A, N) is a weak equivalence. These maps are precisely the weak equivalences in $L_{\mathcal{C}}(\mathcal{M})$.

It is often more convenient to work with left Bousfield localizations that invert a set of cofibrations (i.e. with left derived Bousfield localization). This can always be guaranteed in the following way. For any map f let Qf denote the cofibrant replacement and let $\widetilde{f}$ denote the left factor in the cofibration-trivial fibration factorization of Qf. Then $\widetilde{f}$ is a cofibration between cofibrant objects and we may define $\widetilde{\mathcal{C}} = \{\widetilde{f} \mid f \in \mathcal{C}\}$. Localization with respect to $\widetilde{\mathcal{C}}$ yields the same result as localization with respect to $\mathcal{C}$, so our assumption that the maps in $\mathcal{C}$ are cofibrations between cofibrant objects loses no generality. We thus make the following convention.

Convention 5.1 Throughout this chapter we assume $\mathcal{C}$ is a set of cofibrations between cofibrant objects, and that the model category $L_{\mathcal{C}}(\mathcal{M})$ exists.

The existence of $L_{\mathcal{C}}(\mathcal{M})$ can be guaranteed by assuming $\mathcal{M}$ is left proper and either combinatorial (as discussed in [3]) or cellular (as discussed in [32]). A model category is *left proper* if pushouts of weak equivalences along cofibrations are again weak equivalences. We will make this a standing hypothesis on $\mathcal{M}$. However, as we have not needed the cellularity or combinatoriality assumptions for our work, outside of the existence of $L_{\mathcal{C}}(\mathcal{M})$, we have decided not to assume them. In this way if a Bousfield localization is known to exist for some reason other than the theory in [32] then our results will be applicable.

5.3 General Preservation Result

In this section we provide a general result regarding when Bousfield localization preserves P-algebras. Essentially, this means that (up to weak equivalence) the localization of a P-algebra is again a P-algebra and the localization morphism is a P-algebra homomorphism. We make this precise in Definition 5.2.

Throughout this section, let $\mathcal{M}$ be a monoidal model category and let $\mathcal{C}$ be a set of maps in $\mathcal{M}$ such that Bousfield localization $L_{\mathcal{C}}(\mathcal{M})$ is a also monoidal model category. On the model category level the functor $L_{\mathcal{C}}$ is the identity. So when we write $L_{\mathcal{C}}$ as a functor we shall mean the composition of derived functors $\mathrm{Ho}(\mathcal{M}) \to \mathrm{Ho}(L_{\mathcal{C}}(\mathcal{M})) \to \mathrm{Ho}(\mathcal{M})$, i.e. $E \to L_{\mathcal{C}}(E)$ is the unit map of the adjunction $\mathrm{Ho}(\mathcal{M}) \leftrightarrows \mathrm{Ho}(L_{\mathcal{C}}(\mathcal{M}))$. In particular, for any E in $\mathcal{M}$, $L_{\mathcal{C}}(E)$ is weakly equivalent to $R_{\mathcal{C}}QE$ where $R_{\mathcal{C}}$ is a choice of fibrant replacement in $L_{\mathcal{C}}(\mathcal{M})$ and Q is a cofibrant replacement in $\mathcal{M}$.

Let P be an operad valued in $\mathcal{M}$. Because the objects of $L_{\mathcal{C}}(\mathcal{M})$

are the same as the objects of $\mathcal{M}$, P is also valued in $L_\mathcal{C}(\mathcal{M})$. Thus, we may consider P-algebras in both categories and these classes of objects agree (because the P-algebra action is independent of the model structure). We denote the categories of P-algebras by P-alg$(\mathcal{M})$ and P-alg$(L_\mathcal{C}(\mathcal{M}))$. These are identical as categories, but in a moment they will receive different model structures. Inspired by [14], we make the following definition.

Definition 5.2   Assume that $\mathcal{M}$ and $L_\mathcal{C}(\mathcal{M})$ are monoidal model categories, P is an operad valued in $\mathcal{M}$, and U is the forgetful functor from P-algebras to $\mathcal{M}$. Then $L_\mathcal{C}$ is said to *preserve P-algebras* if the following two properties are satisfied.

1   When E is a P-algebra there is some P-algebra $\widetilde{E}$ such that $U\widetilde{E}$ is weakly equivalent in $\mathcal{M}$ to the localization $L_\mathcal{C}(UE) := R_\mathcal{C}Q(UE)$, where $R_\mathcal{C}$ is fibrant replacement in $L_\mathcal{C}(\mathcal{M})$ and Q is cofibrant replacement in $\mathcal{M}$.

2   In addition, when E is a cofibrant P-algebra, then there is a choice of $\widetilde{E}$ in P-alg$(\mathcal{M})$ with $U\widetilde{E}$ local in $\mathcal{M}$, there is a P-algebra homomorphism $r_E : E \to \widetilde{E}$, and there is a weak equivalence $\beta_E : L_\mathcal{C}(UE) \to U\widetilde{E}$ such that $\beta_E \circ l_{UE} \cong Ur_E$ in Ho$(\mathcal{M})$.

This definition also appears in [6], where it is compared with other notions of preservation, e.g., preservation in the homotopy category. The condition $\beta_E \circ l_{UE} \cong Ur_E$ means that r_E is a lift of the localization map $l_{UE} : UE \to L_\mathcal{C}(UE)$ to the category of algebras, at least up to homotopy.

We are ready to prove the main result of this section. Recall that when we say P-alg$(\mathcal{M})$ *inherits* a model structure from $\mathcal{M}$ we mean that this model structure is transferred by the free-forgetful adjunction. In particular, a map of P-algebras f is a weak equivalence (resp. fibration) if and only if f is a weak equivalence (resp. fibration) in $\mathcal{M}$. A version of this result for semi-model categories will be proven as Corollary 5.5, after semi-model categories are defined. An alternative proof is given in Theorem 5.2 of [6].

Theorem 5.3   *Let $\mathcal{M}$ be a monoidal model category such that the Bousfield localization $L_\mathcal{C}(\mathcal{M})$ exists and is a monoidal model category. Let P be an operad valued in $\mathcal{M}$. If the categories of P-algebras in $\mathcal{M}$ and in $L_\mathcal{C}(\mathcal{M})$ inherit model structures from $\mathcal{M}$ and $L_\mathcal{C}(\mathcal{M})$ then $L_\mathcal{C}$ preserves P-algebras.*

Proof Let R_C denote fibrant replacement in $L_C(\mathcal{M})$, let $R_{C,P}$ denote fibrant replacement in P-alg$(L_C(\mathcal{M}))$, let Q_P denote cofibrant replacement in P-alg$(\mathcal{M})$, and let Q denote cofibrant replacement in $\mathcal{M}$. We will prove the first form of preservation and our method of proof will allow us to deduce the second form of preservation in the special case where E is a cofibrant P-algebra.

Let E be a P-algebra, and define $\tilde{E} = R_{C,P}Q_P(E)$. First, Q is the left derived functor of the identity adjunction between $\mathcal{M}$ and $L_C(\mathcal{M})$, and R_C is the right derived functor of the identity, so $L_C(UE) \simeq R_C Q(UE)$. We must therefore show $R_C Q(UE) \simeq U R_{C,P} Q_P(E)$.

The map $q : Q_P E \to E$ is a trivial fibration in P-alg$(\mathcal{M})$, hence Uq is a trivial fibration in $\mathcal{M}$. The map $QUE \to UE$ is also a weak equivalence in $\mathcal{M}$. Consider the following lifting diagram in $\mathcal{M}$:

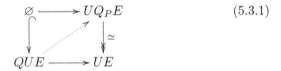

$$\tag{5.3.1}$$

The lifting axiom gives the map $QUE \to UQ_P E$ and it is necessarily a weak equivalence in $\mathcal{M}$ by the two out of three property.

Since $Q_P E$ is a P-algebra in $\mathcal{M}$ it must also be a P-algebra in $L_C(\mathcal{M})$, since the monoidal structure of the two categories is the same. We may therefore apply $R_{C,P}$ to $Q_P E$. We next consider the following lift in $L_C(\mathcal{M})$, which exists because the left vertical map is a trivial cofibration in $L_C(\mathcal{M})$ and $U R_{C,P} Q_P E$ is fibrant in $L_C(\mathcal{M})$:

$$\tag{5.3.2}$$

In this diagram the top horizontal map is U applied to a weak equivalence in P-alg$(L_C(\mathcal{M}))$. Because the model category P-alg$(L_C(\mathcal{M}))$ inherits weak equivalences from $L_C(\mathcal{M})$, this map is a weak equivalence in $L_C(\mathcal{M})$. The left vertical map is also a weak equivalence in $L_C(\mathcal{M})$. Therefore, by the two out of three property, the lift is a weak equivalence in $L_C(\mathcal{M})$. We make use of this map as the horizontal map in the lower right corner of the diagram below.

The top horizontal map $QUE \to UQ_P E$ in the following diagram is the first map we constructed, which was proven to be a weak equivalence in $\mathcal{M}$. The square in the diagram below is then obtained by applying $R_{\mathcal{C}}$ to that map. In particular, $R_{\mathcal{C}}QUE \to R_{\mathcal{C}}UQ_PE$ is a weak equivalence in $L_{\mathcal{C}}(\mathcal{M})$:

$$
\begin{array}{ccc}
QUE & \longrightarrow & UQ_PE \\
\downarrow & & \downarrow \\
R_{\mathcal{C}}QUE & \longrightarrow R_{\mathcal{C}}UQ_PE \longrightarrow & UR_{\mathcal{C},P}Q_PE
\end{array}
\qquad (5.3.3)
$$

We have shown that both of the bottom horizontal maps are weak equivalences in $L_{\mathcal{C}}(\mathcal{M})$. Thus, by the two out of three property, their composite $R_{\mathcal{C}}QUE \to UR_{\mathcal{C},P}Q_PE$ is a weak equivalence in $L_{\mathcal{C}}(\mathcal{M})$. All the objects in the bottom row are fibrant in $L_{\mathcal{C}}(\mathcal{M})$, so these $\mathcal{C}$-local equivalences are actually weak equivalences in $\mathcal{M}$.

As E was a P-algebra and Q_P and $R_{\mathcal{C},P}$ are endofunctors on categories of P-algebras, it is clear that $R_{\mathcal{C},P}Q_PE$ is a P-algebra. We have just shown that $L_{\mathcal{C}}(UE)$ is weakly equivalent to this P-algebra, so we are done.

When E is assumed to be a cofibrant P-algebra. We have seen that there is an $\mathcal{M}$-weak equivalence $R_{\mathcal{C}}QUE \to UR_{\mathcal{C},P}Q_PE$, and above we took $R_{\mathcal{C},P}Q_PE$ in $\mathcal{M}$ as our representative for $L_{\mathcal{C}}(UE)$ in $\mathrm{Ho}(\mathcal{M})$. Since E is a cofibrant P-algebra, there are weak equivalences $E \leftrightarrows Q_P(E)$ in P-$\mathrm{alg}(L_{\mathcal{C}}(\mathcal{M}))$. This is because all cofibrant replacements of a given object are weakly equivalent, e.g. by diagram (5.3.1). So passage to $Q_P(E)$ is unnecessary when E is cofibrant, and we take $\widetilde{E} = R_{\mathcal{C},P}E$ as our representative for $L_{\mathcal{C}}(E)$. Observe that $U\widetilde{E}$ is local as the model structure on P-algebras is transferred. The P-algebra morphism $r_E : E \to \widetilde{E}$ is just the fibrant replacement map $R_{\mathcal{C},P}$, and lifts the localization map $UE \to L_{\mathcal{C}}(UE)$ in $\mathrm{Ho}(\mathcal{M})$. The comparison β_E is the following lift in $L_{\mathcal{C}}(\mathcal{M})$:

$$
\begin{array}{ccc}
UE & \longrightarrow & U\widetilde{E} \\
\scriptstyle{\simeq_{\mathcal{C}}}\downarrow & \nearrow_{\beta_E} & \downarrow \\
L_{\mathcal{C}}(UE) & \longrightarrow & *
\end{array}
\qquad (5.3.4)
$$

The two out of three property guarantees that β_E is a weak equivalence (again using that the model structure on P-algebras is transferred), and the diagram above demonstrates that $\beta_E \circ l_{UE} \cong Ur_E$ in $\mathrm{Ho}(\mathcal{M})$. $\qquad \square$

This theorem alone would not be a satisfactory answer to the question of when L_C preserves P-algebras, because there is no clear way to check the hypotheses. For this reason, in the coming sections we will discuss conditions on $\mathcal{M}$ and P so that P-algebras inherit model structures, and then we will discuss which localizations L_C preserve these conditions (so that P-alg$(L_C(\mathcal{M}))$ inherits a model structure from $L_C(\mathcal{M})$). One such condition on $\mathcal{M}$ is the monoid axiom. In Section 5.8, we discuss which localizations L_C preserve the monoid axiom. However, it will turn out that the monoid axiom is not necessary in order for our preservation results to apply. This is because the work in [33] and [53] produces semi-model structures on P-algebras and these will be enough for our proof above to go through.

Observe that in the proof above we only used formal properties of fibrant and cofibrant replacement functors, and the fact that the model structures on P-algebras were inherited from $\mathcal{M}$ and $L_C(\mathcal{M})$. So the same proof works when P-algebras only form semi-model categories, a notion we define presently. The motivating example is when $\mathcal{D} = P$-alg is obtained from $\mathcal{M}$ via the general transfer principle for transferring a model structure across an adjunction (see Lemma 2.3 in [50] or Theorem 12.1.4 in [23]) when not all the conditions needed to get a full model structure are satisfied. The following definition is taken from [8] and [7], and is distilled from the definitions in [3] and [53]. Recall that, for a set of morphisms S, inj S refers to the class of morphisms having the right lifting property with respect to S.

Definition 5.4 A *semi-model structure* on a category $\mathcal{D}$ consists of classes of weak equivalences $\mathcal{W}$, fibrations $\mathcal{F}$, and cofibrations Ω satisfying the following axioms:

M1 Fibrations are closed under pullback.

M2 The class $\mathcal{W}$ is closed under the two out of three property.

M3 $\mathcal{W}, \mathcal{F}, \Omega$ are all closed under retracts.

M4 i Cofibrations have the left lifting property with respect to trivial fibrations.

ii Trivial cofibrations whose domain is cofibrant have the left lifting property with respect to fibrations.

M5 i Every map in $\mathcal{D}$ can be functorially factored into a cofibration followed by a trivial fibration.

ii Every map whose domain is cofibrant can be functorially factored into a trivial cofibration followed by a fibration.

If, in addition, $\mathcal{D}$ is bicomplete, then we call $\mathcal{D}$ a *semi-model category*. $\mathcal{D}$ is said to be *cofibrantly generated* if there are sets of morphisms I' and J' in $\mathcal{D}$ such that inj I' is the class of trivial fibrations, inj J' is the class of fibrations in $\mathcal{D}$, the domains of I' are small relative to I'-cell, and the domains of J' are small relative to maps in J'-cell whose domain is cofibrant.

In practice, there is often an adjunction $F : \mathcal{M} \leftrightarrows \mathcal{D} : U$ where $\mathcal{M}$ is a model category, U is a forgetful functor, the weak equivalences and fibrations in $\mathcal{D}$ are maps that forget to weak equivalences and fibrations in $\mathcal{M}$, and the generating (trivial) cofibrations of $\mathcal{D}$ are maps of the form $F(I)$ and $F(J)$ where $F : \mathcal{M} \to \mathcal{D}$ is the free algebra functor and I and J are the generating (trivial) cofibrations of $\mathcal{M}$.

Note that the only difference between a semi-model structure and a model structure is that one of the lifting properties and one of the factorization properties requires the domain of the map in question to be cofibrant. Because fibrant and cofibrant replacements are constructed via factorization, (4) implies that every object has a cofibrant replacement and that cofibrant objects have fibrant replacements. So one could construct a fibrant replacement functor that first does cofibrant replacement and then does fibrant replacement. These functors behave as they would in the presence of a full model structure.

We are now prepared to state our preservation result in the presence of only a semi-model structure on P-algebras. When we say P-algebras inherit a semi-model structure we mean with weak equivalences and fibrations reflected and preserved by the forgetful functor. We state this as a corollary because its proof is so similar to that of Theorem 5.3.

Corollary 5.5 *Let $\mathcal{M}$ be a monoidal model category such that the Bousfield localization $L_{\mathcal{C}}(\mathcal{M})$ exists and is a monoidal model category. Let P be an operad valued in $\mathcal{M}$. If the categories of P-algebras in $\mathcal{M}$ and in $L_{\mathcal{C}}(\mathcal{M})$ inherit transferred semi-model structures from $\mathcal{M}$ and $L_{\mathcal{C}}(\mathcal{M})$ then $L_{\mathcal{C}}$ preserves P-algebras.*

Proof The proof proceeds exactly as the proof of Theorem 5.3. We highlight where care must be taken in the presence of semi-model categories. As remarked above, the cofibrant replacement Q_P in the semi-model category P-alg($\mathcal{M}$) exists and the cofibrant replacement map $Q_P E \to E$ is a weak equivalence in P-alg($\mathcal{M}$), hence in $\mathcal{M}$, because the semi-model structure is transferred. Diagram (5.3.1) is a lifting diagram in $\mathcal{M}$, so still yields a weak equivalence $QUE \to UQ_P E$.

Next, the fibrant replacement $R_\mathcal{C}UQ_PE$ is a replacement in the model category $L_\mathcal{C}(\mathcal{M})$. The fibrant replacement $Q_PE \to R_{\mathcal{C},P}Q_PE$ is a fibrant replacement in the semi-model category P-alg$(L_\mathcal{C}(\mathcal{M}))$, and exists because Q_PE is cofibrant in P-alg$(L_\mathcal{C}(\mathcal{M}))$. The resulting object $R_{\mathcal{C},P}Q_PE$ is fibrant in P-alg$(L_\mathcal{C}(\mathcal{M}))$ hence in $L_\mathcal{C}(\mathcal{M})$, since the semi-model structure is transferred. The lift in (5.3.2) is a lift in $L_\mathcal{C}(\mathcal{M})$, and again by the two out of three property in $L_\mathcal{C}(\mathcal{M})$ the diagonal map $R_\mathcal{C}UQ_PE \to UR_{\mathcal{C},P}Q_PE$ is a $\mathcal{C}$-local equivalence.

Next, the map $R_\mathcal{C}QUE \to R_\mathcal{C}UQ_PE$ in (5.3.3) is a fibrant replacement of the map $QUE \to UQ_PE$ in the model category $L_\mathcal{C}(\mathcal{M})$, and so the argument that $R_\mathcal{C}QUE \to R_\mathcal{C}UQ_PE$ is a $\mathcal{C}$-local equivalence remains unchanged. The composite across the bottom of (5.3.3), $R_\mathcal{C}QUE \to UR_{\mathcal{C},P}Q_PE$ is a weak equivalence between fibrant objects in $L_\mathcal{C}(\mathcal{M})$ and so is a weak equivalence in $\mathcal{M}$, as in the proof of Theorem 5.3.

Finally, for the case of E cofibrant in the semi-model category P-alg$(\mathcal{M})$, note that the localization map $E \to L_\mathcal{C}(E)$ is again fibrant replacement $E \to R_{\mathcal{C},P}E$ in P-alg$(L_\mathcal{C}(\mathcal{M}))$. This exists because the domain is cofibrant by assumption. By construction, this map is a P-algebra morphism, as desired. The lift defining β in (5.3.4) occurs in $L_\mathcal{C}(\mathcal{M})$, and the rest of the proof only uses that weak equivalences and fibrations in P-alg$(L_\mathcal{C}(\mathcal{M}))$ forget to weak equivalences and fibrations in $L_\mathcal{C}(\mathcal{M})$. □

Remark 5.6 Corollary 5.5 has been generalized to algebras over colored operads in [61], and to right Bousfield localization in [63]. It has been applied to localizations of Smith ideals in [60].

5.4 Monoidal Bousfield Localizations

In both Theorem 5.3 and Corollary 5.5 we assumed that $L_\mathcal{C}(\mathcal{M})$ is a monoidal model category. In this section we provide conditions on $\mathcal{M}$ and $\mathcal{C}$ so that this occurs. First, we provide an example demonstrating that the pushout product axiom can fail for $L_\mathcal{C}(\mathcal{M})$, even if it holds for $\mathcal{M}$. The author learned this example from Mark Hovey.

Example 5.7 It is not true that every Bousfield localization of a monoidal model category is a monoidal model category. Let $R = \mathbb{F}_2[\Sigma_3]$. An R module is simply an $\mathbb{F}_2$ vector space with an action of the symmetric group Σ_3. Because R is a Frobenius ring, we may pass from R-mod to the *stable module category* $StMod(R)$ by identifying any two morphisms whose difference factors through a projective module.

Section 2.2 of [34] introduces a model category $\mathcal{M}$ of R-modules whose homotopy category is $StMod(R)$, then proves $\mathcal{M}$ is a finitely generated, combinatorial, stable model category in which all objects are cofibrant (hence, $\mathcal{M}$ is also left proper). Proposition 4.2.15 of [34] proves that for $R = \mathbb{F}_2[\Sigma_3]$, this model category is a monoidal model category because R is a Hopf algebra over $\mathbb{F}_2$. The monoidal product of two R-modules is $M \otimes_{\mathbb{F}_2} N$ where R acts via its diagonal $R \to R \otimes_{\mathbb{F}_2} R$.

We now check that cofibrant objects are flat in $\mathcal{M}$. By the pushout product axiom, $X \otimes -$ is left Quillen. Since all objects are cofibrant, all weak equivalences are weak equivalences between cofibrant objects. So Ken Brown's lemma implies $X \otimes -$ preserves weak equivalences.

Let $f : 0 \to \mathbb{F}_2$, where the codomain has the trivial Σ_3 action. We'll show that the Bousfield localization with respect to f cannot be a monoidal Bousfield localization. First observe that if an object is f-locally trivial then it has no Σ_3-fixed points, i.e., fails to admit Σ_3-equivariant maps from $\mathbb{F}_2$ (the non-identity element would need to be taken to a Σ_3-fixed point because the Σ_3-action on $\mathbb{F}_2$ is trivial).

If the pushout product axiom held in $L_f(\mathcal{M})$ then the pushout product of two f-locally trivial cofibrations g, h would have to be f-locally trivial. We will now demonstrate an f-locally trivial object N for which $N \otimes_{\mathbb{F}_2} N$ is not f-locally trivial, so $(\varnothing \to N) \square (\varnothing \to N)$ is not a trivial cofibration in $L_f(\mathcal{M})$.

Define $N \cong \mathbb{F}_2 \oplus \mathbb{F}_2$ where the element (12) sends $a = (1, 0)$ to $b = (0, 1)$ and the element (123) sends a to b and b to $c = a + b$. The reader can check that $(12)(123)$ acts the same as $(123)^2(12)$, so that this is a well-defined Σ_3-action. This object N is f-locally trivial, since localization by f kills both factors of $\mathbb{F}_2$. However, $N \otimes_{\mathbb{F}_2} N$ is not f-locally trivial because $N \otimes_{\mathbb{F}_2} N$ does admit any map from $\mathbb{F}_2$ taking the non-identity element of $\mathbb{F}_2$ to the Σ_3-invariant element $a \otimes a + b \otimes b + c \otimes c$. Thus, $L_f(\mathcal{M})$ is not a monoidal model category.

There are two ways to get around examples such as the above. One can change the mapping space used to define the localization, e.g., using the derived internal hom rather than a homotopy function complex (as in [3, Definition 4.45]), or one can place hypotheses on the set of morphisms $\mathcal{C}$ that we are inverting, to ensure that the $\mathcal{C}$-local equivalences play nicely with the monoidal product. These two approaches are, in fact, equivalent.

A similar program, which amounts to a condition on the set of morphisms $\mathcal{C}$, was conducted in [14], in order to guarantee that localizations of stable model categories commute with suspension. Similarly, a condition on a stable localization to ensure that it is additionally monoidal was given in Definition 6.2 of [2] and the same condition appeared in Theorem 4.46 of [3]. This condition states that $\mathcal{C} \,\square\, I$ is contained in the $\mathcal{C}$-local equivalences.

Remark 5.8 The counterexample above fails to satisfy the condition that $\mathcal{C} \,\square\, I$ is contained in the $\mathcal{C}$-local equivalences. If this condition were satisfied then I would be contained in the f-local equivalences and this would imply all cofibrant objects (hence all objects) are f-locally trivial. But $0 \to N \otimes_{\mathbb{F}_2} N$ is not f-locally trivial. Thus, this counterexample has no bearing on the work of [2] or [3].

Remark 5.9 The counterexample demonstrates a general principle that we now highlight. In any G-equivariant world, there are multiple spheres due to the different group actions. In the example above, one can suspend by representations of the symmetric group Σ_n, i.e. copies of $\mathbb{F}_2$ on which Σ_n acts. The 1-point compactification of such an object is a sphere S^n on which Σ_n acts. A localization that kills a representation sphere should not be expected to respect the monoidal structure, because not all acyclic cofibrant objects can be built from one of the representation spheres alone. In particular, $N \otimes N$ will not be in the smallest thick subcategory generated by $\mathbb{F}_2$. The point is that the homotopy categories of stable model categories in an equivariant context are not monogenic axiomatic stable homotopy categories in the sense of [37].

Note that this example also demonstrates that the monoid axiom can fail on $L_{\mathcal{C}}(\mathcal{M})$. The author does not know an example of a model category satisfying the pushout product axiom but failing the monoid axiom.

In our applications we will need to know that $L_{\mathcal{C}}(\mathcal{M})$ satisfies the pushout product axiom, the unit axiom, and the axiom that cofibrant objects are flat. We therefore give a name to such localizations, and then we characterize them. The reader is advised to keep Convention 5.1 in mind.

Definition 5.10 A Bousfield localization $L_{\mathcal{C}}$ is said to be a *monoidal Bousfield localization* if $L_{\mathcal{C}}(\mathcal{M})$ satisfies the pushout product axiom, the unit axiom, and the axiom that cofibrant objects are flat.

Theorem 5.11 *Suppose that $\mathcal{M}$ is a cofibrantly generated monoidal model category in which cofibrant objects are flat and the domains of* *the generating cofibrations are cofibrant. Let I denote the generating cofibrations of $\mathcal{M}$. Then $L_{\mathcal{C}}$ is a monoidal Bousfield localization if and only if every map of the form $f \otimes id_K$, where f is in $\mathcal{C}$ and K is a domain or codomain of a map in I, is a $\mathcal{C}$-local equivalence.*

Theorem 5.12 *Suppose $\mathcal{M}$ is a cofibrantly generated monoidal model category in which cofibrant objects are flat. Then $L_{\mathcal{C}}$ is a monoidal Bousfield localization if and only if every map of the form $f \otimes id_K$, where f is in $\mathcal{C}$ and K is cofibrant, is a $\mathcal{C}$-local equivalence.*

Note that the condition $\mathcal{C} \,\square\, I \subset \mathcal{C}$-local equivalences, from [2, 3], implies the condition from these theorems. In fact, one can prove it is equivalent to $L_{\mathcal{C}}(\mathcal{M})$ being a monoidal model category, because $\mathcal{C}$ can be taken to be a set of $\mathcal{C}$-local trivial cofibrations. However, the condition stated in the theorems above is easier to check. We shall prove Theorem 5.11 in Subsection 5.4.1 and we shall prove Theorem 5.12 in Subsection 5.4.2. These theorems demonstrate precisely what must be done if one wishes to invert a given set of morphisms $\mathcal{C}$ and ensure that the resulting model structure is a monoidal model structure.

Definition 5.13 Suppose $\mathcal{M}$ is left proper, is either cellular or combinatorial, and that the domains of the generating cofibrations are cofibrant. The *smallest monoidal Bousfield localization* which inverts a given set of morphisms $\mathcal{C}$ is the Bousfield localization with respect to the set $\mathcal{C}' = \{\mathcal{C} \otimes id_K\}$ where K runs through the domains and codomains of the generating cofibrations I.

This notion has already been used in [39]. The reason for the hypothesis on the domains of the generating cofibrations is to ensure that $\mathcal{C}'$ is a set. Requiring left properness and either cellularity or combinatoriality ensures that $L_{\mathcal{C}'}$ exists. The smallest Bousfield localization has a universal property, that we now highlight.

Proposition 5.14 *Suppose $\mathcal{C}'$ is the smallest monoidal Bousfield localization inverting $\mathcal{C}$, and let $j : \mathcal{M} \to L_{\mathcal{C}'}(\mathcal{M})$ be the left Quillen functor realizing the localization. Suppose $\mathcal{N}$ is a monoidal model category with cofibrant objects flat. Suppose $F : \mathcal{M} \to \mathcal{N}$ is a monoidal left Quillen functor such that $\mathbb{L}F$ takes the images of $\mathcal{C}$ in $\mathrm{Ho}(\mathcal{M})$ to isomorphisms in $\mathrm{Ho}(\mathcal{N})$. Then there is a unique monoidal left Quillen functor $\delta : L_{\mathcal{C}'}\mathcal{M} \to \mathcal{N}$ such that $\delta j = F$.*

Proof Suppose $F : \mathcal{M} \to \mathcal{N}$ is a monoidal left Quillen functor, that $\mathcal{N}$ has cofibrant objects flat, and that $\mathbb{L}F$ takes the images of $\mathcal{C}$ in $\mathrm{Ho}(\mathcal{M})$

to isomorphisms in Ho($\mathcal{N}$). Then F also takes the images of maps in $\mathcal{C}'$ to isomorphisms in Ho($\mathcal{N}$), because for any $f \in \mathcal{C}$ and any cofibrant K, $F(f \otimes K) \cong F(f) \otimes F(K)$ is a weak equivalence in $\mathcal{N}$. This is because $F(K)$ is cofibrant in $\mathcal{N}$ (as F is left Quillen), cofibrant objects are flat in $\mathcal{N}$, and $F(f)$ is a weak equivalence in $\mathcal{N}$ by hypothesis.

The universal property of the localization $L_{\mathcal{C}'}$ then provides a unique left Quillen functor $\delta : L_{\mathcal{C}'}\mathcal{M} \to \mathcal{N}$ that is the same as F on objects and morphisms (Theorem 3.3.18 and Theorem 3.3.19 in [32]). In particular, δ is a monoidal functor and $\delta q = Fq : F(QS) \to F(S)$ is a weak equivalence in $\mathcal{N}$ because the cofibrant replacement $QS \to S$ is the same in $L_{\mathcal{C}'}(\mathcal{M})$ as in $\mathcal{M}$. So δ is a unique monoidal left Quillen functor as required, and the commutativity $\delta j = F$ follows immediately from the definition of δ. $\qquad\square$

5.4.1 Proof of Theorem 5.11

In this section we will prove Theorem 5.11. We first prove that under the hypotheses of Theorem 5.11, cofibrant objects are flat in $L_{\mathcal{C}}(\mathcal{M})$.

Proposition 5.15 *Let $\mathcal{M}$ be a cofibrantly generated monoidal model category in which cofibrant objects are flat and the domains of the generating cofibrations are cofibrant. Let I denote the generating cofibrations of $\mathcal{M}$. Suppose that every map of the form $f \otimes id_K$, where f is in $\mathcal{C}$ and K is a domain or codomain of a map in I, is a $\mathcal{C}$-local equivalence. Then cofibrant objects are flat in $L_{\mathcal{C}}(\mathcal{M})$.*

Proof We must prove that the class of maps $\{g \otimes X \mid g$ is a $\mathcal{C}$-local equivalence and X is a cofibrant object$\}$ is contained in the $\mathcal{C}$-local equivalences. Let X be a cofibrant object in $L_{\mathcal{C}}(\mathcal{M})$ (equivalently, in $\mathcal{M}$). Let $g : A \to B$ be a $\mathcal{C}$-local equivalence. To prove $- \otimes X$ preserves $\mathcal{C}$-local equivalences, it suffices to show that it takes $L_{\mathcal{C}}(\mathcal{M})$ trivial cofibrations between cofibrant objects to weak equivalences. This is because we can always do cofibrant replacement on g to get $Qg : QA \to QB$. While Qg need not be a cofibration in general, we can always factor it into $QA \hookrightarrow Z \xrightarrow{\simeq} QB$. By abuse of notation we will continue to use the symbol QB to denote Z, and we will rename the cofibration $QA \to Z$ as Qg since Z is cofibrant and maps via a trivial fibration to B. Smashing with

X gives:

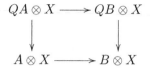

If we prove that $Qg \otimes X$ is a $\mathcal{C}$-local equivalence, then $g \otimes X$ must also be by the two out of three property, since the vertical maps are weak equivalences in $\mathcal{M}$ due to X being cofibrant and cofibrant objects being flat in M. So we may assume that g is an $L_{\mathcal{C}}(\mathcal{M})$ trivial cofibration between cofibrant objects. Since X is built as a transfinite composition of pushouts of maps in I, we proceed by transfinite induction. For the rest of the proof, let K, K_1, and K_2 denote domains/codomains of maps in I. These objects are cofibrant in $\mathcal{M}$ by hypothesis, so they are also cofibrant in $L_{\mathcal{C}}(\mathcal{M})$.

For the base case $X = K$ we appeal to Theorem 3.3.18 in [32]. The composition $F = id \circ K \otimes - : \mathcal{M} \to \mathcal{M} \to L_{\mathcal{C}}(\mathcal{M})$ is left Quillen because K is cofibrant. F takes maps in $\mathcal{C}$ to weak equivalences by hypothesis. So Theorem 3.3.18 implies F induces a left Quillen functor $K \otimes - : L_{\mathcal{C}}(\mathcal{M}) \to L_{\mathcal{C}}(\mathcal{M})$. Thus, $K \otimes -$ takes $\mathcal{C}$-local equivalences between cofibrant objects to $\mathcal{C}$-local equivalences and in particular takes Qg to a $\mathcal{C}$-local equivalence. Note that this is the key place in this proof where we use the hypothesis that $L_{\mathcal{C}}$ is a monoidal Bousfield localization. This theorem is the primary tool when one wishes to get from a statement about $\mathcal{C}$ to a statement about all $\mathcal{C}$-local equivalences.

For the successor case, suppose X_α is built from K as above and is flat in $L_{\mathcal{C}}(\mathcal{M})$. Suppose $X_{\alpha+1}$ is built from X_α and a map in I via a pushout diagram:

$$
\begin{array}{ccc}
K_1 & \overset{i}{\hookrightarrow} & K_2 \\
\downarrow & \searrow & \downarrow \\
X_\alpha & \longrightarrow & X_{\alpha+1}
\end{array}
$$

We smash this diagram with $g : A \to B$ and note that smashing a pushout square with an object yields a pushout square.

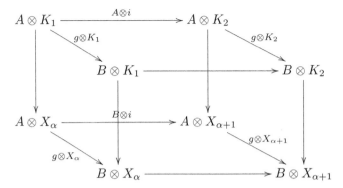

Because g is a cofibration of cofibrant objects, A and B are cofibrant. Because pushouts of cofibrations are cofibrations, $X_\alpha \hookrightarrow X_{\alpha+1}$ for all α. Because X_0 is cofibrant, X_α is cofibrant for all α. So all objects above are cofibrant. Furthermore, $g \otimes K_i = g \,\square\, (0 \hookrightarrow K_i)$. Thus, by the pushout product axiom on $\mathcal{M}$ and the fact that cofibrations in $\mathcal{M}$ match those in $L_\mathcal{C}(\mathcal{M})$, these maps are cofibrations.

Finally, the maps $g \otimes K_i$ are weak equivalences in $L_\mathcal{C}(\mathcal{M})$ by the base case above, while $g \otimes X_\alpha$ is a weak equivalence in $L_\mathcal{C}(\mathcal{M})$ by the inductive hypothesis. Thus, by the Cube Lemma (Lemma 5.2.6 in [34]), the map $g \otimes X_{\alpha+1}$ is a weak equivalence in $L_\mathcal{C}(\mathcal{M})$.

For the limit case, suppose $X = \lim_{\substack{\to \\ \alpha < \beta}} X_\alpha$ where each X_α is cofibrant and flat in $L_\mathcal{C}(\mathcal{M})$. Because each X_α is cofibrant, $g \otimes X_\alpha = g \,\square\, (0 \hookrightarrow X_\alpha)$ is still a cofibration. By the inductive hypothesis, each $g \otimes X_\alpha$ is also a $\mathcal{C}$-local equivalence, hence a trivial cofibration in $L_\mathcal{C}(\mathcal{M})$. Since trivial cofibrations are always closed under transfinite composition, $g \otimes X = g \otimes \lim_{\to} X_\alpha = \lim_{\to}(g \otimes X_\alpha)$ is also a trivial cofibration in $L_\mathcal{C}(\mathcal{M})$. $\square$

We now pause for a moment to extract the key point in the proof above, where we applied the universal property of Bousfield localization. This is a reformulation Theorem 3.3.18 in [32] that we will need below.

Lemma 5.16 *A left Quillen functor $F : \mathcal{M} \to \mathcal{M}$ induces a left Quillen functor $L_\mathcal{C} F : L_\mathcal{C}(\mathcal{M}) \to L_\mathcal{C}(\mathcal{M})$ if and only if for all $f \in \mathcal{C}$, $F(f)$ is $\mathcal{C}$-local equivalence.*

We turn now to the unit axiom.

Proposition 5.17 *If $\mathcal{M}$ satisfies the unit axiom then any Bousfield localization $L_\mathcal{C}(\mathcal{M})$ satisfies the unit axiom. If cofibrant objects are flat in*

$\mathcal{M}$ then the map $QS \otimes Y \to Y$, induced by cofibrant replacement $QS \to S$, is a weak equivalence for all Y, not just cofibrant Y. Furthermore, for any weak equivalence $f : K \to L$ between cofibrant objects, $f \otimes Y$ is a weak equivalence.

Proof Since $L_\mathcal{C}(\mathcal{M})$ has the same cofibrations as $\mathcal{M}$, it must also have the same trivial fibrations. Thus, it has the same cofibrant replacement functor and the same cofibrant objects. Thus, the unit axiom on $L_\mathcal{C}(\mathcal{M})$ follows directly from the unit axiom on $\mathcal{M}$, because a weak equivalence in $\mathcal{M}$ is in particular a $\mathcal{C}$-local equivalence.

We now assume cofibrant objects are flat and that Y is an object of $\mathcal{M}$. Consider the following diagram:

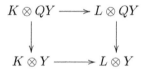

The top map is a weak equivalence by the unit axiom for the cofibrant object QY. The left vertical map is a weak equivalence because cofibrant objects are flat and QS is cofibrant. The right vertical is a weak equivalence by definition of QY. Thus, the bottom arrow is a weak equivalence by the two out of three property.

For the final statement we again apply cofibrant replacement to Y and we get

$$
\begin{array}{ccc}
K \otimes QY & \longrightarrow & L \otimes QY \\
\downarrow & & \downarrow \\
K \otimes Y & \longrightarrow & L \otimes Y
\end{array}
$$

Again the top horizontal map and the vertical maps are weak equivalences because cofibrant objects are flat (for the first use that QX is cofibrant, for the second use that K and L are cofibrant). $\square$

We turn now to proving Theorem 5.11. As mentioned in the proof of Proposition 5.15, if h and g are $L_\mathcal{C}(\mathcal{M})$-cofibrations then they are cofibrations in $\mathcal{M}$ and so $h \square g$ is a cofibration in $\mathcal{M}$ (hence in $L_\mathcal{C}(\mathcal{M})$) by the pushout product axiom on $\mathcal{M}$. To verify the rest of the pushout product axiom on $L_\mathcal{C}(\mathcal{M})$ we must prove that if h is a trivial cofibration in $L_\mathcal{C}(\mathcal{M})$ and g is a cofibration in $L_\mathcal{C}(\mathcal{M})$ then $h \square g$ is a weak equivalence in $L_\mathcal{C}(\mathcal{M})$.

Proposition 5.18 *Let $\mathcal{M}$ be a cofibrantly generated monoidal model category in which cofibrant objects are flat and the domains of the generating cofibrations are cofibrant. Let I denote the generating cofibrations of $\mathcal{M}$. Suppose that every map of the form $f \otimes id_K$, where f is in $\mathcal{C}$ and K is a domain or codomain of a map in I, is a $\mathcal{C}$-local equivalence. Then $L_{\mathcal{C}}(\mathcal{M})$ satisfies the pushout product axiom.*

Proof We have already remarked that the cofibration part of the pushout product axiom on $L_{\mathcal{C}}(\mathcal{M})$ follows from the pushout product axiom on $\mathcal{M}$, since the two model categories have the same cofibrations. By Proposition 4.2.5 of [34] it is sufficient to check the pushout product axiom on generating (trivial) cofibrations. So suppose $h : X \to Y$ is an $L_{\mathcal{C}}(\mathcal{M})$ trivial cofibration and $g : K \to L$ is a generating cofibration in $L_{\mathcal{C}}(\mathcal{M})$ (equivalently, in $\mathcal{M}$). Then we must show $h \,\square\, g$ is an $L_{\mathcal{C}}(\mathcal{M})$ trivial cofibration

By hypothesis on $\mathcal{M}$, K and L are cofibrant. Because h is a cofibration, $K \otimes h$ and $L \otimes h$ are cofibrations by the pushout product axiom on $\mathcal{M}$ (because $K \otimes h = (\varnothing \hookrightarrow K) \,\square\, h$). By Proposition 5.15, cofibrant objects are flat in $L_{\mathcal{C}}(\mathcal{M})$. So $K \otimes h$ and $L \otimes h$ are also weak equivalences. In particular, $K \otimes -$ and $L \otimes -$ are left Quillen functors. Consider the following diagram:

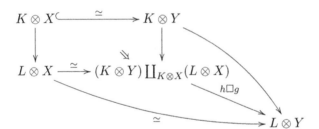

The map $L \otimes X \to (K \otimes Y) \coprod_{K \otimes X} (L \otimes X)$ is a trivial cofibration because it is the pushout of a trivial cofibration. Thus, by the two out of three property for the lower triangle, $h \,\square\, g$ is a weak equivalence. Since we already knew it was a cofibration (because it is so in $\mathcal{M}$), this means it is a trivial cofibration. $\square$

We are now ready to complete the proof of Theorem 5.11.

Proof of Theorem 5.11 We begin with the forwards direction. Suppose $L_{\mathcal{C}}(\mathcal{M})$ satisfies the pushout product axiom and has cofibrant objects flat. Let f be any map in $\mathcal{C}$. Note that in particular, f is a $\mathcal{C}$-local equivalence.

Because cofibrant objects are flat, the map $f \otimes K$ is a $\mathcal{C}$-local equivalence for any cofibrant K. So the collection $\mathcal{C} \otimes K$ is contained in the $\mathcal{C}$-local equivalences, where K runs through the class of cofibrant objects, i.e. $L_{\mathcal{C}}$ is a monoidal Bousfield localization.

For the converse, we apply our three previous propositions. That cofibrant objects are flat in $L_{\mathcal{C}}(\mathcal{M})$ is the content of Proposition 5.15. The unit axiom on $L_{\mathcal{C}}(\mathcal{M})$ follows from Proposition 5.17 applied to $L_{\mathcal{C}}(\mathcal{M})$. That the pushout product axiom holds on $L_{\mathcal{C}}(\mathcal{M})$ is Proposition 5.18.   □

5.4.2 Proof of Theorem 5.12

We will now prove Theorem 5.12, following the outline above. The proof that cofibrant objects are flat in $L_{\mathcal{C}}(\mathcal{M})$ will proceed just as it did in Proposition 5.15. Proposition 5.17 again implies the unit axiom in $L_{\mathcal{C}}(\mathcal{M})$. Deducing the pushout product axiom on $L_{\mathcal{C}}(\mathcal{M})$ will be more complicated without the assumption on the domains of I. For this reason, we need the following lemma. First, let I' be obtained from the generating cofibrations I by applying any cofibrant replacement Q to all $i \in I$ and then taking the left factor in the cofibration-trivial fibration factorization of Qi. So I' consists of cofibrations between cofibrant objects.

Lemma 5.19 *Suppose $\mathcal{M}$ is a left proper model category cofibrantly generated by sets I and J in which the domains of maps in J are small relative to I-cell. Then the sets $I' \cup J$ and J cofibrantly generate $\mathcal{M}$.*

Proof We verify the conditions given in Definition 11.1.2 of [32]. We have not changed J, so the fibrations are still precisely the maps satisfying the right lifting property with respect to J and the maps in J still permit the small object argument because the domains are small relative to J-cell.

Any map that has the right lifting property with respect to all maps in I is a trivial fibration, so will in particular have the right lifting property with respect to all cofibrations, hence with respect to maps in $I' \cup J$. Conversely, suppose p has the right lifting property with respect to all maps in $I' \cup J$. We are faced with the following lifting problem:

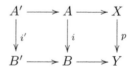

Because p has lifting with respect to $I' \cup J$, it has the right lifting property with respect to J. This guarantees us that p is a fibration. Now because $\mathcal{M}$ is left proper, Proposition 13.2.1 in [32] applies to solve the lifting diagram above. In particular, because p has the right lifting property with respect to I', p must have the right lifting property with respect to I. Thus, p is a trivial fibration as desired.

We now turn to smallness. Any domain of a map in J is small relative to J-cell, but in general this would not imply smallness relative to I-cell. We have assumed the domains of maps in J are small relative to I-cell, so they are small relative to $(J \cup I')$-cell because $J \cup I'$ is contained in I-cell.

Any domain of a map in I' is of the form QA for A a domain of a map in I. We will show QA is small relative to I-cell. As $J \cup I'$ is contained in I-cell this will show QA is small relative to $J \cup I'$. Consider the construction of QA as the left factor in

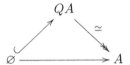

The map $\varnothing \to QA$ is in I-cell, so QA is a colimit of cells (let us say κ_A many), each of which is κ-small where κ is the regular cardinal associated to I by Proposition 11.2.5 of [32]. So for any λ greater than the cofinality of $\max(\kappa, \kappa_A)$, a map from QA to a λ-filtered colimit of maps in I-cell must factor through some stage of the colimit because all the cells making up QA will factor in this way. One can find a uniform λ for all objects QA by an appeal to Lemma 10.4.6 of [32]. $\qquad\square$

Remark 5.20   In a combinatorial model category no smallness hypothesis needs to be made because all objects are small. In a cellular model category, the assumption that the domains of J are small relative to cofibrations is included. As these hypotheses are standard when working with left Bousfield localization, we shall say no more about the additional smallness hypothesis placed on J above.

Corollary 5.21   *Suppose $\mathcal{M}$ is a left proper model category cofibrantly generated by sets I and J in which the domains of maps in J are small relative to I-cell and are cofibrant. Then there exist a set of generating cofibrations I' with cofibrant domains.*

Remark 5.22 Note that this corollary does not say that any left proper, cofibrantly generated model category has generating sets I and J with cofibrant domains. There is an example due to Carlos Simpson (found on page 199 of [52]) of a left proper, combinatorial model category that has no such sets I and J. In this example the cofibrations and trivial cofibrations are the same, so cannot be leveraged against one another in the way we have done above.

We are now prepared to prove Theorem 5.12.

Proof of Theorem 5.12 First, if $L_\mathcal{C}$ is a monoidal Bousfield localization then every map of the form $f \otimes id_K$, where $f \in \mathcal{C}$ and K is cofibrant, is a $\mathcal{C}$-local equivalence. This is because f is a $\mathcal{C}$-local equivalence and cofibrant objects are flat in $L_\mathcal{C}(\mathcal{M})$. We turn now to the converse.

Assume every map of the form $f \otimes id_K$, where $f \in \mathcal{C}$ and K is cofibrant, is a $\mathcal{C}$-local equivalence. Then cofibrant objects are flat in $L_\mathcal{C}(\mathcal{M})$. To see this, let X be cofibrant and define $F(-) = X \otimes -$. Then Lemma 5.16 implies F is left Quillen when viewed as a functor from $L_\mathcal{C}(\mathcal{M})$ to $L_\mathcal{C}(\mathcal{M})$. So F takes $\mathcal{C}$-local equivalences between cofibrant objects to $\mathcal{C}$-local equivalences. By the reduction at the beginning of the proof of Proposition 5.15, this implies F takes all $\mathcal{C}$-local equivalences to $\mathcal{C}$-local equivalences.

Next, the unit axiom on $L_\mathcal{C}(\mathcal{M})$ follows from the unit axiom on $\mathcal{M}$, by Proposition 5.17. Finally, we must prove the pushout product axiom holds on $L_\mathcal{C}(\mathcal{M})$. As in the proof of Proposition 5.18, Proposition 4.2.5 of [34] reduces the problem to checking the pushout product axiom on a set of generating (trivial) cofibrations. We apply Lemma 5.19 to $\mathcal{M}$ and check the pushout product axiom with respect to this set of generating maps.

As in the case of Theorem 5.11, let $h : X \to Y$ be a trivial cofibration in $L_\mathcal{C}(\mathcal{M})$ and let $g : K \to L$ be a generating cofibration. By the lemma, the map g is either a cofibration between cofibrant objects or a trivial cofibration in $\mathcal{M}$. If the former, then the proof of Proposition 5.18 goes through verbatim and proves that $h \square g$ is an $L_\mathcal{C}(\mathcal{M})$-trivial cofibration, since cofibrant objects are flat in $L_\mathcal{C}(\mathcal{M})$. If the latter, then because g is a trivial cofibration in $\mathcal{M}$ and h is a cofibration in $\mathcal{M}$ we may apply the pushout product axiom on $\mathcal{M}$ to see that $h \square g$ is a trivial cofibration in $\mathcal{M}$ (hence in $L_\mathcal{C}(\mathcal{M})$ too). This completes the proof of the pushout product axiom on $L_\mathcal{C}(\mathcal{M})$. $\square$

Remark 5.23 The use of the lemma demonstrates that this proposition proves something slightly more general. Namely, if $\mathcal{M}$ is cofibrantly generated, left proper, has cofibrant objects flat, and the class of cofibrations is closed under pushout product then $\mathcal{M}$ satisfies the pushout product axiom.

Additionally, one could also prove the forwards direction in the theorem using only that $L_{\mathcal{C}}(\mathcal{M})$ satisfies the pushout product axiom. For any cofibrant K we have a cofibration $\phi_K : \varnothing \hookrightarrow K$. Note that for any $f \in \mathcal{C}$, $f \otimes K = f \,\square\, \phi_K \subset \mathcal{C}$-local equivalences, because f is a trivial cofibration in $L_{\mathcal{C}}(\mathcal{M})$.

We record this remark because in the future we hope to better understand the connection between monoidal Bousfield localizations and the closed localizations that appeared in [14], and this remark may be useful.

5.5 Preservation of algebras over Σ-cofibrant operads

In this section we will provide several applications of the results in the previous section. We remind the reader that for operads valued in $\mathcal{M}$, a map of operads $A \to B$ is said to be a trivial fibration if $A_n \to B_n$ is a trivial fibration in $\mathcal{M}$ for all n. An operad P is said to be *cofibrant* if the map from the initial operad into P has the left lifting property in the category of operads with respect to all trivial fibrations of operads. An operad P is said to be Σ-*cofibrant* if it has this left lifting property only in the category of symmetric sequences. The E_{∞}-operads considered in [44] are Σ-cofibrant precisely because the nth space is assumed to be an $E\Sigma_n$ space, where Σ_n is the symmetric group.

We begin with a theorem due to Markus Spitzweck, proven as Theorem 5 in [53] and as Theorem A.8 in [26], that makes it clear that the hypotheses of Corollary 5.5 are satisfied when $L_{\mathcal{C}}$ is a monoidal Bousfield localization and P is a cofibrant operad.

Theorem 5.24 *Suppose P is a Σ-cofibrant operad and $\mathcal{M}$ is a monoidal model category. Then P-alg is a semi-model category.*

This theorem, applied to both $\mathcal{M}$ and $L_{\mathcal{C}}(\mathcal{M})$ (if the localization is monoidal), endows the categories of P-algebras in $\mathcal{M}$ and $L_{\mathcal{C}}(\mathcal{M})$ with inherited semi-model structures. By Corollary 5.5, monoidal Bousfield localizations preserve algebras over Σ-cofibrant operads. In particular, monoidal localizations preserve A_∞ and E_∞-algebras in $\mathcal{M}$, since these algebras are encoded by A_∞ and E_∞-operads P that are Σ-cofibrant (and weakly equivalent to Ass and Com respectively in the category $Coll(\mathcal{M})$). A definition of A_∞ and E_∞ operads over general model categories $\mathcal{M}$ can be found in [53, Section 8], among other places. When $\mathcal{M}$ is a category of spectra we are free to work with operads valued in spaces because the Σ^∞ functor will take a (Σ-cofibrant) space-valued operad to a (Σ-cofibrant) spectrum-valued operad with the same algebras.

5.5.1 Spaces and Spectra

We now provide examples demonstrating the power of Theorem 5.3. For topological spaces the situation is especially nice. We will always work in the context of pointed spaces, with the Quillen model structure.

Proposition 5.25 *Let $\mathcal{M}$ be the model category of (pointed) simplicial sets or k-spaces. Every Bousfield localization of $\mathcal{M}$ is a monoidal Bousfield localization.*

Proof For a review of the monoidal model structures on spaces and simplicial sets see Chapter 4 of [34]. Both are cellular, left proper, monoidal model categories with cofibrant objects flat and the domains of the generating cofibrations cofibrant.

For $\mathcal{M} = sSet$ or $sSet_*$, we can simply rely on Theorem 4.1.1 of [32], which guarantees that $L_\mathcal{C}(\mathcal{M})$ is a simplicial model category. The pushout product axiom is equivalent to the SM7 axiom for $sSet$, so this proves $L_\mathcal{C}(\mathcal{M})$ is a monoidal model category and hence that $L_\mathcal{C}$ is monoidal. There is also an elementary proof of this fact, obtained from the proof below by replacing $F(-, -)$ everywhere by $\mathrm{map}(-, -)$.

We turn now to $\mathcal{M} = Top_*$. By definition, any Bousfield localization $L_\mathcal{C}$ will be a monoidal Bousfield localization as soon as we show $\mathcal{C} \wedge S^n_+$ is contained in the $\mathcal{C}$-local equivalences (the codomains of the generating cofibrations are contractible, so do not matter). As remarked in the discussion below Definition 4.1 in [39], for topological model categories Bousfield localization with respect to a set of cofibrations can be defined using topological mapping spaces rather than simplicial mapping spaces (at least when all maps in $\mathcal{C}$ are cofibrations). Let $F(X, Y)$ denote the space of based maps $X \to Y$.

We will make use of Proposition 3.2 in [35], a version of which states that because Top is left proper and cofibrantly generated, a map f is a weak local equivalence if and only if $F(T, f)$ is a weak equivalence of topological spaces for all T in the (co)domains of the generating cofibrations I in Top_*.

Now consider the following equivalent statements, where T runs through the domains and codomains of generating cofibrations.

f is a C-local equivalence

iff $F(f, Z)$ is a weak equivalence for all C-local Z

iff $F(T, F(f, Z))$ is a weak equivalence for all C-local Z and all T

iff $F(T \wedge f, Z)$ is a weak equivalence for all C-local Z (by adjointness)

iff $T \wedge f$ is a C-local equivalence

This proves that the class of C-local equivalences is closed under smashing with a domain or codomain of a generating cofibration, so L_C is a monoidal Bousfield localization. An analogous proof works for $\mathcal{M} = Top$. $\qquad\square$

The reader may wonder whether all Bousfield localizations preserve algebras over cofibrant operads in general model categories $\mathcal{M}$, i.e. whether all Bousfield localizations are monoidal. This is false, as demonstrated by the following example, from Section 6 in [14].

Example 5.26 Let $\mathcal{M}$ be symmetric spectra, S-modules or orthogonal spectra. Recall that in pointed spaces, the nth Postnikov section functor P_n is the Bousfield localization L_f corresponding to $f : S^{n+1} \to *$, for $n \geq 0$. Applying Σ^∞ gives a map of spectra and we again denote by P_n the Bousfield localization with respect to this map. The Bousfield localization P_{-1} on $\mathcal{M}$ does not preserve A_∞-algebras. If R is a non-connective A_∞-algebra then the unit map $\nu : S \to P_{-1}R$ is null because $\pi_0(P_{-1}R) = 0$. Thus, $P_{-1}R$ cannot admit a ring spectrum structure (not even up to homotopy) because $S \wedge P_{-1}R \to P_{-1}R \wedge P_{-1}R \to P_{-1}R$ is not a homotopy equivalence as it would have to be for $P_{-1}R$ to be a homotopy ring. It follows that the model category $P_{-1}\mathcal{M}$ fails the pushout product axiom, because if $P_{-1}\mathcal{M}$ satisfied the pushout product axiom, then A_∞-algebras in $P_{-1}\mathcal{M}$ would inherit a transferred semi-model structure, by Theorem 5.24, and Corollary 5.5 would imply that P_{-1} preserves A_∞-algebras.

In [14], examples of the sort above are prohibited by assuming that L-equivalences are closed under the monoidal product. It is then shown in Theorem 6.5 that for symmetric spectra this property is implied if the localization is *stable*, i.e. $L \circ \Sigma \simeq \Sigma \circ L$. We now compare our requirement that L_C be a monoidal Bousfield localization to existing results regarding preservation of monoidal structure.

Proposition 5.27 *Let $\mathcal{M}$ be a stable model category. Then every monoidal Bousfield localization is stable. In a monogenic setting such as spectra, every stable localization is monoidal.*

This is clear, since suspending is the same as smashing with the suspension of the unit sphere. The Postnikov section is clearly not stable, and indeed the counterexample above hinges on the fact that the section has truncated the spectrum by making trivial the degree in which the unit must live. Stable localizations preserve cofiber sequences, but P_{-1} does not. Under the hypothesis that localization respects the monoidal product, Theorem 6.1 of [14] proves that cofibrant algebras over a cofibrant colored operad valued in $sSet_*$ or Top_* are preserved. Theorem 5.3 recovers this result in the case of operads, and improves on it by extending the class of operads so that they do not need to be valued in $sSet_*$ or Top_*, by discussing preservation of non-cofibrant algebras, by weakening the cofibrancy required of the operad to Σ-cofibrancy (using Theorem 5.24 above), and by potentially weakening the hypothesis on the localization. A different generalization of [14] has been given in [26].

Proposition 5.28 *Every Bousfield localization for which the local equivalences are closed under $\otimes$ is a monoidal Bousfield localization, but the converse fails.*

Proof To see why this is true, consider the maps id_K as L-equivalences when testing whether or not $id_K \otimes \mathcal{C}$ is a $\mathcal{C}$-local equivalence. To see that the converse fails, take $\mathcal{C}$ to be the generating trivial cofibrations of any cofibrantly generated model category in which the weak equivalences are not closed under $\otimes$. $\square$

Thus, our hypothesis on a monoidal Bousfield localization is strictly weaker than requiring L-equivalences to be closed under $\otimes$. Theorems 5.11 and 5.12 demonstrate that the hypothesis that $\mathcal{C} \otimes id_K$ is contained in the $\mathcal{C}$-local equivalences is best-possible, since $L_{\mathcal{C}}$ is a monoidal Bousfield localization if and only if this property holds, and without the pushout product axiom on $L_{\mathcal{C}}(\mathcal{M})$ the question of preservation of algebras under localization is not even well-posed. Note that cofibrant objects are flat for symmetric spectra by 5.3.10 in [38].

Remark 5.29 In light of the Postnikov Section example, the argument of Proposition 5.25 must break down for spectra. The precise place where the argument fails is the passage through $\mathrm{map}(T, \mathrm{map}(f, Z))$. In spectra, this expression has no meaning, because T is a spectrum but $\mathrm{map}(f, Z)$

is a space. So the argument of Proposition 5.25 relies precisely on the fact that $\mathcal{M} = sSet$ (or $\mathcal{M} = Top$ in the topological case), so that the SM7 axiom for $\mathcal{M}$ is precisely the same as the pushout product axiom.

Theorem 5.3 and Theorem 5.24 combine to prove that any monoidal Bousfield localization of spectra preserves A_∞ and E_∞-algebras. In particular, A_∞ and E_∞-algebras are preserved by stable Bousfield localizations such as L_E where E is a homology theory. So our results recover Theorems VIII.2.1 and VIII.2.2 of [19].

5.5.2 Equivariant Spectra

In order to specialize Corollary 5.5 to the case of G-equivariant spectra, where G is a compact Lie group, we must first understand the generating cofibrations. For Top^G, the (co)domains of maps in I take the form $((G/H) \times S^{n-1})_+$ and $((G/H) \times D^n)_+$ for H a closed subgroup of G, by Definition 1.1 in [42]. For G-spectra, the situation is more complicated. We first fix a G-universe, i.e., a set $\mathcal{U}$ of orthogonal G-representations closed under direct sums and summands, containing the trivial one-dimensional representation. For any finite dimensional orthogonal G-representation W there is an evaluation functor $Ev_W : \mathcal{S}^G \to Top^G$. This functor has a left adjoint F_W (see Proposition 3.1 in [39] for more details). The (co)domains of maps in I take the form $F_W((G/H)_+ \wedge S_+^{n-1})$ and $F_W((G/H)_+ \wedge D_+^n)$ by Definition 1.11 in [42], where W runs through our fixed G-universe. The latter are contractible, and so smashing with them does not make a difference. Observe that the domains of the generating cofibrations are cofibrant, since on the space level they are G-CW complexes, and on the spectra level they are created by the left Quillen functors F_W.

Left Bousfield localization yields the *stable model structure*, which we denote $\mathcal{S}^G$. That $\mathcal{S}^G$ is a monoidal model category with cofibrant objects flat is verified in Proposition III.7.3 of [42], and may also be deduced from Corollary 4.4 in [39]. We could also work with the *positive stable model structure* $\mathcal{S}_+^G$, which has the same weak equivalences as $\mathcal{S}^G$, but cofibrations defined by functors F_W where $W^G \neq 0$. The proof that these model structures are left proper and cellular can be found in the appendix of [27]. For $\mathcal{M} = \mathcal{S}^G$ or $\mathcal{S}_+^G$, our preservation result (Corollary 5.5 together with Theorem 5.11) becomes:

Theorem 5.30 *Let G be a compact Lie group. In $\mathcal{S}^G$ (resp. $\mathcal{S}_+^G$), a Bousfield localizations $L_\mathcal{C}$ is monoidal if and only if each morphism in the set $\mathcal{C} \wedge F_W((G/H)_+ \wedge S_+^{n-1})$ is a $\mathcal{C}$-local equivalence for all closed subgroups*

H of G, for all W in the universe (resp. all W such that $W^G \neq 0$), and for all n (resp. $n > 0$). Furthermore, such localizations preserve P-algebra structures for any Σ-cofibrant P, including any equivariant E_∞-operad P.

Here a G-operad P is called *equivariant E_∞* if it is Σ-free, the spaces $P(n)$ are G-CW complexes, and $P(n)^H \simeq *$ for all closed subgroups H of G. These operads are Σ-cofibrant with respect to the model structure on G-operads transferred from the model structure on G-collections $\prod_{n \geq 0} (Top^G)^{\Sigma_n}$. Here Top^G has the usual G-equivariant model structure, and $(Top^G)^{\Sigma_n}$ has the projective model structure, where Σ_n is the symmetric group. Hence, a morphism $f = (f_n)$ is a weak equivalence (resp. fibration) if f_n^H is a weak equivalence (resp. fibration) in Top for every closed subgroup H of G and every n. Note that these operads do *not* encode genuine equivariant commutativity. To do that, subgroups of $G \times \Sigma_n$ would need to be considered. In particular, there is not a Quillen equivalence between algebras over an E_∞-operad and commutative equivariant ring spectra. The N_∞-operads of [10] were introduced to encode genuine commutativity in a homotopy coherent way (relative to a choice of a collection of families of subgroups of $G \times \Sigma_n$ for $n \geq 0$) and were constructed in [27] as cofibrant replacements of the operad Com in various model structures on the category of G-operads, corresponding to the choice of a collection of families of subgroups of $G \times \Sigma_n$.

Ignoring suspensions, Theorem 5.30 demonstrates that monoidal Bousfield localizations are precisely the ones for which L_C respects smashing with $(G/H)_+$ for all subgroups H. In this light, Theorem 5.30 can be seen as a generalization of Proposition 5.27, saying that if L_C respects stabilization with respect to all the objects $F_W((G/H)_+ \wedge S_+^{n-1})$ then L_C is monoidal. We think of these monoidal localizations as the ones that can 'see' the information of all subgroups. A natural question is: what if L_C can only 'see' the information of some subgroups H? To answer this question, we must consider the following model structures, from Theorem 6.3 in [42] (on spectra either the stable or positive stable model structure can be used):

Definition 5.31 Let $\mathcal{F}$ be a family of closed subgroups of G, i.e. a nonempty set of subgroups closed under conjugation and taking subgroups. Then the *$\mathcal{F}$-fixed point model structure* on pointed G-spaces is a cofibrantly generated model structure in which a map f is a weak equivalence (resp. fibration) if and only if f^H is a weak equivalence (resp. fibration) in

Top for all $H \in \mathcal{F}$. We will denote this model structure by $Top^{\mathcal{F}}$. The generating (trivial) cofibrations are $(G/H \times g)_+$, where g is a generating (trivial) cofibration of topological spaces, and $H \in \mathcal{F}$.

The corresponding cofibrantly generated model structure on G-spectra will be denoted $\mathcal{S}^{\mathcal{F}}$. Again, weak equivalences (resp. fibrations) are maps f such that f^H is a weak equivalence (resp. fibration) of orthogonal spectra for all $H \in \mathcal{F}$. The generating (trivial) cofibrations are $F_W((G/H)_+ \wedge g)$ as H runs through $\mathcal{F}$, g runs through the generating (trivial) cofibrations of spaces, and W runs through our fixed G-universe $\mathcal{U}$.

With the generating cofibrations in hand, Theorem 5.11 implies that monoidal Bousfield localizations in $\mathcal{S}^{\mathcal{F}}$ are characterized by the property that $\mathcal{C} \wedge (G/H)_+$ is a $\mathcal{C}$-local equivalence for all $H \in \mathcal{F}$ (again, ignoring suspensions). One can also define $\mathcal{F}$-fixed point semi-model structures $Oper^{\mathcal{F}}$ on the category of G-operads by applying the general machinery of Theorem 12.2.A in [23]. Indeed, [27] demonstrates how to define full model structures on these categories of operads, for even more general families of subgroups.

Definition 5.32 Let $E_\infty^{\mathcal{F}}$ be the cofibrant replacement for the operad Com in the $\mathcal{F}$-fixed point semi-model structure on G-operads.

These operads form a lattice (ordered by family inclusion) interpolating between non-equivariant E_∞ (corresponding to the family $\mathcal{F} = \{e\}$) and equivariant E_∞ (corresponding to the family $\mathcal{F} = \{All\}$ and denoted E_∞^G). To understand the algebraic structure encoded by $E_\infty^{\mathcal{F}}$, we pause to introduce some new terminology.

Given a G-space X and a closed subgroup H, one may restrict the G action to H and obtain an H-space denoted $res_H(X)$. This association is functorial and lifts to a functor $res_H : \mathcal{S}^G \to \mathcal{S}^H$. This *restriction functor* has a left adjoint $G_+ \wedge_H (-)$, the *induction functor*. We refer the reader to Section 2.2.4 of [31] for more details. If one shifts focus to commutative monoids $Comm_G$ in $\mathcal{S}^G$ (equivalently to genuine E_∞-algebras) then there is again a restriction functor $res_H : Comm_G \to Comm_H$ and it again has a left adjoint functor $N_H^G(-)$ called the *norm*. This functor is discussed in Section 2.3.2 of [31].

An $E_\infty^{\mathcal{F}}$-algebra X has a multiplicative structure on $res_H(X)$ (compatible with the transfers) for every $H \in \mathcal{F}$. However, $N_H^G(res_H(X))$ need not have a multiplicative structure. This is related to Example 5.37 below. These operads $E_\infty^{\mathcal{F}}$ have been generalized and further studied in [27], which includes a comparison between these operads and the

N_∞-operads of [10], results about transferred model structures, and rectification results. For now we will focus on how $E_\infty^{\mathcal{F}}$-algebra structure interacts with Bousfield localization. First, observe that both $Oper^{\mathcal{F}}$ and $\mathcal{S}^{\mathcal{F}}$ are $Top^{\mathcal{F}}$-model structures (in the sense of Definition 4.2.18 in [34]) and the cofibrancy of $E_\infty^{\mathcal{F}}$ is relative to the $\mathcal{F}$-model structure. Thus, from a model category theoretic standpoint, $E_\infty^{\mathcal{F}}$-algebras are best viewed in $\mathcal{S}^{\mathcal{F}}$. The following two theorems also have formulations for the positive stable model structure, in analogy with Theorem 5.30, that we leave to the reader.

Theorem 5.33   Let $\mathcal{M} = \mathcal{S}^G$ and let $\mathcal{F}$ be a family of closed subgroups of G. Assume $F_W((G/H)_+ \wedge S_+^{n-1}) \wedge \mathcal{C}$ is contained in the $\mathcal{C}$-local equivalences for all $H \in \mathcal{F}$, for all n, and for all W in the universe. Then $L_\mathcal{C}$ preserves $E_\infty^{\mathcal{F}}$-structure.

Localizations of the form above are $\mathcal{F}$-monoidal but not necessarily G-monoidal. For this reason, when $X \in E_\infty^G$-alg, $L_\mathcal{C}(X)$ has $E_\infty^{\mathcal{F}}$-algebra structure but may not have E_∞^G-algebra structure. More generally, we have the following result, encoding the fact that if we work in $\mathcal{S}^{\mathcal{K}}$ rather than $\mathcal{S}^G$, then localizations are compatible with both $\mathcal{K}$ and $\mathcal{F}$. Because there are now two families involved, the localization will preserve algebraic structure corresponding to the meet of these two families in the lattice of families.

Theorem 5.34   Let $\mathcal{M}$ be the $\mathcal{K}$-fixed point model structure on the category of G-spectra and let $\mathcal{K}'$ be a subfamily of $\mathcal{K}$. Assume the morphisms $F_W((G/H)_+ \wedge S_+^{n-1}) \wedge \mathcal{C}$ are contained in the $\mathcal{C}$-local equivalences for all $H \in \mathcal{K}'$, for all n, and for all W in the universe. Then $L_\mathcal{C}$ takes any $E_\infty^{\mathcal{F}}$-algebra to an $E_\infty^{\mathcal{F} \cap \mathcal{K}'}$-algebra.

Proof   In order to apply Corollary 5.5, first forget to the model structure $\mathcal{S}^{\mathcal{F} \cap \mathcal{K}'}$ and observe that any $E_\infty^{\mathcal{F}}$-algebra is sent to a $E_\infty^{\mathcal{F} \cap \mathcal{K}'}$-algebra. The hypothesis on $L_\mathcal{C}$ guarantees that $L_\mathcal{C}$ is a monoidal Bousfield localization with respect to the $\mathcal{F} \cap \mathcal{K}'$ model structure, and so $E_\infty^{\mathcal{F} \cap \mathcal{K}'}$ is preserved. $\square$

Remark 5.35   It is easy to produce examples of localizations $L_\mathcal{C}$ that reduce $E_\infty^{\mathcal{F}}$-algebra structure to $E_\infty^{\mathcal{F}'}$-algebra structure for any families $\mathcal{F}' \subsetneq \mathcal{F}$ of closed subgroups of G, by generalizing the Postnikov section 5.26. For every closed subgroup $H \in \mathcal{F} \setminus \mathcal{F}'$, consider a truncation of the spectrum $F_W((G/H)_+ \wedge S_+^{n-1})$. Localizing with respect to the wedge of these truncation maps will take $E_\infty^{\mathcal{F}}$-algebras to $E_\infty^{\mathcal{F}'}$-algebras, by Theorem 5.34.

Together, Theorem 5.34 and Remark 5.35 resolve the preservation question for operads in the lattice $E_\infty^{\mathcal{F}}$. As expected, preservation of lesser algebraic structure comes down to requiring a less stringent condition on the Bousfield localization. The least stringent condition is for $\mathcal{F} = \{e\}$ and recovers the notion of a stable localization (i.e. one which is monoidal on the category of spectra after forgetting the G-action). However, because none of the operads $E_\infty^{\mathcal{F}}$ rectify with respect to the Com operad, we do not have preservation results for commutative equivariant ring spectra. For the remainder of the section, we discuss localizations that preserve $E_\infty^{\mathcal{F}}$-algebra structure but fail to preserve commutative structure. We begin with the example that motivated this chapter, which the author learned from a talk given by Mike Hill at Oberwolfach (the proceedings can be found in [29]). A similar example appeared in [45]. This example will be generalized in Example 5.37 below.

Example 5.36 There are localizations that destroy genuine commutative structure but that preserve equivariant E_∞-algebra structure. For this example, let G be a (non-trivial) finite group. Consider the reduced real regular representation $\bar{\rho}$ obtained by taking the quotient of the real regular representation ρ by the trivial representation. We write $\bar{\rho}_G = \rho_G - 1$ where 1 means the trivial representation $\mathbb{R}[e]$. Taking the one-point compactification of this representation yields a representation sphere $S^{\bar{\rho}}$. There is a natural inclusion $a_{\bar{\rho}} : S^0 \to S^{\bar{\rho}}$ induced by the inclusion of the trivial representation into $\bar{\rho}$. Consider the spectrum $E = \mathbb{S}[a_{\bar{\rho}}^{-1}]$ obtained from the unit $\mathbb{S}$ (certainly a commutative algebra in $\mathbb{S}^G$) by localization with respect to $a_{\bar{\rho}}$. We will show that this spectrum cannot be commutative.

First, for any proper $H < G$, the restriction $\rho_G|_H$ is $[G : H]\rho_H$, so $\bar{\rho}_G|_H = [G : H]\bar{\rho}_H + ([G : H]1 - 1)$. Let $k = [G : H] - 1$. Observe that $res_H S^{\bar{\rho}_G} = (S^{\bar{\rho}_H})^{\#[G:H]} \wedge S^k$. It follows that $res_H(E)$ is contractible, because $k > 0$.

If E were a commutative equivariant ring spectrum, then the counit of the norm-restriction adjunction would provide a ring homomorphism $N_H^G res_H(E) \to E$. But the domain is contractible for every proper subgroup H because $res_H(E)$ is contractible. This cannot be a ring map unless E to be contractible, and we know E is not contractible, because its G-fixed points are not contractible.

Justin Noel has pointed out that the localization in Example 5.36 is smashing, hence monoidal (this is clear from the reformulation in Example 5.37, but was not clear to the author from the formulation above). It follows that the localization preserves E_∞-algebra structure, by Theorem 5.30, and hence takes commutative monoids to E_∞-algebras. Because *any* such H will lead to a failure of $L(\mathcal{S}) = E$ to be commutative, Example 5.36 is in some sense maximally bad. We now leverage this observation to generalize Example 5.36.

Recall from Definition 5.7 of [27] that an $\mathcal{F}$-N_∞-operad P is a generalization of an N_∞-operad, where P-algebras have multiplicative norms for all $H \in \mathcal{F}$. The formulation of Example 5.37 matches the presentation from [29] for the case when G is finite, $\mathcal{F}$ is the family of all subgroups of G, and $\mathcal{P}$ the family of proper subgroups.

Example 5.37   Let G be a compact Lie group and $\mathcal{F}$ a family of closed subgroups of G. If X is an algebra over an $\mathcal{F}$-N_∞-operad then there is a localization L sending X to an $E_\infty^{\mathcal{F}}$-algebra. Consider the cofiber sequence $E\mathcal{P}_+ \to S^0 \to \widetilde{E}\mathcal{P}$ for any family $\mathcal{P}$ properly contained in $\mathcal{F}$. Recall the fixed-point property of the space $E\mathcal{P}$ (discussed very nicely in Section 7 of [49]) and deduce:

$$(E\mathcal{P}_+)^H \simeq \begin{cases} *_+ = S^0 & \text{if } H \in \mathcal{P} \\ \varnothing_+ = * & \text{if } H \notin \mathcal{P} \end{cases}$$

For all H, the H-fixed points of S^0 are S^0, so that the cofiber obtained by mapping this space into S^0 satisfies the following fixed-point property

$$(\widetilde{E}\mathcal{P})^H \simeq \begin{cases} * & \text{if } H \in \mathcal{P} \\ S^0 & \text{if } H \notin \mathcal{P} \end{cases}$$

Now apply Σ_+^∞ to the map $S^0 \to \widetilde{E}\mathcal{P}$. If G is a finite group, and $\mathcal{F}$ is the family of all subgroups of G, then the resulting map $\mathcal{S} \to E$ is the same localization map considered in Example 5.36 (see Section 7 of [49]). Returning to the general case, note that E is not contractible because $E\mathcal{P}_+$ is not homotopy equivalent to S^0 (since $\mathcal{P}$ is properly contained in $\mathcal{F}$), though $res_H(E\mathcal{P}_+)$ *is* homotopy equivalent to $res_H(S^0)$ for any $H \in \mathcal{P}$. In this formulation it is clear that the map $\mathcal{S} \to E$ is a nullification that kills all maps out of the induced cells $G_+ \wedge_H (H/K)_+ \cong (G/H)_+$ for all $H \in \mathcal{P}$.

This localization is monoidal with respect to the $\mathcal{F}$-model structure, so E is still an $E_\infty^{\mathcal{F}}$-algebra by Theorem 5.34. When G is finite, Example 5.37 makes it clear that the localization is simply inverting a homotopy element (namely: the Euler class $a_{\overline{\rho}}$ discussed in Section 2.6.3 of [31]). The presentation in Example 5.37 has several benefits of its own: it generalizes to compact Lie groups G, it demonstrates that a smaller localization than Example 5.36 is needed to destroy $\mathcal{F}\text{-}N_\infty$-algebra structure rather than N_∞-algebra structure, and it shows how localization can reduce one's place in the lattice of $\mathcal{F}\text{-}N_\infty$-algebras without reducing it all the way down to $E_\infty^{\mathcal{F}}$. To see this, observe that the localization E can still admit some multiplicative norms, for subgroups $H \in \mathcal{F} \setminus \mathcal{P}$. Hence, E can still be a $\mathcal{K}\text{-}N_\infty$-algebra if the family $\mathcal{K}$ only intersects $\mathcal{P}$ in the trivial subgroup (using the fact that $\mathbb{S}$ is an N_∞-algebra for all choices of families of subgroups). Examples such as that of II.2.3 [40] can be used for this purpose.

Motivated by Example 5.37, we devote the next two sections to determining when a left Bousfield localization must preserve commutative structure. We will see that the key compatibility condition required is that the maps in $\mathcal{C}$ respect the free commutative monoid functor. In the case of equivariant spectra, this will imply that $\mathcal{C}$ respects the multiplicative norm functors.

5.6 Bousfield Localization and Commutative Monoids

In this section we turn to the interplay between monoidal Bousfield localizations and commutative monoids, i.e. algebras over the (non-cofibrant) operad Com. In [58], the following theory is developed as Definition 3.1, Theorem 3.2, and Corollary 3.8. Here Σ_n is the symmetric group.

Definition 5.38   A monoidal model category $\mathcal{M}$ is said to satisfy the *commutative monoid axiom* if whenever h is a trivial cofibration in $\mathcal{M}$ then $h^{\square n}/\Sigma_n$ is a trivial cofibration in $\mathcal{M}$ for all $n > 0$.

If, in addition, the class of cofibrations is closed under the operation $(-)^{\square n}/\Sigma_n$ then $\mathcal{M}$ is said to satisfy the *strong commutative monoid axiom*.

Theorem 5.39   *Let $\mathcal{M}$ be a cofibrantly generated monoidal model category satisfying the commutative monoid axiom and the monoid axiom, and assume that the domains of the generating maps I (resp. J) are small relative to $(I \otimes \mathcal{M})$-cell (resp. $(J \otimes \mathcal{M})$-cell). Let R be a commutative monoid in $\mathcal{M}$. Then the category $CAlg(R)$ of commutative R-algebras is a cofibrantly generated model category in which a map is a weak equiva-*

lence or fibration if and only if it is so in $\mathcal{M}$*. In particular, when* $R = S$ *this gives a model structure on commutative monoids in* $\mathcal{M}$*.*

Corollary 5.40   *Let* $\mathcal{M}$ *be a cofibrantly generated monoidal model category satisfying the commutative monoid axiom, and assume that the domains of the generating maps* I *(resp.* J*) are small relative to* $(I \otimes \mathcal{M})$*-cell (resp.* $(J \otimes \mathcal{M})$*-cell). Then for any commutative monoid* R*, the category of commutative* R*-algebras is a cofibrantly generated semi-model category in which a map is a weak equivalence or fibration if and only if it is so in* $\mathcal{M}$*.*

While these results only make use of the commutative monoid axiom, in practice we usually desire the strong commutative monoid axiom so that in the category of commutative R-algebras cofibrations with cofibrant domains forget to cofibrations in $\mathcal{M}$. This is discussed further in [58] and numerous examples of model categories satisfying these axioms are given.

In order to apply the corollary above to verify the hypotheses of Corollary 5.5 we must give conditions on the maps $\mathcal{C}$ so that if $\mathcal{M}$ satisfies the commutative monoid axiom then so does $L_\mathcal{C}(\mathcal{M})$. As for the pushout product axiom, our method will be to apply Lemma 5.16, which is just the universal property of Bousfield localization. However, $(-)^{\Box n}/\Sigma_n$ is not a functor on $\mathcal{M}$, but rather on $\mathrm{Arr}(\mathcal{M})$. The following lemma lets us instead work with the functor $\mathrm{Sym}^n : \mathcal{M} \to \mathcal{M}$ defined by $\mathrm{Sym}^n(X) = X^{\otimes n}/\Sigma_n$. This lemma is proved in Appendix A of [58], and appears in [25].

Lemma 5.41   *Assume that for every* $g \in I$*,* $g^{\Box n}/\Sigma_n$ *is a cofibration. Suppose* f *is a trivial cofibration between cofibrant objects and* $f^{\Box n}/\Sigma_n$ *is a cofibration for all* n*. Then* $f^{\Box n}/\Sigma_n$ *is a trivial cofibration for all* n *if and only if* $\mathrm{Sym}^n(f)$ *is a trivial cofibration for all* n*.*

With this lemma in hand, we are ready to prove the main result of this section, regarding preservation of the commutative monoid axiom by Bousfield localization.

Theorem 5.42   *Assume* $\mathcal{M}$ *is a cofibrantly generated monoidal model category satisfying the strong commutative monoid axiom and with domains of the generating cofibrations cofibrant. Suppose that* $L_\mathcal{C}(\mathcal{M})$ *is a monoidal Bousfield localization with generating trivial cofibrations* $J_\mathcal{C}$*. If* $\mathrm{Sym}^n(f)$ *is a* $\mathcal{C}$*-local equivalence for all* $n \in \mathbb{N}$ *and for all* $f \in J_\mathcal{C}$*, then* $L_\mathcal{C}(\mathcal{M})$ *satisfies the strong commutative monoid axiom. In particular,* the category of commutative monoids inherits a transferred semi-model structure from $L_\mathcal{C}(\mathcal{M})$.

Remark 5.43 The condition of Theorem 5.42, that for all $n \in \mathbb{N}$ and for all $f \in J_\mathcal{C}$, $\mathrm{Sym}^n(f)$ is a $\mathcal{C}$-local equivalence is equivalent to the condition that $\mathrm{Sym}(-)$ preserves $\mathcal{C}$-local equivalences between cofibrant objects. The latter condition implies the former because $\mathrm{Sym}^n(f)$ is a retract of $\mathrm{Sym}(f)$, and the maps in $J_\mathcal{C}$ may be assumed to have cofibrant domains, as shown in [36, Proposition 4.3]. That the former implies the latter follows from Theorem 5.6 of [6], which shows that the existence of a transferred semi-model structure on commutative monoids in $L_\mathcal{C}(\mathcal{M})$ implies $\mathrm{Sym}(-)$ preserves $\mathcal{C}$-local equivalences between cofibrant objects. This result, together with Theorem 5.42, implies that both conditions are equivalent to existence of a transferred semi-model structure on commutative monoids in $L_\mathcal{C}(\mathcal{M})$. Furthermore, the existence of this semi-model structure is equivalent to $L_\mathcal{C}(\mathcal{M})$ satisfying the strong commutative monoid axiom, since the existence of the semi-model structure implies $\mathrm{Sym}(-)$ preserves $\mathcal{C}$-local equivalences, which implies the strong commutative monoid axiom by Theorem 5.42. We have refrained from stating Theorem 5.42 as an 'if and only if' to match the discussion in [6] where the converse was first noticed.

We turn now to the proof of Theorem 5.42, and to several related results. All of these results are meant to find the easiest possible condition to check on $\mathcal{C}$ so that $L_\mathcal{C}(\mathcal{M})$ satisfies the commutative monoid axiom. Theorem 5.42 reduces the problem from having to check the class of all $\mathcal{C}$-local equivalences to only having to check the set $J_\mathcal{C}$ (which, unfortunately, is often mysterious in practice). It is tempting to try to prove Theorem 5.42 using Lemma 5.16, as we did in Theorem 5.11, since this would reduce the problem to checking the set $J \cup \mathcal{C}$ (which is much less mysterious than $J_\mathcal{C}$). However, Sym^n is not a left adjoint. One could attempt to get around this by applying Lemma 5.16 with the functor $\mathrm{Sym} : \mathcal{M} \to \mathrm{CMon}(\mathcal{M})$, but this would require the existence of a model structure on $\mathrm{CMon}(\mathcal{M})$ in which the weak equivalences are $\mathcal{C}$-local equivalences. As this is what we're trying to prove by obtaining the commutative monoid axiom on $L_\mathcal{C}(\mathcal{M})$, this approach is doomed to fail. Instead, we opt for a more technical argument, following the techniques of [58].

Proof of Theorem 5.42 By Appendix A of [58], if $(-)^{\Box n}/\Sigma_n$ takes generating (trivial) cofibrations to (trivial) cofibrations, then it takes all (trivial) cofibrations to (trivial) cofibrations. The generating cofibrations of $L_{\mathcal{C}}(\mathcal{M})$ are the same as those in $\mathcal{M}$ and $\mathcal{M}$ satisfies the strong commutative monoid axiom, so the class of cofibrations of $L_{\mathcal{C}}(\mathcal{M})$ is closed under the operation $(-)^{\Box n}/\Sigma_n$.

Suppose, for every generating trivial cofibration $f : X \to Y$ of $L_{\mathcal{C}}(\mathcal{M})$, that $\mathrm{Sym}^n(f)$ is a $\mathcal{C}$-local equivalence. Because the domains of the generating cofibrations in $\mathcal{M}$ are cofibrant, the same is true in $L_{\mathcal{C}}(\mathcal{M})$ (see Proposition 4.3 in [36]), so we may assume f has cofibrant domain and codomain. In particular, the proof of Lemma 5.41 implies $\mathrm{Sym}^n(f)$ is a cofibration, because $f^{\Box k}/\Sigma_k$ is a cofibration for all k and the domain X of f is cofibrant.

By hypothesis, $\mathrm{Sym}^n(f)$ is a trivial cofibration of $L_{\mathcal{C}}(\mathcal{M})$ for all n. We are therefore in the situation of Lemma 5.41 and may conclude that $f^{\Box n}/\Sigma_n$ is a trivial cofibration for all n. We now apply the result from Appendix A of [58] to conclude that all trivial cofibrations of $L_{\mathcal{C}}(\mathcal{M})$ are closed under the operation $(-)^{\Box n}/\Sigma_n$. The main theorem of [58] produces the resulting transferred semi-model structure on commutative monoids. $\qquad\square$

If we know more about $\mathcal{M}$ in the statement of Theorem 5.42, then we can in fact get a sharper condition regarding the generating trivial cofibrations $J_{\mathcal{C}}$. One way to better understand the trivial cofibrations in $L_{\mathcal{C}}(\mathcal{M})$ is via the theory of framings. Briefly, the idea here is to cofibrantly replace morphisms in $\mathcal{C}$ in an appropriate simplicial model category. One starts by shifting focus from $\mathcal{M}$ to the category of cosimplicial objects $\mathcal{M}^\Delta$, i.e., functors from the simplex category Δ to $\mathcal{M}$. For any model category $\mathcal{M}$, $\mathcal{M}^\Delta$ carries the Reedy model structure [34, Theorem 5.2.5]. For an object X of $\mathcal{M}$, a *cosimplicial resolution* is a cofibrant replacement, in the Reedy model structure, of the constant cosimplicial object valued at X. As cofibrant replacement is functorial, this also defines the notion of a cosimplicial resolution of a morphism f in $\mathcal{M}$. For a class of morphisms $\mathcal{C}$ in $\mathcal{M}$, Definition 4.2.1 of [32] defines the *full class of horns on $\mathcal{C}$* to be the class

$$\Lambda(\mathcal{C}) = \{\widetilde{f} \Box i_n \mid f \in \mathcal{C}, n \geq 0\}$$

where $i_n : \partial\Delta[n] \to \Delta[n]$ is the natural inclusion, and $\widetilde{f} : \widetilde{A} \to \widetilde{B}$ is a Reedy cofibration between cosimplicial resolutions, obtained by factoring the cosimplicial resolution of f into a Reedy cofibration followed by a Reedy trivial fibration. In the case where $\mathcal{C}$ is a set and $\mathcal{M}$ is cofibrantly generated, Definition 4.2.2 of [32] defines an *augmented set of $\mathcal{C}$-horns* to be $\overline{\Lambda(\mathcal{C})} = \Lambda(\mathcal{C}) \cup J$. Finally, 4.2.5 of [32] defines a set $\widehat{\Lambda(\mathcal{C})}$ to be a set of relative I-cell complexes with cofibrant domains obtained from $\overline{\Lambda(\mathcal{C})}$ via cofibrant replacement. Note that, according to the erratum to [32], we do not know that the domains of maps in $\widetilde{\Lambda(\mathcal{C})}$ are cofibrant, but we do know that they are small relative to I.

We now advertise the surprising and powerful Theorem 4.11 in [2]. This result states that if $\mathcal{M}$ is proper and stable, if the $\mathcal{C}$-local objects are closed under Σ (such $L_{\mathcal{C}}$ are called *stable*), and if $\mathcal{C}$ consists of cofibrations between cofibrant objects then $J_{\mathcal{C}}$ is $J \cup \Lambda(\mathcal{C})$. The last hypothesis is a standing assumption for this chapter. The key input to [2, Theorem 4.11] is the observation that for such $\mathcal{M}$, a map is a $\mathcal{C}$-fibration if and only if its fiber is $\mathcal{C}$-fibrant.

Corollary 5.44 *Suppose $\mathcal{M}$ is a stable, proper, simplicial model category satisfying the strong commutative monoid axiom. Suppose that $L_{\mathcal{C}}$ is a stable and monoidal Bousfield localization such that for all $n \in \mathbb{N}$ and $f \in \mathcal{C}$, $\mathrm{Sym}^n(f)$ is a $\mathcal{C}$-local equivalence. Then $L_{\mathcal{C}}(\mathcal{M})$ satisfies the strong commutative monoid axiom. In particular, the category of commutative monoids inherits a transferred semi-model structure from $L_{\mathcal{C}}(\mathcal{M})$.*

Proof By Theorem 5.42 we must only check that Sym^n takes maps in $J_{\mathcal{C}} = J \cup \Lambda(\mathcal{C})$ to $\mathcal{C}$-local equivalences. By the commutative monoid axiom on $\mathcal{M}$, maps in J are taken to weak equivalences, so we must only consider maps in $\Lambda(\mathcal{C})$.

The reason for the hypothesis that $\mathcal{M}$ is simplicial is Remark 5.2.10 in [34], which states that the functor $\tilde{A}^m = A \otimes \Delta[m]$ is a cosimplicial resolution of A (at least, when A is cofibrant), in both $\mathcal{M}$ and $L_{\mathcal{C}}(\mathcal{M})$. Thus, we may take our map in $\Lambda(\mathcal{C})$ to be of the form $(f \otimes \Delta[m]) \square i_n$ where $f : A \to B$ is in $\mathcal{C}$, since $\tilde{f}^m = f \otimes \Delta[m]$.

The map $(f \otimes \Delta[m]) \square i_n$ can be realized as the corner map in the diagram

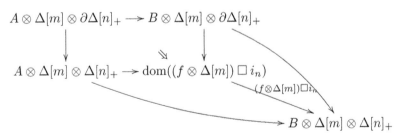

If we can prove that $(g \otimes K)^{\square n}/\Sigma_n$ is a $\mathcal{C}$-local trivial cofibration for any $\mathcal{C}$-local trivial cofibration g between cofibrant objects then we can

apply the same reasoning from the proof of Proposition 5.18 to deduce that the corner map becomes a $\mathcal{C}$-local trivial cofibration after applying $(-)^{\square n}/\Sigma_n$. This reasoning goes by proving that after applying $(-)^{\square n}/\Sigma_n$ the lower curved map and the top horizontal map are $\mathcal{C}$-local trivial cofibrations, so the bottom horizontal map is as well (because it is a pushout), and hence the corner map is a weak equivalence by the two out of three property. This reasoning works because whenever f is a pushout of g then $f^{\square n}/\Sigma_n$ is a pushout of $g^{\square n}/\Sigma_n$ as shown in Appendix A of [58].

Because $g \otimes K$ is a $\mathcal{C}$-local trivial cofibration between cofibrant objects, we may apply Lemma 5.41 to reduce the final step to checking that if $\mathrm{Sym}^n(g)$ is a $\mathcal{C}$-local trivial cofibration for all n then so is $\mathrm{Sym}^n(g \otimes K)$. This is proven in [25]. $\qquad\square$

When the hypotheses of stability and properness are dropped one can no longer easily write down the set $J_\mathcal{C}$. However, Theorem 4.1.1 (and its proof, notably 4.3.1) in [32] demonstrate that the class of maps $X \to L_\mathcal{C}(X)$ are contained in $\widetilde{\Lambda(\mathcal{C})}$-cell. Given a $\mathcal{C}$-local trivial cofibration $g : X_1 \to X_2$ between cofibrant objects, applying fibrant replacement $L_\mathcal{C}$ results in a map $L_\mathcal{C}(g)$, that is a weak equivalence between cofibrant objects. An appeal to Ken Brown's lemma on the functor Sym^n and to the two out of three property reduces the verification that $(-)^{\square n}/\Sigma_n$ takes g to a $\mathcal{C}$-local equivalence to verifying that $(-)^{\square n}/\Sigma_n$ takes $X_i \to L_\mathcal{C}(X_i)$ to $\mathcal{C}$-local equivalences.

Since such maps are in $\widetilde{\Lambda(\mathcal{C})}$-cell, by Appendix A of [58] one must only show that maps in $\widetilde{\Lambda(\mathcal{C})}$ are taken to $\mathcal{C}$-local equivalences by $(-)^{\square n}/\Sigma_n$ (that they are taken to cofibrations is immediate by the strong commutative monoid axiom on $\mathcal{M}$). This observation leads to the following result, which we have recently learned was independently discovered in [25].

Theorem 5.45 *Suppose $\mathcal{M}$ is a cofibrantly generated, simplicial model category satisfying the strong commutative monoid axiom and with domains of the generating cofibrations cofibrant. Suppose that for all $n \in \mathbb{N}$ and $f \in \mathcal{C}$, $\mathrm{Sym}^n(f)$ is a $\mathcal{C}$-local equivalence. Then $L_\mathcal{C}(\mathcal{M})$ satisfies the strong commutative monoid axiom. In particular, the category of commutative monoids inherits a transferred semi-model structure from $L_\mathcal{C}(\mathcal{M})$.*

As the proof of this Theorem appears in [25], we will content ourselves with the sketch of the proof given above and we refer the interested reader to [25] for details. With a careful analysis of $\widetilde{\Lambda(\mathcal{C})}$ the author believes one could remove the need for $\mathcal{M}$ to be simplicial. However, lacking equations of the sort found in Remark 5.2.10 of [34], he does not know how to proceed.

Remark 5.46 The commutative monoid axiom has a natural generalization to an arbitrary operad P. The proof of Proposition 7.6 in [28] demonstrates a precise hypothesis on $\mathcal{M}$ so that P-algebras inherit a model structure, namely that for all $A \in P$-alg and for all n, $P_A[n] \otimes_{\Sigma_n} (-)^{\Box n}$ preserves trivial cofibrations (where P_A is the enveloping operad). If these hypotheses are only satisfied for cofibrant A then P-alg inherits a semi-model structure. We hope in the future to study the types of localizations that preserve these axioms, so that Corollary 5.5 may be applied to deduce preservation results for arbitrary operads P. We conjecture that the correct condition on a localization is that for all $f \in \mathcal{C}$, for all $A \in P$-alg, and for all n, then $P_A[n] \otimes_{\Sigma_n} f^{\Box n}$ is contained in the $\mathcal{C}$-local equivalences. Assuming a P-algebra analogue of Lemma 5.41, the proof of Corollary 5.44 will go through, if we assume $P_A[n] \otimes_{\Sigma_n} f^{\Box n}$ is contained in the $\mathcal{C}$-local equivalences, for all f of the form $g \otimes K$ where $g \in \mathcal{C}$ and K is a simplicial set.

Remark 5.47 Theorem 5.42 also has a converse, that the author discovered in joint work with Michael Batanin [6] (Theorem 5.6 and Example 5.9). For nicely behaved model categories, including all examples considered in this chapter, the following are equivalent:

1 $L_{\mathcal{C}}$ preserves P-algebras,
2 P-alg$(L_{\mathcal{C}}(\mathcal{M}))$ admits a transferred semi-model structure from $L_{\mathcal{C}}(\mathcal{M})$,
3 $L_{\mathcal{C}}$ lifts to a localization of P-algebras (inverting the maps $P(\mathcal{C})$),
4 U preserves local equivalences,

and any of these statements implies $P(-)$ preserves $\mathcal{C}$-local equivalences between cofibrant objects. A dual result, for the situation of right Bousfield localization, appears in [59]. It follows that, for any of the situations from Theorem 5.42, 5.44, or 5.45, $L_{\mathcal{C}}$ preserves commutative monoids if and only if $\mathrm{Sym}(-)$ preserves $\mathcal{C}$-local equivalences between cofibrant objects. Note that the condition that the objects be cofibrant is no obstacle, in any model category satisfying the strong commutative monoid axiom, since $\mathcal{C}$ can be taken to be a set of cofibrations between cofibrant objects, and $\mathrm{Sym}^n(X)$ is cofibrant whenever X is cofibrant, if the commutative monoid axiom is satisfied. This follows from the filtration on $\mathrm{Sym}^n(\varnothing) \to \mathrm{Sym}^n(X)$ from Lemma A.3 of [58]. Although the positive stable model structure only satisfies the (weak) commutative monoid axiom, one can use the positive flat stable model structure to prove all statements needed for the positive stable model structure, as demonstrated in [58].

5.7 Preservation of Commutative Monoids

We turn now to the question of preservation under Bousfield localization of commutative monoids. We will be applying Theorem 5.42 and Corollary 5.5 for this purpose in a moment, but we first remark on a simpler case where the hypotheses of Theorem 5.42 are not necessary.

5.7.1 Spectra

Preservation of commutative monoids by monoidal Bousfield localizations is easy in certain categories of spectra, because of the property that for all cofibrant X in $\mathcal{M}$, the map $(E\Sigma_n)_+ \wedge_{\Sigma_n} X^{\wedge n} \to X^{\wedge n}/\Sigma_n$ is a weak equivalence, where the symmetric group Σ_n acts by permuting copies of X. This property was first noticed in [19], and we will now discuss it more generally.

Recall that, given a morphism of operads $\alpha : O \to P$, the two operads O and P are said to satisfy *rectification* if the induced adjunction between P-alg and O-alg is a Quillen equivalence. In [58], we introduced the *rectification axiom*, which states that if $Q_{\Sigma_n} S \to S$ is cofibrant replacement for the unit S in the projective model structure $\mathcal{M}^{\Sigma_n}$ of Σ_n-objects in $\mathcal{M}$, then for all cofibrant X in $\mathcal{M}$, the map $Q_{\Sigma_n} S \otimes_{\Sigma_n} X^{\otimes n} \to X^{\otimes n}/\Sigma_n$ is a weak equivalence (this is the natural generalization of the property from [19] mentioned above, and was further generalized in [62]). Observe that this property automatically holds on $L_{\mathcal{C}}(\mathcal{M})$ if it holds on $\mathcal{M}$, because the cofibrant objects are the same and the weak equivalences are contained in the $\mathcal{C}$-local equivalences. We now prove that in the presence of the rectification axiom, preservation results for commutative monoids are particularly nice.

Theorem 5.48 *Let $QCom$ denote a Σ-cofibrant replacement of Com in $\mathcal{M}$. Let $\mathcal{M}$ be a monoidal model category satisfying the conditions of Theorem 4.6 of [58], so that the rectification axiom implies that $QCom$ and Com rectify. Let $L_{\mathcal{C}}$ be a monoidal Bousfield localization. Then $L_{\mathcal{C}}$ preserves commutative monoids. In particular:*

- *For positive (flat) symmetric spectra, positive (flat) orthogonal spectra, or* S-*modules, QCom is* E_∞ *and any monoidal Bousfield localization preserves strict commutative ring spectra.*

- *For positive (flat) G-equivariant orthogonal spectra, QCom is* E_∞^G *and any monoidal Bousfield localization preserves strict commutative equivariant ring spectra.*

Proof Let E be a commutative monoid, so in particular E is a $QCom$ algebra via the map $QCom \to Com$. Because $QCom$ is Σ-cofibrant, $QCom$-algebras in both $\mathcal{M}$ and $L_\mathcal{C}(\mathcal{M})$ inherit semi-model structures. Corollary 5.5 implies $L_\mathcal{C}(E)$ is weakly equivalent to some $QCom$-algebra E_Q. The rectification axiom in $L_\mathcal{C}(\mathcal{M})$ now implies E_Q is weakly equivalent to a commutative monoid $\widehat{E}$. □

Currently, this result is only known to apply to the categories of spectra listed in the statement of the theorem. We conjectured in [58] that the rectification axiom implies rectification between $QCom$ and Com for general $\mathcal{M}$. If this conjecture is proven then the theorem will apply to all $\mathcal{M}$ satisfying the rectification axiom. Even if the conjecture is false, the following proposition demonstrates that when $\mathcal{M}$ satisfies the rectification axiom then the conditions of Theorem 5.42 are satisfied and so any monoidal localization preserves commutative monoids.

Proposition 5.49 *Suppose* $\mathcal{N}$ *is a monoidal model category satisfying the rectification axiom. Then* $\mathrm{Sym}^n(-)$ *takes trivial cofibrations between cofibrant objects to weak equivalences.*

In particular, if $L_\mathcal{C}(\mathcal{M})$ *is a monoidal Bousfield localization and* $\mathcal{M}$ *satisfies the rectification axiom, then* $L_\mathcal{C}$ *preserves commutative monoids.*

Proof The first part is proven as Proposition 4.6 in [58], and we refer the reader there for a proof. For the second part, we apply the first part with $\mathcal{N} = L_\mathcal{C}(\mathcal{M})$, using our observation that the rectification axiom holds on $L_\mathcal{C}(\mathcal{M})$ whenever it holds on $\mathcal{M}$. Thus, $\mathrm{Sym}^n : L_\mathcal{C}(\mathcal{M}) \to L_\mathcal{C}(\mathcal{M})$ takes $\mathcal{C}$-local trivial cofibrations between cofibrant objects to $\mathcal{C}$-local equivalences. In particular, the hypotheses of Theorem 5.42 are satisfied and we may deduce from Corollary 5.5 that $L_\mathcal{C}$ preserves commutative monoids. □

5.7.2 Spaces

We turn our attention now to simplicial sets and topological spaces. Rectification is known to fail (see Example 4.4 in [58]), so even though all localizations are monoidal we may not apply the result above. For spaces the path connected commutative monoids are weakly equivalent to generalized Eilenberg-Mac Lane spaces, i.e. products of Eilenberg-Mac Lane spaces. Preservation of commutative monoids has been proven for pointed CW complexes as Theorem 1.4 in [15].

Theorem 5.50 *Let $\mathcal{M}$ be the category of pointed CW complexes. Let $\mathcal{C}$ be any set of maps. Then $\mathrm{Sym}(-)$ preserves $\mathcal{C}$-local equivalences and $L_{\mathcal{C}}$ sends GEMs to GEMs.*

As a special case of this theorem, we recover classical results of Bousfield, e.g. parts of Theorem 5.1 and Lemma 9.8 from [13]. The proof of Theorem 5.50 is based on work of Dror Farjoun, Chapter 4 of [21], in the setting of $\mathcal{M} = sSet$. That work is generalized in [58] to hold for the category of k-spaces. So we may extend the theorem above to k-spaces as well. Observe that the theorem above implies both $sSet$ and k-spaces satisfy the conditions of Theorem 5.42 because Sym^n is a retract of Sym.

Theorem 5.51 *Let $\mathcal{M}$ be either simplicial sets or k-spaces. Then every Bousfield localization preserves GEMs.*

Thus, we have extended the result above and Theorem 4.B.4 in [21] to a wider class of topological spaces than spaces having the homotopy type of a CW complex.

5.7.3 Chain Complexes

When k is a field of characteristic zero, there are model structures on $Ch(k)_{\geq 0}$ and $Ch(k)$ where the weak equivalences are quasi-isomorphisms, the fibrations are morphisms that are degreewise surjections in positive degree, and the cofibrations are the degreewise split monomorphisms [34, Theorem 2.3.11]. All operads are Σ-cofibrant, hence all operad-algebras are preserved by any monoidal Bousfield localization. That cofibrant objects are flat is an easy exercise, using the observation that, for every cofibrant A, the map $A \otimes QX \to A \otimes X$ induced by cofibrant replacement (i.e. projective resolution) is a quasi-isomorphism.

This model structure is cofibrantly generated by morphisms of the form $I := \{i_n : S^{n-1} \to D^n \mid n \in \mathbb{N}\}$ and $J := \{j_n : 0 \to D^n \mid n \in \mathbb{N}\}$ where the *sphere* S^{n-1} is the chain complex that is R in degree $n-1$ (and 0 elsewhere), the *disk* D^n is the chain complex that is R in degrees $n-1$ and n (and 0 elsewhere) with the only non-zero differential being the identity on R, i_n is the natural inclusion, and j_n is the unique morphism from the zero module.

Proposition 5.52 *Let k be a field of characteristic zero. The only Bousfield localizations of $Ch(k)_{\geq 0}$ are truncations.*

Proof Since k is a field, all modules are free. Hence, all chain complexes are wedges of spheres, and morphisms between them are generated by zero maps and identities on copies of R. Inverting an identity map changes nothing. Inverting a zero map has the effect of killing some object, also known as a nullification.

Furthermore, killing the wedge sum k^2 in degree n is the same as killing k in degree n, and this also kills copies of k in higher degrees, because differentials are identity maps. Thus, the localization is completely determined by the lowest dimension in which the first nullification occurs. The localization is therefore equivalent to $0 \to V$ where V is the sphere on k in that dimension. $\square$

Corollary 5.53 *All Bousfield localizations of $Ch(k)_{\geq 0}$ are monoidal and hence preserve algebras over any operad P.*

Remark 5.54 For unbounded chain complexes, truncations need not preserve algebraic structure. For example, if $f : S^{-2} \to D^{-3}$ gets inverted then just as with the Postnikov Section, an algebra will be taken to an object with no unit.

Quillen proved in Proposition 2.1 of Appendix B of [47] that bounded chain complexes over a field of characteristic zero satisfies the commutative monoid axiom. The proof that all quasi-isomorphisms are closed under Sym^n goes via the cofiber and the 5-lemma on homology groups. The key observation is that $\mathrm{Sym}^n(-)$ commutes with homology. The same proof demonstrates that Sym^n preserves $\mathcal{C}$-local equivalences for all $L_\mathcal{C}$ as above. Hence, all Bousfield localizations of $Ch(k)_{\geq 0}$ preserve commutative differential graded algebras. Of course, this can also be seen directly from the description of $L_\mathcal{C}$ as a truncation.

5.7.4 Equivariant Spectra

We conclude this section by returning to our motivating example, Example 5.36. Throughout this section, G is a finite group, since otherwise we do not know how to transfer a semi-model structure to commutative equivariant ring spectra. Note that when it was discovered, Example 5.36 represented a potential gap in the proof of the Kervaire Invariant One Theorem (because the spectrum $\Omega = D^{-1}MU^{(4)}$ needed to be commutative for the computations in [31]). Thankfully, the following theorem from [30] demonstrates that Ω was indeed commutative.

Theorem 5.55 *Let G be a finite group. Let L be a localization of equivariant spectra. If for all L-acyclics Z and for all subgroups H, $N_H^G Z$ is L-acyclic, then for all commutative G-ring spectra R, $L(R)$ is a commutative G-ring spectrum.*

The hypothesis in this theorem is designed so that the proof in [19] regarding preservation of E_∞-structure under localization (i.e. via the skeletal filtration) may go through. We now specialize our preservation result to the context of G-spectra, by combining Theorem 5.3, Theorem 5.45, and Remark 5.47. Recall from Proposition 5.27 that monoidal localizations are stable. In order to have a transferred semi-model structure on commutative monoids, we need to work with either the positive stable model structure on G-spectra (of Theorem 14.2 of [43]), the positive flat stable model structure (of Theorem 2.3.27 of [54]), or the positive complete stable model structure (of Proposition B.4.1 of [31]). The proof that these model structures are left proper and cellular can be found in the appendix of [27].

Theorem 5.56 *Let $\mathcal{M}$ denote any of the positive model structure on G-spectra discussed above. Suppose $L_{\mathcal{C}}$ is a monoidal left Bousfield localization. Then the following are equivalent:*

1 $L_{\mathcal{C}}$ preserves commutative equivariant ring spectra,

2 $\mathrm{Sym}^n(-)$ preserves local equivalences between cofibrant objects, for all n,

3 $\mathrm{Sym}^n(-)$ takes maps in $\mathcal{C}$ to local equivalences, and

4 $\mathrm{Sym}^n(-)$ preserves L-acyclicity for all n.

That preservation of L-acyclics is the same as preservation of L-local equivalences as can be seen via the rectification axiom and the property that cofibrant objects are flat, but it is easier to observe that this equivalence holds for any stable localization in any stable model category (because consideration of cofibers allows one to reduce to the study of nullifications). In [30], several equivalent conditions are given in order for a localization to preserve commutative structure. Condition (4) above is equivalent to the condition that, for all L-acyclics Z and for all subgroups H, $N_H^G Z$ is L-acyclic. Hence, Theorem 5.56 sharpens Theorem 5.55 to make it an 'if and only if' result.

Another equivalent formulation states that preservation occurs whenever the functors $(E_G \Sigma_n)_+ \wedge_{\Sigma_n} (-)^{\wedge n}$ preserve L-acyclicity. This condition can be verified via the skeletal filtration of $E_G \Sigma_n$ into a homotopy colimit of Σ_n-free $G \times \Sigma_n$ sets T of the form $(G \times \Sigma_n)/\Gamma$ where Γ is the graph of a subgroup. This formulation of what is required for L to preserve commutativity is at the heart of the arguments in [31] and [10] and allows for the preservation machinery to be extended to N_∞-operads in [27]. The condition is analogous to the non-equivariant condition that functors $(E\Sigma_n)_+ \wedge_{\Sigma_n} (-)^{\wedge n}$ preserve L-acyclicity.

By Theorem 6.3 of [27], for any complete N_∞-operad P whose spaces have the homotopy type of $G \times \Sigma_n$-CW complexes, P-algebras are Quillen equivalent to commutative monoids. Hence, Theorem 5.48 implies the same conditions from Theorem 5.56 are equivalent to preservation of P-algebras for any (hence all) complete N_∞-operad P whose spaces have the homotopy type of $G \times \Sigma_n$-CW complexes. Preservation results for non-complete N_∞-operads and for $\mathcal{F}$-N_∞-operads can be found in Section 7 of [27].

5.8 Bousfield localization and the monoid axiom

Recall that the monoid axiom is required to transfer a full model structure to the category of monoids in a monoidal model category [50]. However, Theorem 5.24 demonstrates that there is a transferred semi-model structure even if the monoid axiom is not satisfied. It follows from Corollary 5.5, that our preservation results do not require $L_{\mathcal{C}}(\mathcal{M})$ to satisfy the monoid axiom. However, the monoid axiom is an important part of the study of monoidal model categories, with many applications beyond the ability to transfer a model structure to monoids, and in this section we provide a result that guarantees it holds on $L_{\mathcal{C}}(\mathcal{M})$.

We remark that Proposition 3.8 of [1] proves that $L_{\mathcal{C}}(\mathcal{M})$ inherits the monoid axiom from $\mathcal{M}$ if $L_{\mathcal{C}}$ takes a special form similar to localization at a homology theory. In contrast, our result will place no hypothesis on the maps in $\mathcal{C}$ at all, beyond our standing hypothesis that these maps are cofibrations. We additionally remark that [46] has independently considered the question of when Bousfield localization preserves the monoid axiom, towards the goal of rectification results in general categories of spectra.

In order to understand when Bousfield localization will preserve the monoid axiom we must introduce a definition, taken from [5]. Note that this is a different usage of the term h-cofibration than the usage in [19] where it means 'Hurewicz cofibration.' The meaning here is for 'homotopical cofibration' for reasons which will become clear.

Definition 5.57 A map $f : X \to Y$ is called an *h-cofibration* if the functor $f_! : X/\mathcal{M} \to Y/\mathcal{M}$ given by cobase change along f preserves weak equivalences. Formally, this means that in any diagram as below, in which both squares are pushout squares and w is weak equivalence, then w' is also a weak equivalence:

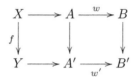

It is clear that any trivial cofibration is an h-cofibration, by the two out of three property. If $\mathcal{M}$ is left proper then any cofibration is an h-cofibration (because $A \to A'$ is automatically a cofibration if f is). In fact, the converse holds as well [5, Lemma 1.2] and [5, Lemma 1.3] proves that h-cofibrations are closed under composition, pushout, and finite coproduct.

If $\mathcal{M}$ is left proper. Proposition 1.5 in [5] proves that an h-cofibration is the same as a map f such that every pushout along f is a homotopy pushout (this version of the definition above was independently discovered in [56]) and also that f is an h-cofibration if and only if there is a factorization of f into a cofibration followed by a *cofiber equivalence* $w : W \to Y$. This means, for any map $g : W \to K$, the right-hand vertical map in the following pushout diagram is a weak equivalence:

$$
\begin{array}{ccc}
W & \longrightarrow & K \\
{\scriptstyle w}\downarrow & \searrow & \downarrow \\
X & \longrightarrow & T
\end{array}
$$

We will make use of these various properties of h-cofibrations in this section. The purpose for introducing h-cofibrations is to make the following definition, which should be thought of as saying that the cofibrations in $\mathcal{M}$ behave like inclusions of closed neighborhood deformation retracts of topological spaces.

Definition 5.58 $\mathcal{M}$ is said to be *h-monoidal* if for each (trivial) cofibration f and each object Z, $f \otimes Z$ is a (trivial) h-cofibration.

We will find conditions so that Bousfield localization preserves the property of being *h*-monoidal, and we will then use this to deduce when Bousfield localization preserves the monoid axiom. In [5], *h*-monoidality is verified for the model categories of topological spaces, simplicial sets, equivariant spaces, chain complexes over a field (with the projective model structure), symmetric spectra (with the stable projective model structure), and several other model categories not considered in this chapter. We now verify *h*-monoidality for the remaining model structures of interest in this chapter. We remind the reader that an *injective model structure* has weak equivalences and cofibrations defined levelwise, and fibrations defined by the right lifting property.

Proposition 5.59 *The following eight model structures on symmetric spectra are h-monoidal (4 stable and 4 unstable model structures):*

1 *The levelwise projective (stable) model structure (of Theorem 5.1.2 in [38], see also Proposition 1.14 of [5]).*
2 *The positive (stable) model structure (of Theorem 14.1 and 14.2 in [43]).*
3 *The flat (stable) model structure (of Proposition 2.2 and Theorem 2.4 in [51], there called the S-model structure).*
4 *The positive flat (stable) model structure (obtained by redefining the cofibrations from the model structure above to be isomorphisms in level 0, see Proposition 3.1 in [51]).*

Proof We appeal to Proposition 1.9 in [5], and make use of the injective (or injective stable for (5)-(8)) model structure on symmetric spectra, introduced in Definition 5.1.1 (resp. after Definition 5.3.6) of [38]. The references above prove that all eight of the model structures above are monoidal and that both injective model structures are left proper (e.g. because all objects are cofibrant). The final condition in Proposition 1.9 is that for any (trivial) cofibration f and any object X, the map $f \otimes X$ is a (trivial) cofibration in the corresponding injective model structure. The cofibration part of this is Proposition 4.15(i) in version 3 of Stefan Schwede's book project [48], since for all eight of the model structures above the cofibrations are contained in the flat cofibrations and for any X the map $\varnothing \to X$ is an injective (a.k.a. levelwise) cofibration. The trivial cofibration part is Proposition 4.15(iv) in [48], which includes statements for both levelwise and stable weak equivalences. □

We turn now to orthogonal and equivariant orthogonal spectra. To mimic constructions from the theory of symmetric spectra built on simplicial sets, we must build orthogonal spectra on the category Top_Δ of Δ-generated spaces (pointed, of course), which we now briefly discuss. The category of Δ-generated spaces is a locally presentable category of topological spaces that admits a combinatorial model structure [17, 20]. Essentially, these are spaces that are colimits of simplices, and the category of Δ-generated spaces is the final closure of the subcategory Δ of spaces consisting of the topological simplices Δ^n. The construction of the category of Δ-generated spaces is analogous to Vogt's construction of the category of compactly generated spaces as the colimit-closure of the category of compact Hausdorff spaces [55], and so proofs set in compactly generated spaces have analogues in Δ-generated spaces.

In both settings, care must be taken, as certain common constructions can take you out of the subcategory. For example, when computing limits, one needs Vogt's k-ification functor to get back into the category of compactly generated spaces, and there is an analogous "delta-fication" functor to get back into the category of Δ-generated spaces, used when taking limits or passing to subspaces takes you out of the category of Δ-generated spaces. For examples where this kind of care is taken, see [4, 17, 24].

For the two proofs below, these kinds of issues did not arise, for several reasons. First, [38] works at the level of simplices, and thus translates immediately to the setting of Δ-generated spaces. Secondly, Δ-generated spaces are closed under the smash product, fixed points commute with pushouts and filtered colimits along closed inclusions, and fixed points commute with the smash product. We note that in general, closed subspaces of Δ-generated spaces need not be Δ-generated, though this can be arranged by enlarging Δ. Nevertheless, the H-fixed points of a space in Top_Δ^G (or in Top_Δ^H) is a Δ-generated space, since the H-fixed points functor is a right adjoint (by the special adjoint functor theorem) to the functor $Top_\Delta \to Top_\Delta^H$ taking any space to the H-space with trivial H-action.

Let Sp_Δ^O denote orthogonal spectra built on Δ-generated spaces, i.e. where each space in the spectrum is a Δ-generated space. Let G be a compact Lie group and let GSp_Δ^O denote G-equivariant orthogonal spectra built on Δ-generated spaces. We first need a lemma regarding the existence of injective model structures.

Lemma 5.60 *The following model structures exist and are left proper and combinatorial: the levelwise injective model structure on Sp_Δ^O, the* stable injective model structure on Sp^O_Δ, the levelwise injective model structure on GSp^O_Δ, and the stable injective model structure on GSp^O_Δ.

Proof Left properness will be inherited from Δ-generated spaces, where it is verified just as for topological spaces [17]. For the existence of these model structures, we proceed as in Theorem 5.1.2 and Lemma 5.1.4 of [38], starting with the level model structures. One chooses the set of generating cofibrations C to consist of one representative from each isomorphism class of cofibrations $i : X \to Y$ where Y is a cell spectrum built with countably many cells. This size restriction is required to ensure that C is a *set*. One similarly defines a set tC of generating trivial cofibrations.

As usual for injective model structures, it is difficult to precisely describe the morphisms in C or tC, but one can prove that morphisms with the right lifting property with respect to C are injective fibrations and level equivalences, just as in [38, Lemma 5.1.4]. The idea is to construct a lift bit-by-bit (that is, cell-by-cell), and use of Zorn's Lemma to construct a full lift. In the case of symmetric spectra, this involves looking at individual simplicies not yet covered by the partial lift, and generating a countable subspectrum from such a simplex.

In the Δ-generated case, one does the same thing with individual cells, using that the category of Δ-generated spaces is the final closure of Δ, and hence that arguments at the level of simplices translate directly to arguments about the corresponding Δ-generated spaces. The rest of Lemma 5.1.4 goes through mutatis mutandis, using formal properties of model categories, using properties of topological fibrations, and using Lemma 12.2 in [43] when checking that injective cofibrations are closed under smashing with an arbitrary object.

Together with the fact that a category of spectra built on a locally presentable category is again locally presentable, this proves the level injective model structures are combinatorial. The stable injective structures are obtained by Bousfield localization in the usual way, which exists because the levelwise structures are left proper and combinatorial. □

Proposition 5.61 *Let G be a compact Lie group and fix a universe $\mathcal{U}$, that we take to mean a G-universe when working equivariantly. Assume all spectra are built on Δ-generated spaces. The following eight model structures (4 stable and 4 unstable) are h-monoidal:*

1 The levelwise projective (stable) model structure on G-equivariant orthogonal spectra (of Theorem III.2.4 and III.4.2 in [42]).

2 The positive (stable) model structure on G-equivariant orthogonal spectra (of Theorem III.2.10 and III.5.3 in [42]).

3 The flat (stable) model structure on G-equivariant orthogonal spectra (of Theorem 2.3.13 of in [54]).

4 The positive flat (stable) model structure on G-equivariant orthogonal spectra (obtained by redefining the cofibrations from the model structure above to be isomorphisms in level 0, of Theorem 2.3.27 in [54]).

Taking G to be the trivial group yields eight model structures on orthogonal spectra, that this proposition proves are h-monoidal. They are the levelwise projective (stable) model structure (of Theorem 6.5 and Theorem 9.2 in [43]), the positive (stable) model structure on orthogonal spectra (of Theorem 14.1 and 14.2 in [43]), the flat (stable) model structure on orthogonal spectra (of Proposition 1.3.5 and 2.3.27 in [54]), and the positive flat (stable) model structure on orthogonal spectra (of Proposition 1.3.10 and Theorem 2.3.27 in [54]).

Proof of Proposition 5.61   The proof proceeds just as it does for Proposition 5.59, i.e. by comparison to the injective (stable) model structures in each of these settings. For the statement that for any cofibration f and any object X, the map $f \otimes X$ is a cofibration in the corresponding injective model structure, we appeal to Lemma 12.2 of [43] (which works equally well in the equivariant context). Finally, we turn to the statement that for any trivial cofibration f and any object X, the map $f \otimes X$ is a weak equivalence in the corresponding injective model structure. For the levelwise model structures above this property is inherited from spaces, e.g. by Lemma 12.2 in [43]. For the stable model structures we appeal to the monoid axiom on all of the model structures in the theorem and to the fact that projective (stable) equivalences are the same as injective (stable) equivalences. The monoid axiom has been verified in [54] for all these model structures by Theorems 1.2.54 and 1.2.57 (both originally proven in [43]), 1.3.10, 2.2.46 and 2.2.50 (both originally from [42]), and 2.3.27. $\square$

As previously mentioned, for the two proofs above, no constructions appeared that could take us out of the category of Δ-generated spaces. However, to prove other properties of Top_{Δ}^{G} and spectra built on them, such constructions do appear. The author intends to write a separate manuscript gathering together the required properties needed to work with (G-equivariant) Δ-generated spaces, in much the same way that May [44] and Vogt [55] did for compactly generated and compactly generated weak Hausdorff spaces, and to prove further properties of spectra built on them.

We return now to the question of the monoid axiom. It is proven in Proposition 2.5 of [5] that if $\mathcal{M}$ is left proper, h-monoidal, and the weak equivalences in $(\mathcal{M} \otimes I)$-cell are closed under transfinite composition, then $\mathcal{M}$ satisfies the monoid axiom. We will use this to find conditions on $\mathcal{M}$ so that $L_{\mathcal{C}}(\mathcal{M})$ satisfies the monoid axiom. First, we improve Proposition 2.5 from [5] by replacing the third condition with the hypothesis that the (co)domains of I are finite relative to the class of h-cofibrations (in the sense of Section 7.4 of [34]).

Proposition 5.62 *Suppose $\mathcal{M}$ is cofibrantly generated, left proper, h-monoidal, and the (co)domains of I are finite relative to the class of h-cofibrations. Then $\mathcal{M}$ satisfies the monoid axiom.*

Proof We follow the proof of Proposition 2.5 in [5]. Consider the class $\{f \otimes Z \mid Z \in \mathcal{M}, f \in J\}$. As $\mathcal{M}$ is h-monoidal, this is a class of trivial h-cofibrations. By Lemma 1.3 in [5], h-cofibrations are closed under pushout. By Lemma 1.6 in [5], because $\mathcal{M}$ is left proper, trivial h-cofibrations are closed under pushouts (e.g. because weak equivalences are closed under homotopy pushout). In order to prove $\{f \otimes Z \mid Z \in \mathcal{M}, f \in J\}$-cell is contained in the weak equivalences of $\mathcal{M}$ we must only prove that transfinite compositions of trivial h-cofibrations are weak equivalences.

Consider a λ-sequence $A_0 \to A_1 \to \cdots \to A_\lambda$ of trivial h-cofibrations. Let j_β denote the map $A_\beta \to A_{\beta+1}$ in this λ-sequence. As in Proposition 17.9.4 of [32] we may construct a diagram

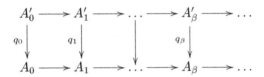

in which each A'_β is cofibrant, all the maps $A'_\beta \to A_\beta$ are trivial fibrations, and all the maps $A'_\beta \to A'_{\beta+1}$ are trivial cofibrations. Construction of this diagram proceeds by applying the cofibration-trivial fibration factorization iteratively to every composition $j_\beta \circ q_\beta : A'_\beta \to A_\beta \to A_{\beta+1}$ in order to construct $A'_{\beta+1}$. As j_β and q_β are both weak equivalences, so is their composite and so the cofibration $A'_\beta \to A'_{\beta+1}$ produced by the cofibration-trivial fibration factorization is a weak equivalence by the two out of three property.

We now show that the map $q_\lambda : A'_\lambda \to A_\lambda$ is a weak equivalence, following the approach of Lemma 7.4.1 in [34]. Consider the lifting problem

Where f is in the set I of generating cofibrations. Because the domains and codomains of maps in I are finitely presented we know that the map $X \to A'_\lambda$ factors through some finite stage A'_n. Similarly, $Y \to A_\lambda$ factors through some finite stage A_m. Let $k = \max(n, m)$. The map $A'_k \to A_k$ is a trivial fibration so there is a lift $g : Y \to A'_k$. Define $h : Y \to A'_\lambda$ as the composite with $A'_k \to A'_\lambda$.

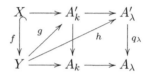

Both triangles in the left-hand square commute by definition of lift. The triangle featuring g and h commutes because it is a composition. So the triangle featuring f and h commutes. The right-hand square commutes by construction of A'_λ and A_λ, so the trapezoid containing g and q_λ commutes. Thus, the triangle featuring h and q_λ commutes.

The existence of this lift h for all $f \in I$ proves that $A'_\lambda \to A_\lambda$ is a trivial fibration. Now consider that transfinite compositions of trivial cofibrations are always trivial cofibrations, so $A'_0 \to A'_\lambda$ is a weak equivalence. Furthermore, the vertical maps $q_0 : A'_0 \to A_0$ and $q_\lambda : A'_\lambda \to A_\lambda$ are trivial fibrations. So by the two out of three property, the map $A_0 \to A_\lambda$ is a weak equivalence as required.   □

It is shown in [5] that the compactness hypothesis of the proposition is satisfied for topological spaces, simplicial sets, equivariant and motivic spaces, and chain complexes. Similarly, it holds for all our categories of structured spectra because the sphere spectrum is $\aleph_0$-compact as a spectrum. Lastly, it holds for all the stable analogues of these structures because the compactness hypothesis is automatically preserved by any Bousfield localization (the set of generating cofibrations of $L_\mathcal{C}(\mathcal{M})$ is simply I again).

Remark 5.63 The proof above only uses the fact that the maps j_β were h-cofibrations in order to factor $Y \to A_\lambda$ through some finite stage. So if the (co)domains of I are finite relative to the class of weak equivalences then the proof above demonstrates that weak equivalences are preserved under transfinite composition [34, Corollary 7.4.2]. This property was already known classically for $sSet$ and $Ch(k)$, but need not hold for a general model category.

Proposition 5.64   *Suppose $\mathcal{M}$ a cofibrantly generated, h-monoidal, left proper, and that the (co)domains of I are cofibrant and are finite relative to the class of h-cofibrations. Suppose cofibrant objects are flat. Let $L_\mathcal{C}$ be a monoidal Bousfield localization. Then $L_\mathcal{C}(\mathcal{M})$ is h-monoidal.*

Proof   Suppose $f : A \to B$ is a cofibration in $L_\mathcal{C}(\mathcal{M})$ and Z is any object of $L_\mathcal{C}(\mathcal{M})$. We must show $f \otimes Z$ is an h-cofibration in $L_\mathcal{C}(\mathcal{M})$. Because $L_\mathcal{C}(\mathcal{M})$ is left proper, Proposition 1.5 in [5] reduces us to proving that there is a factorization of $f \otimes Z$ into a cofibration followed by a cofiber equivalence $w : X \to B \otimes Z$, i.e. for any map $g : X \to K$ the right-hand vertical map in the following pushout diagram is a $\mathcal{C}$-local equivalence:

Because f is a cofibration in $\mathcal{M}$, the h-monoidality of $\mathcal{M}$ guarantees us that $f \otimes Z$ is an h-cofibration in $\mathcal{M}$. Apply the cofibration-trivial fibration factorization in $\mathcal{M}$. Note that this is also a cofibration-trivial fibration factorization of $f \otimes Z$ in $L_\mathcal{C}(\mathcal{M})$ because cofibrations and trivial fibrations in the two model categories agree. The resulting $w : X \to B \otimes Z$ is a trivial fibration in either model structure. Because $\mathcal{M}$ is left proper we know that the map w is a cofiber equivalence in $\mathcal{M}$ by Proposition 1.5 in [5] applied to the h-cofibration $f \otimes Z$. So in any pushout diagram as above the map $K \to T$ is a weak equivalence in $\mathcal{M}$, hence in $L_\mathcal{C}(\mathcal{M})$. Thus, w is a cofiber equivalence in $L_\mathcal{C}(\mathcal{M})$ and its existence proves $f \otimes Z$ is an h-cofibration in $L_\mathcal{C}(\mathcal{M})$.

Now suppose f were a trivial cofibration in $L_\mathcal{C}(\mathcal{M})$ to start. We must show that $f \otimes Z$ is a $\mathcal{C}$-local equivalence. We do this first in the case where f is a generating trivial cofibration. By hypothesis, A and B are cofibrant. Apply cofibrant replacement to Z:

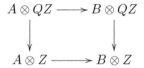

The fact that cofibrant objects are flat in $L_\mathcal{C}(\mathcal{M})$ implies the vertical maps are $\mathcal{C}$-local equivalences (because A and B are cofibrant) and that the top horizontal map is a $\mathcal{C}$-local equivalence (because QZ is cofibrant). By the two out of three property the bottom horizontal map is a $\mathcal{C}$-local equivalence.

By Lemma 1.3 in [5], the class of h-cofibrations is closed under cobase change and retracts. By Lemma 1.6, the class of trivial h-cofibrations is closed under cobase change (because $L_\mathcal{C}(\mathcal{M})$ is left proper). Weak equivalences are always closed under retract. Finally, by Proposition 5.62 the class of trivial h-cofibrations is closed under transfinite composition by our compactness hypothesis on $\mathcal{M}$ (equivalently, on $L_\mathcal{C}(\mathcal{M})$). So for a general f in the trivial cofibrations of $L_\mathcal{C}(\mathcal{M})$, realize f as a retract of $g \in J_\mathcal{C}$-cell, so that $g \otimes Z$ is a transfinite composite of pushouts of maps of the form $j \otimes Z$ for $j \in J_\mathcal{C}$. We have just proven that all $j \otimes Z$ are trivial h-cofibrations and closure properties imply $g \otimes Z$ and hence $f \otimes Z$ are trivial h-cofibrations as well.   □

Theorem 5.65 *Suppose $\mathcal{M}$ is a cofibrantly generated, left proper, h-monoidal model category such that the (co)domains of I are cofibrant and are finite relative to the h-cofibrations and cofibrant objects are flat. Then for any monoidal Bousfield localization $L_\mathcal{C}$, the model category $L_\mathcal{C}(\mathcal{M})$ satisfies the monoid axiom.*

Proof Apply Proposition 5.62 to the category $L_\mathcal{C}(\mathcal{M})$. By Proposition 5.64, $L_\mathcal{C}(\mathcal{M})$ is h-monoidal. It is left proper because $\mathcal{M}$ is left proper. To verify the monoid axiom, consider a λ-sequence of maps that are pushouts of maps in $\{f \otimes Z \mid f$ is a trivial cofibration in $L_\mathcal{C}(\mathcal{M})\}$. Such maps are h-cofibrations in $\mathcal{M}$ because $\mathcal{M}$ is h-monoidal, f is a cofibration in $\mathcal{M}$, and h-cofibrations are closed under pushout. Thus, the hypothesis that the (co)domains of I are finite relative to the h-cofibrations in $\mathcal{M}$ is sufficient to construct the lift in Proposition 5.62 and to prove the transfinite composition part of the proof of the monoid axiom.   □

References

[1] D. Barnes. Splitting monoidal stable model categories. *J. Pure Appl. Algebra*, 213(5):846–856, 2009.

[2] D. Barnes and C. Roitzheim. Stable left and right Bousfield localisations. *Glasg. Math. J.*, 56(1):13–42, 2014.

[3] C. Barwick. On left and right model categories and left and right Bousfield localizations. *Homology Homotopy Appl.*, 12(2):245–320, 2010.

[4] C. Barwick. The fundamental groupoid and the Postnikov tower. Available from https://www.maths.ed.ac.uk/~cbarwick/papers/post-notes.pdf, 2014.

[5] M. A. Batanin and C. Berger. Homotopy theory for algebras over polynomial monads. *Theory Appl. Categ.*, 32:Paper No. 6, 148–253, 2017.

[6] M. A. Batanin and D. White. Left Bousfield localization and Eilenberg-Moore categories. available as arXiv:1606.01537, to appear in *Homology, Homotopy Appl.*, 2016.

[7] M. A. Batanin and D. White. Homotopy theory of algebras of substitudes and their localisation. available as arXiv:2001.05432, 2020.

[8] M. A. Batanin and D. White. Left Bousfield localization without left properness. available as arXiv:2001.03764, 2020.

[9] C. Berger and I. Moerdijk. Axiomatic homotopy theory for operads. *Comment. Math. Helv.*, 78(4):805–831, 2003.

[10] A. J. Blumberg and M. A. Hill. Operadic multiplications in equivariant spectra, norms, and transfers. *Adv. Math.*, 285:658–708, 2015.

[11] A. K. Bousfield. The localization of spaces with respect to homology. *Topology*, 14:133–150, 1975.

[12] A. K. Bousfield. The localization of spectra with respect to homology. *Topology*, 18(4):257–281, 1979.

[13] A. K. Bousfield. Homotopical localizations of spaces. *Amer. J. Math.*, 119(6):1321–1354, 1997.

[14] C. Casacuberta, J. J. Gutiérrez, I. Moerdijk, and R. M. Vogt. Localization of algebras over coloured operads. *Proc. Lond. Math. Soc. (3)*, 101(1):105–136, 2010.

[15] C. Casacuberta, J. L. Rodríguez, and J. Tai. Localizations of abelian Eilenberg–Mac Lane spaces of finite type. *Algebr. Geom. Topol.*, 16(4):2379–2420, 2016.

[16] D. Dugger. Combinatorial model categories have presentations. *Adv. Math.*, 164(1):177–201, 2001.

[17] D. Dugger. Notes on delta-generated spaces. preprint available electronically from http://math.uoregon.edu/ ddugger/delta.html, 2003.

[18] W. G. Dwyer and J. Spaliński. Homotopy theories and model categories. In *Handbook of algebraic topology*, pages 73–126. North-Holland, Amsterdam, 1995.

[19] A. D. Elmendorf, I. Kriz, M. A. Mandell, and J. P. May. *Rings, modules, and algebras in stable homotopy theory*, volume 47 of *Mathematical Surveys and Monographs*. American Mathematical Society, Providence, RI, 1997. With an appendix by M. Cole.

[20] L. Fajstrup and J. Rosický. A convenient category for directed homotopy. *Theory Appl. Categ.*, 21:No. 1, 7–20, 2008.

[21] E. D. Farjoun. *Cellular spaces, null spaces and homotopy localization*, volume 1622 of *Lecture Notes in Mathematics*. Springer-Verlag, Berlin, 1996.

[22] H. Fausk and D. C. Isaksen. t-model structures. *Homology Homotopy Appl.*, 9(1):399–438, 2007.

[23] B. Fresse. *Modules over operads and functors*, volume 1967 of *Lecture Notes in Mathematics*. Springer-Verlag, Berlin, 2009.

[24] P. Gaucher. Homotopical interpretation of globular complex by multi-pointed d-space. *Theory Appl. Categ.*, 22:588–621, 2009.

[25] S. Gorchinskiy and V. Guletskiĭ. Symmetric powers in abstract homotopy categories. *Adv. Math.*, 292:707–754, 2016.

[26] J. J. Gutiérrez, O. Röndigs, M. Spitzweck, and P. A. Ø stvær. Motivic slices and coloured operads. *J. Topol.*, 5(3):727–755, 2012.

[27] J. J. Gutiérrez and D. White. Encoding equivariant commutativity via operads. *Algebr. Geom. Topol.*, 18(5):2919–2962, 2018.

[28] J. E. Harper. Homotopy theory of modules over operads and non-Σ operads in monoidal model categories. *J. Pure Appl. Algebra*, 214(8):1407–1434, 2010.

[29] M. A. Hill and M. J. Hopkins. Localizations of equivariant commutative rings. Oberwolfach Reports 46/2011, 2640–2643, 2011.

[30] M. A. Hill and M. J. Hopkins. Equivariant multiplicative closure. In *Algebraic topology: applications and new directions*, volume 620 of *Contemp. Math.*, pages 183–199. Amer. Math. Soc., Providence, RI, 2014.

[31] M. A. Hill, M. J. Hopkins, and D. C. Ravenel. On the nonexistence of elements of Kervaire invariant one. *Ann. of Math. (2)*, 184(1):1–262, 2016.

[32] P. S. Hirschhorn. *Model categories and their localizations*, volume 99 of *Mathematical Surveys and Monographs*. American Mathematical Society, Providence, RI, 2003.

[33] M. Hovey. Monoidal model categories. reprint available electronically from http://arxiv.org/abs/math/9803002, 1998.

[34] M. Hovey. *Model categories*, volume 63 of *Mathematical Surveys and Monographs*. American Mathematical Society, Providence, RI, 1999.

[35] M. Hovey. Spectra and symmetric spectra in general model categories. *J. Pure Appl. Algebra*, 165(1):63–127, 2001.

[36] M. Hovey. Homotopy theory of comodules over a Hopf algebroid. In *Homotopy theory: relations with algebraic geometry, group cohomology, and algebraic K-theory*, volume 346 of *Contemp. Math.*, pages 261–304. Amer. Math. Soc., Providence, RI, 2004.

[37] M. Hovey, J. H. Palmieri, and N. P. Strickland. Axiomatic stable homotopy theory. *Mem. Amer. Math. Soc.*, 128(610):x+114, 1997.

[38] M. Hovey, B. Shipley, and J. Smith. Symmetric spectra. *J. Amer. Math. Soc.*, 13(1):149–208, 2000.

[39] M. Hovey and D. White. An alternative approach to equivariant stable homotopy theory. *Tbilisi Math. Journal*, Special Issue on Homotopy Theory, Spectra, and Structured Ring Spectra, 51–69, 2020.

[40] L. G. Lewis, Jr., J. P. May, M. Steinberger, and J. E. McClure. *Equivariant stable homotopy theory*, volume 1213 of *Lecture Notes in Mathematics*. Springer-Verlag, Berlin, 1986. With contributions by J. E. McClure.

[41] C. Malkiewich, M. Merling, D. White, F. L. Wolcott, and C. Yarnall. The user's guide project: Giving experiential context to research papers. *Journal of Humanistic Mathematics*, 5(2):186–188, 2015.

[42] M. A. Mandell and J. P. May. Equivariant orthogonal spectra and *S*-modules. *Mem. Amer. Math. Soc.*, 159(755):x+108, 2002.

[43] M. A. Mandell, J. P. May, S. Schwede, and B. Shipley. Model categories of diagram spectra. *Proc. London Math. Soc. (3)*, 82(2):441–512, 2001.

[44] J. P. May. *The geometry of iterated loop spaces*. Springer-Verlag, Berlin-New York, 1972. Lectures Notes in Mathematics, Vol. 271.

[45] J. E. McClure. E_∞-ring structures for Tate spectra. *Proc. Amer. Math. Soc.*, 124(6):1917–1922, 1996.

[46] D. Pavlov and J. Scholbach. Homotopy theory of symmetric powers. *Homology Homotopy Appl.*, 20(1):359–397, 2018.

[47] D. Quillen. Rational homotopy theory. *Ann. of Math. (2)*, 90:205–295, 1969.

[48] S. Schwede. An untitled book project about symmetric spectra. preprint available electronically from http://www.math.uni-bonn.de/people/schwede/SymSpec-v3.pdf, 2012.

[49] S. Schwede. Lectures on equivariant stable homotopy theory. preprint available electronically from http://www.math.uni-bonn.de/people/schwede/equivariant.pdf, 2020.

[50] S. Schwede and B. E. Shipley. Algebras and modules in monoidal model categories. *Proc. London Math. Soc. (3)*, 80(2):491–511, 2000.

[51] B. E. Shipley. A convenient model category for commutative ring spectra. In *Homotopy theory: relations with algebraic geometry, group cohomology, and algebraic K-theory*, volume 346 of *Contemp. Math.*, pages 473–483. Amer. Math. Soc., Providence, RI, 2004.

[52] C. Simpson. *Homotopy theory of higher categories*, volume 19 of *New Mathematical Monographs*. Cambridge University Press, Cambridge, 2012.

[53] M. Spitzweck. Operads, algebras and modules in general model categories. preprint available electronically from http://arxiv.org/abs/math/0101102, 2001.

[54] M. Stolz. Equivariant structure on smash powers of commutative ring spectra. Ph.D. thesis available electronically from https://web.math.rochester.edu/people/faculty/doug/otherpapers/mstolz.pdf, 2011.

[55] Rainer M. Vogt. Convenient categories of topological spaces for homotopy theory. *Arch. Math. (Basel)*, 22:545–555, 1971.

[56] D. White. Monoidal bousfield localizations and algebras over operads. Ph.D. thesis – Wesleyan University, 2014.

[57] D. White. A user's guide: Monoidal bousfield localizations and algebras over operads. *Enchiridion: Mathematical User's Guides*, 1, 2015.

[58] D. White. Model structures on commutative monoids in general model categories. *J. Pure Appl. Algebra*, 221(12):3124–3168, 2017.

[59] D. White and D. Yau. Right Bousfield localization and Eilenberg-Moore categories. available as arXiv:1609.03635, 2016.

[60] D. White and D. Yau. Smith ideals of operadic algebras in monoidal model categories. available as arXiv:1703.05377, 2017.

[61] D. White and D. Yau. Bousfield localization and algebras over colored operads. *Appl. Categ. Structures*, 26(1):153–203, 2018.

[62] D. White and D. Yau. Homotopical adjoint lifting theorem. *Appl. Categ. Structures*, 27(4):385–426, 2019.

[63] D. White and D. Yau. Right Bousfield localization and operadic algebras. *Tbilisi Math. Journal*, Special Issue on Homotopy Theory, Spectra, and Structured Ring Spectra, 71–118, 2020.

6
Stratification and Duality for Unipotent Finite Supergroup Schemes

Dave Benson[a]

Srikanth B. Iyengar[b]

Henning Krause[c]

Julia Pevtsova[d]

Abstract

We survey some methods developed in a series of papers, for classifying localising subcategories of tensor triangulated categories. We illustrate these methods by proving a new theorem, providing such a classification in the case of the stable module category of a unipotent finite supergroup scheme.

To John Greenlees on his 60th birthday.

6.1 Introduction

John Greenlees' influence on mathematics is reflected throughout this volume, and our own work has benefitted enormously from his insights. We should like to mention in particular his collaboration with May [50] and Dwyer [41] on derived completions and local cohomology; on the local cohomology spectral sequence in the context of group cohomology [48] and his subsequent work with Lyubeznik [49]; and on duality in algebra and topology; in particular, his work with Benson [12, 13, 14], and with Dwyer and Iyengar [42].

A broad framework for understanding localisation and duality was developed in a series of papers [17]–[26]. Originally geared towards applications to representation theory of finite groups and finite group schemes,

[a] Institute of Mathematics, University of Aberdeen
[b] Department of Mathematics, University of Utah
[c] Fakultät für Mathematik, Universität Bielefeld
[d] Department of Mathematics, University of Washington

the framework has been applied in a number of other areas, such as commutative algebra [35], ring spectra [6], modules over algebras of cochains [15, 18], and equivariant KK-theory [34]. This theory is closely related to that of Balmer [3, 4, 5], but the point of view is different.

The purpose of this chapter is to give an outline of the theory, by way of explaining how it applies in the case of the stable module category of a unipotent finite supergroup scheme. In particular, we shall prove the following theorem.

Theorem 6.1   *Let G be a unipotent finite supergroup scheme over a field k of positive characteristic $p \geqslant 3$. The theory of support gives a one to one correspondence between the localising subcategories of $\mathrm{StMod}(kG)$ and subsets of $\mathrm{Proj}\, H^{*,*}(G, k)$.*

Here $\mathrm{StMod}(kG)$ denotes the stable module category of kG-modules, $H^{*,*}(G, k)$ the cohomology ring of G, and $\mathrm{Proj}\, H^{*,*}(G, k)$ its projective spectrum. As a consequence, we get a classification of the thick subcategories of $\mathrm{stmod}(kG)$, the stable category of finitely generated kG-modules; they are in one to one correspondence with the specialisation closed subsets of $\mathrm{Proj}\, H^{*,*}(G, k)$.

Supergroup schemes are introduced Section 6.6 and the result above is proved in Section 6.10. The statement above speaks of "the" theory of support, but in fact there are at least two, quite distinct, notions of support in this context. One is obtained from specialising the support theory introduced in [17], which is based on cohomology and applies to any compactly generated triangulated category with a ring action; in this case $\mathrm{StMod}(kG)$ with the canonical action of $H^{*,*}(G, k)$. The other notion of support is inspired by the theory of π-points introduced by Friedlander and Pevtsova [46] in the context of finite group schemes. That these two notions of support coincide for modules over finite supergroup schemes is one of the crucial steps in proving the theorem above.

Theorem 6.1 was first proved, with no restrictions on the prime p, for finite groups [19], using ideas from commutative algebra, mainly the Bernstein-Gelfand-Gelfand correspondence. A new proof, from a more representation theoretic perspective, was given in [22], exploiting ideas from [46] and a concept of cosupport for representations, which is related to derived completions introduced by Greenlees and May [50]. These ideas also played a pivotal role in our proof of Theorem 6.1 for the case of finite group schemes [23].

Theorem 6.1 is a culmination of our work reported in [24, 26, 28]. We sketch the details of how to put these pieces together in Section 6.9.

This work takes as a starting point the applications of the theory of super polynomial functors to representations and cohomology of finite supergroup schemes, developed in a series of papers by Drupieski and Kujawa [37, 38, 40]. Their work [36] represents a development parallel to our own. The main difference is that we treat both connected and non-connected cases of unipotent supergroup schemes and are interested in infinite dimensional kG-modules. In [36] one can find many intricate results, examples and calculations for finite dimensional modules over connected group schemes.

Finally, in Section 6.11 we give a brief overview of the local duality for StMod(kG), summarised in the following result.

Theorem 6.2 *Let G be a finite supergroup scheme over a field k of positive characteristic $p \geqslant 3$. For each $\mathfrak{p} \in \operatorname{Proj} H^{*,*}(G, k)$, the corresponding localising subcategory $\Gamma_{\mathfrak{p}} \operatorname{StMod}(kG)$ satisfies local duality.*

Observe that, in contrast with Theorem 6.1, here G is not restricted to be unipotent. For finite groups, such a statement was conjectured by Benson [9] and proved by Benson and Greenlees [14]; see also [10]. In [25] we gave a new proof of their result covering also the case of finite group schemes. It turns out that local duality is a general feature of Gorenstein algebras, and can be viewed as an extension of a duality theorem due to Auslander [1] and Buchweitz [30] generalising Tate duality for finite groups. All this is explained in [27]. One can view the result above as a variation, dealing with graded Gorenstein algebras.

In what follows, we have tried to present the development of the theory from elementary abelian p-groups to finite groups to finite group schemes to finite supergroups schemes. We focus mostly on stratification, and even then cannot hope to present a complete account of the theory in a short survey. Thus the aim is to give just a flavour of numerous techniques and developments that happened in the theory of support varieties in (relatively) recent years many of which were influenced by John Greenlees' work.

Acknowledgements

It is a pleasure to acknowledge the support provided by the American Institute of Mathematics in San Jose, California, through their "SQuaREs" program. We also acknowledge the National Science Foundation under Grant No. DMS-1440140 which supported three of the authors (DB, SBI, JP) while they were in residence at the Mathematical Sciences Research

Institute in Berkeley, California, during the Spring 2018 semester. Finally, two of the authors (DB, JP) are grateful for hospitality provided by City, University of London.

SBI was partly supported by NSF grant DMS-2001368. JP was partly supported by NSF grants DMS-1501146, DMS-1901854, and a Brian and Tiffinie Pang faculty fellowship.

6.2 Local cohomology and support

Let T be a triangulated category, with shift Σ, admitting all small coproducts, and compactly generated. The examples we have in mind come from algebra, topology and geometry. The *centre* $Z^*(\mathsf{T})$ of T is the graded ring whose degree n part consists of the natural transformations η from the identity functor to Σ^n satisfying $\eta\Sigma = (-1)^n \Sigma\eta$. These form a graded commutative ring which is usually not Noetherian. We assume that we are given a graded commutative Noetherian ring R together with a homomorphism of rings $R \to Z^*(\mathsf{T})$. This is called an *action* of R on T. For X, Y in T set

$$\operatorname{Hom}^*_{\mathsf{T}}(X, Y) := \bigoplus_{n \in \mathbb{Z}} \operatorname{Hom}_{\mathsf{T}}(X, \Sigma^n Y) \,;$$

this has a structure of an R-module, compatible with morphisms in T.

A subset V of a topological space X is *specialisation closed* if V contains the closure of its points. We write $\operatorname{Spec} R$ for the homogeneous primes ideals of R, with the Zariski topology. For each specialisation closed subset V of $\operatorname{Spec} R$, the subcategory T_V of V-*torsion* objects in T is the full subcategory consisting of the objects X such that $\operatorname{Hom}^*_{\mathsf{T}}(C, X)_{\mathfrak{p}} = 0$ for each compact object C in T and each $\mathfrak{p} \notin V$. For X in T there is a functorial triangle

$$\Gamma_V X \to X \to L_V X$$

where $\Gamma_V X$ is in T_V, and $L_V X$ admits no non-zero maps from any object in T_V.

Definition 6.3   Given a $\mathfrak{p}$ in $\operatorname{Spec} R$ choose specialisation closed subsets V and W with $V \not\subseteq W$ and $V \subseteq W \cup \{\mathfrak{p}\}$, and define

$$\Gamma_{\mathfrak{p}} X := \Gamma_V L_W X = L_W \Gamma_V X.$$

This turns out to be independent of choice of V and W satisfying these conditions, and defines an idempotent functor $\Gamma_{\mathfrak{p}} : \mathsf{T} \to \mathsf{T}$ which may be thought of as isolating the layer of T corresponding to the prime $\mathfrak{p}$, and consisting of the objects "supported at $\mathfrak{p}$." The functor $\Gamma_{\mathfrak{p}}$ has a right adjoint, which we denote $\Lambda^{\mathfrak{p}}$; it has to do with completions along $\mathfrak{p}$, and plays an equally important role in the theory; for details, see [17, 16].

Definition 6.4 The *support* and *cosupport* of an object X in T are the subsets

$$\mathrm{supp}_R(X) := \{\mathfrak{p} \in \mathrm{Spec}\, R \mid \Gamma_{\mathfrak{p}} X \neq 0\}$$
$$\mathrm{cosupp}_R(X) := \{\mathfrak{p} \in \mathrm{Spec}\, R \mid \Lambda^{\mathfrak{p}} X \neq 0\}$$

It is not hard to prove that both support and cosupport detect zero objects: an object X in T is 0 if and only if $\mathrm{supp}_R(X) = \varnothing$, if and only if $\mathrm{cosupp}_R(X) = \varnothing$.

We focus mostly on support, but cosupport resurfaces in Section 6.10. We write $\mathrm{supp}_R(\mathsf{T})$ for $\bigcup \mathrm{supp}_R(X)$, where X ranges over the objects in T.

A full triangulated subcategory of T is called *localising* if it is closed under arbitrary coproducts. Under mild conditions, such as R having finite Krull dimension, the localising subcategory $\mathrm{Loc}_{\mathsf{T}}(X)$ of T generated by an object X is equal to the localising subcategory

$$\mathrm{Loc}_{\mathsf{T}}(\{\Gamma_{\mathfrak{p}} X \mid \mathfrak{p} \in \mathrm{supp}_R(\mathsf{T})\})$$

generated by the objects $\Gamma_{\mathfrak{p}} X$. If this is the case, we say that the *local-global principle* holds. Then there is a one to one correspondence between localising subcategories of T and functions assigning to each $\mathfrak{p}$ in $\mathrm{supp}_R(\mathsf{T})$ a localising subcategory of $\Gamma_{\mathfrak{p}} \mathsf{T}$. The function corresponding to a localising subcategory S sends $\mathfrak{p}$ to $\mathsf{S} \cap \Gamma_{\mathfrak{p}} \mathsf{T}$. This is described in detail in [20].

We say that an action of a graded commutative ring R on T *stratifies* T if the local-global principle holds and for each $\mathfrak{p}$ in $\mathrm{supp}_R(\mathsf{T})$, the subcategory $\Gamma_{\mathfrak{p}} \mathsf{T}$ is minimal with respect to inclusion, among localising subcategories of T. Under these circumstances, there is one to one correspondence between localising subcategories of T and the subsets of $\mathrm{supp}_R(\mathsf{T})$. The subset corresponding to a localising subcategory S is its support, $\mathrm{supp}_R \mathsf{S}$, by which we mean the set of primes $\mathfrak{p}$ for which $\Gamma_{\mathfrak{p}}$ is not the zero functor on S.

Example 6.5 Let G be a finite group and k a field of characteristic p dividing the order of G. The stable module category $\mathrm{StMod}(kG)$ is a compactly generated triangulated category, with compact objects the finite dimensional kG-modules, and there is an action of the cohomology ring $H^*(G, k)$. In this case, we write $\mathrm{supp}_G(M)$ for $\mathrm{supp}_{H^*(G,k)}(M)$.

The support of $\mathrm{StMod}(kG)$ is $\mathrm{Proj}\, H^*(G, k)$, which is defined to be the set $\mathrm{Spec}\, H^*(G, k)$ excluding the maximal ideal consisting of all elements of positive degree. When M is finite dimensional, $\mathrm{supp}_G(M)$ is the closed subset of $\mathrm{Proj}\, H^*(G, k)$ defined by $\mathrm{Ann}_{H^*(G,k)}\, \mathrm{Ext}^{*,*}(M, M)$.

It is proved in [19] that when G is a p-group this action stratifies $\mathrm{StMod}(kG)$.

Such a stratification does not hold for general finite groups, because the localising subcategories $\Gamma_{\mathfrak{p}}\, \mathrm{StMod}(kG)$ are *tensor ideal*, namely they are closed under tensor products with all modules, whereas there may be localising subcategories that are not. For a finite p-group, the only simple module is the trivial module k, so the localising subcategory it generates is the whole of $\mathrm{StMod}(kG)$, and hence all localising subcategories are tensor ideal. To address arbitrary finite groups, we must take the tensor structure into account.

Tensor triangulated categories

Suppose now that T is a compactly generated triangulated category that comes with a symmetric monoidal tensor product $\otimes \colon \mathsf{T} \times \mathsf{T} \to \mathsf{T}$, which is exact in each variable, preserves small coproducts, and with unit $\mathbb{1}$. In this situation, we say that T is a *tensor triangulated category*. There is a particularly good kind of action of a graded commutative ring R on such a T; namely, one that factors as $R \to \mathrm{End}_{\mathsf{T}}^*(\mathbb{1})$ followed by the natural map $\mathrm{End}_{\mathsf{T}}^*(\mathbb{1}) \to Z^*(\mathsf{T})$ induced by the tensor product. We say that such an action is *canonical*.

Given a canonical action of R on T, the functor $\Gamma_{\mathfrak{p}}$ is naturally isomorphic to $\Gamma_{\mathfrak{p}}\mathbb{1} \otimes -$. The subcategory $\Gamma_{\mathfrak{p}}\mathsf{T}$ is *tensor ideal*, meaning that it is closed under tensoring with arbitrary objects in T. Denote by $\mathrm{Loc}_{\mathsf{T}}^{\otimes}(X)$ the tensor ideal localising subcategory of T generated by an object X. The tensor version of the local-global principle holds in this situation, and says that $\mathrm{Loc}_{\mathsf{T}}^{\otimes}(X)$ is equal to the tensor ideal localising subcategory $\mathrm{Loc}_{\mathsf{T}}^{\otimes}(\{\Gamma_{\mathfrak{p}}X \mid \mathfrak{p} \in \mathrm{supp}_R(\mathsf{T})\})$. So we get a one to one correspondence between tensor ideal localising subcategories of T and functions assigning to each $\mathfrak{p}$ in $\mathrm{supp}_R(\mathsf{T})$ a tensor ideal localising subcategory of $\Gamma_{\mathfrak{p}}\mathsf{T}$, see [20, Theorem 3.6].

We say that a canonical action of a graded commutative ring R *stratifies* a tensor triangulated category T if for each $\mathfrak{p} \in \mathrm{supp}_R(\mathsf{T})$, the subcategory

$\Gamma_{\mathfrak{p}}\mathsf{T}$ is minimal as a tensor ideal localising subcategory. See also the work of Hovey, Palmieri, and Strickland [51] where such a minimality plays a prominent role.

Example 6.6 Let G be a finite group. The stable module category of kG-modules is a tensor triangulated category, with tensor unit the trivial module k. The Tate cohomology ring $\widehat{H}^*(G, k)$ is none other than the graded ring $\operatorname{End}^*_{\operatorname{StMod}(kG)}(k)$, and so has a canonical action on $\operatorname{StMod}(kG)$. While $\widehat{H}^*(G, k)$ is usually not Noetherian, its subring $H^*(G, k)$ is, which then inherits the canonical action on $\operatorname{StMod}(kG)$.

The following theorem is proved in [19].

Theorem 6.7 *Let G be a finite group and k a field of characteristic p. As a tensor triangulated category, the action of $H^*(G, k)$ stratifies $\operatorname{StMod}(kG)$. In particular, there is a bijection, defined by $\operatorname{supp}_G(-)$, between tensor ideal localising subcategories of $\operatorname{StMod}(kG)$ and subsets of $\operatorname{Proj} H^*(G, k)$.* $\square$

A key step in the proof is a reduction to elementary abelian groups, which depends on the Quillen stratification theorem [54], and work of Chouinard, recalled below.

6.3 The rank variety

For an elementary abelian group E, there is another description of support that is much better suited to computation, the *rank variety*. In this section we recall the construction and essential properties of rank varieties, referring the reader to [8, Section 5.8] for more details.

Chouinard [32] proved that over an arbitrary commutative ring of coefficients k, a kG-module is projective if and only if its restriction to every elementary abelian subgroup E of G is a projective kE-module. If k is a field of characteristic p, we only need to consider elementary abelian p-subgroups. In this case, when k is algebraically closed, Dade [33] proved that a finite dimensional kE-module is projective if and only if its restriction to each cyclic shifted subgroup is projective. The precise statement is as follows. If $E := \langle g_1, \ldots, g_r \rangle \cong (\mathbb{Z}/p)^r$, is an elementary abelian p-group, set $X_i = g_i - 1$ in kE for $1 \leqslant i \leqslant r$. Then kE is a truncated polynomial ring:

$$kE \cong \frac{k[X_1, \ldots, X_r]}{(X_1^p, \ldots, X_r^p)}.$$

Let $J(kE)$ be the radical, $(X_1, \ldots, X_r)$, of kE. For $\lambda = (\lambda_1, \ldots, \lambda_r)$ in $\mathbb{A}^r(k)$ set

$$X_\lambda := \lambda_1 X_1 + \cdots + \lambda_r X_r \quad \text{in } kE$$

and $\alpha_\lambda \colon k[t]/(t^p) \to kE$ for the homomorphism of k-algebras sending t to X_λ. Given a kE-module M, we write $\alpha_\lambda^*(M)$ for the $k[t]/(t^p)$-module obtained by restriction along α_λ.

Theorem 6.8 (Dade [33])   *Let k be an algebraically closed field of characteristic p, and E an elementary abelian p-group. A finite dimensional dimensional kE-module M is projective if and only if $\alpha_\lambda^*(M)$ is projective for all $0 \neq \lambda \in \mathbb{A}^r(k)$.* $\square$

Based on this, Carlson [31] introduced the notion of rank variety.

Definition 6.9   Let k be an algebraically closed field of characteristic p, and E an elementary abelian p-group. The *rank variety* of a finitely generated kE-module M is the subset

$$V_E^r(M) := \{\lambda \in \mathbb{A}^r(k) \mid \alpha_\lambda^*(M) \text{ is not projective}\}$$

of $\mathbb{A}^r(k)$. We write $V_{kE}^r(M)$ if the field of coefficients needs to be specified. Observe that $V_E^r(M)$ contains 0 and is homogenous. It is also closed, for projectivity over $k[t]/(t^p)$ is detected by a rank condition on the operator representing the action of t. This also gives a method for calculating equations defining $V_E^r(M)$. Ostensibly, this depends on the choice of generators for E as an elementary abelian group. However, if $\alpha, \beta \colon k[t]/(t^p) \to kE$ are maps such that $\alpha(t)$ and $\beta(t)$ have the same image in $J(kE)/J^2(kE)$ then $\alpha^*(M)$ is projective if and only if $\beta^*(M)$ is projective. So it makes sense to think of the ambient affine space $\mathbb{A}^r(k)$ as identified with $J(kE)/J^2(kE)$. We return to this ambiguity in Section 6.6 when we discuss π-points.

The cohomology ring of E is well-known and easy to compute:

$$H^*(E, k) = \begin{cases} k[y_1, \ldots, y_r] & \text{if } p = 2 \\ k[x_1, \ldots, x_r] \otimes \Lambda(y_1, \ldots, y_r) & \text{for } p \text{ odd} \end{cases}$$

where y_i is in degree one the x_i is in degree two, and is the Bockstein $\beta(y_i)$ of y_i.

Let $k[Y_1, \ldots, Y_r]$ be the coordinate ring of $\mathbb{A}^r(k)$. If $p = 2$, this can be identified with $H^*(E, k)$ in such a way that the Y_i correspond to the y_i. If p is odd, there is a twist: we have to identify x_i with Y_i^p, so that there is a Frobenius twist involved in the identification of $\operatorname{Spec} H^*(E, k)$ with

$\operatorname{Spec} k[Y_1, \ldots, Y_r]$. Carlson conjectured that this identifies support and rank variety for a finite dimensional kE-module.

Theorem 6.10 (Avrunin, Scott, [2, Theorem 1.1]) *Let k be an algebraically closed field. Under the usual identification of radical homogeneous ideals in $\operatorname{Spec} H^*(E, k)$ with affine cones in $\mathbb{A}^r(k)$, the radical ideal defining the subset $\operatorname{supp}_E(M)$ corresponds to $V_E^r(M)$ for any finite-dimensional kE-module M.*

Given an extension of fields $k \subseteq K$ and a kE-module M, let M_K be the KE-module $K \otimes_k M$. Benson, Carlson and Rickard [11, Theorem 5.2] extended Dade's theorem to cover infinitely generated modules, as follows:

Theorem 6.11 *A kE-module M is projective if and only if $\alpha_\lambda^*(M_K)$ is projective for all extension fields K of k and all $0 \neq \lambda$ in $\mathbb{A}^r(K)$.*

For $0 \neq \lambda \in \mathbb{A}^r(K)$ let $\mathfrak{p} \subseteq k[Y_1, \ldots, Y_r]$ be the ideal consisting of homogeneous polynomials that vanish at λ. This is a prime ideal and λ is a *generic* point for $\mathfrak{p}$. Each $\mathfrak{p}$ in $\operatorname{Proj} k[Y_1, \ldots, Y_r]$ occurs this way. If $\lambda \in \mathbb{A}^r(K)$ and $\lambda' \in \mathbb{A}^r(K')$ are generic points for the same prime ideal, then $\alpha_\lambda^*(M_K)$ is projective if and only if $\alpha_{\lambda'}^*(M_{K'})$ is projective. So instead of a rank variety, we assign to M a subset of $\operatorname{Proj} k[Y_1, \ldots, Y_r]$.

Definition 6.12 Set $\mathcal{V}_E^r(M)$ to be the set of $\mathfrak{p} \in \operatorname{Proj} k[Y_1, \ldots, Y_r]$ such that if $\lambda \in \mathbb{A}^r(K)$ is generic for $\mathfrak{p}$ then $\alpha_\lambda^*(M_K)$ is not projective.

Historically the rank variety was defined as a subset of $\mathbb{A}^r(k)$, but we switch to considering projective varieties since all our constructions are "invariant" under scalar multiplication. It follows from [11, Theorem 10.5] that the subset $\operatorname{supp}_E(M) \subseteq \operatorname{Proj} H^*(E, k)$ corresponds to $\mathcal{V}_E^r(M)$ for *any* kE-module M, once the appropriate identification, involving Frobenius for $p > 2$, of $\operatorname{Proj} H^*(E, k)$ and $\mathbb{P}^{r-1}$ is made.

We shall see in Section 6.5 that the appropriate generalisation of these concepts to finite group schemes leads to the theory of π-points.

6.4 Finite group schemes

In this section, we introduce finite group schemes, following the approach given in [52, Chapter 1]. Throughout k will be a field.

An *affine scheme* over k is a representable functor from the category $\mathsf{CAlg}(k)$ of commutative algebras over k to sets. The representing object of an affine scheme S, denoted $k[S]$, is called its *coordinate ring*. Thus $S \colon \mathsf{CAlg}(k) \to \mathsf{Set}$ takes the form $\mathrm{Hom}_{\mathsf{CAlg}(k)}(k[S], -)$.

An *affine group scheme* is a functor $G \colon \mathsf{CAlg}(k) \to \mathsf{Grp}$ whose composite with the forgetful functor $\mathsf{Grp} \to \mathsf{Set}$ is representable. By Yoneda's lemma, the natural transformations given by the group operations give rise to a structure on $k[G]$ of commutative Hopf algebra. This gives a contravariant equivalence of categories from affine group schemes to commutative Hopf algebras, sending G to $k[G]$.

A *finite group scheme* is an affine group scheme G with the property that the coordinate ring $k[G]$ is finite dimensional over k. In this case, we may dualise to get the *group algebra* $kG := \mathrm{Hom}_k(k[G], k)$, which has the structure of cocommutative Hopf algebra. This gives a covariant equivalence of categories from finite group schemes to finite dimensional cocommutative Hopf algebras, sending G to kG.

Finite groups are examples of finite group schemes, but there are many more, including p-restricted Lie algebras, and Frobenius kernels of affine group schemes. See, for example, [22, §1] for examples and an explanation of how finite groups fit into the context.

Friedlander and Suslin [47, Theorem 1.1] proved that for any finite group scheme G, the k-algebra $H^*(G, k)$ is Noetherian; moreover, the $H^*(G, k)$-module $H^*(G, M)$ is finitely generated for any finite dimensional kG-module M. This opened the door for the development of support theories for finite group schemes. Another landmark development in this area was the theory of π-points.

6.5   The theory of π-points

The theory of π-points for finite group schemes was initiated by Friedlander and Pevtsova [45, 46], and generalises rank varieties discussed in Section 6.3. This approach does not rely on the choice of generators needed to define cyclic shifted subgroups for elementary abelian p-groups which makes it applicable in a much greater generality.

Let G be a finite group scheme over a field k of positive characteristic p. A *π-point* α of G is given as follows. We choose an extension field K of k, a unipotent abelian subgroup scheme E of G_K, and a flat map

$$\alpha \colon K[t]/(t^p) \to KE \subseteq KG_K.$$

Note that E does not have to come from a subgroup scheme of G by extension of scalars. The flatness condition is equivalent to the statement that the image of t is in $J(KE)$ but not in $J^2(KE)$.

Given such an α consider the composite

$$H^*(G, k) \subseteq K \otimes_k H^*(G, k) \cong H^*(G_K, K)$$

$$\cong \operatorname{Ext}^*_{KG_K}(K, K) \xrightarrow{\alpha^*} \operatorname{Ext}^*_{K[t]/(t^p)}(K, K).$$

The ring $\operatorname{Ext}^*_{K[t]/(t^p)}(K, K)$ is isomorphic to $K[v]$ with $|v| = 1$ if $p = 2$, and to $K[u, v]/(u^2)$ with $|u| = 1$ and $|v| = 2$ if $p \neq 2$. The nil radical is zero in the first case, and the ideal (u) in the second case. We define $\mathfrak{p}(\alpha)$ to be the inverse image in $H^*(G, k)$ of the nil radical of $\operatorname{Ext}^*_{K[t]/(t^p)}(K, K)$ under the above map. This is a homogeneous prime ideal in $H^*(G, k)$. The following theorem is due to Friedlander and Pevtsova [46, Corollary 2.11, Theorem 4.6].

Theorem 6.13 *If $\alpha\colon K[t]/(t^p) \to KG_K$ and $\beta\colon L[t]/(t^p) \to LG_L$ are π-points, the following conditions are equivalent.*

(i) $\mathfrak{p}(\alpha) = \mathfrak{p}(\beta)$;
(ii) $\alpha^*(M_K)$ is projective if and only if $\beta^*(M_L)$ is projective for a kG-module M;
(iii) $\alpha^*(M_K)$ is projective if and only if $\beta^*(M_L)$ is projective for any finite dimensional kG-module M. $\square$

We therefore put an equivalence relation on the set of π-points of G, where $\alpha \sim \beta$ if and only if the equivalent conditions of the theorem hold. The map sending the equivalence class of a π-point α to the prime $\mathfrak{p}(\alpha)$ gives a bijection between the equivalence classes of π-points in G and the set $\operatorname{Proj} H^*(G, k)$.

With this equivalence relation, every π-point is equivalent to one that factors through a subgroup scheme which is not only abelian unipotent, but elementary. Let $\mathbb{G}_a$ be the additive group scheme and for each integer $r \geqslant 0$, let $\mathbb{G}_{a(r)}$ denote its rth Frobenius kernel; see [52, Chapter 9].

Definition 6.14 A finite group scheme is *elementary* if it is isomorphic to the group scheme $\mathbb{G}_{a(r)} \times (\mathbb{Z}/p)^s$ with $r, s \geqslant 0$.

This terminology was introduced by Bendel [7, Definition 1.2]. For a finite group scheme cohomology is detected, modulo nilpotents, on elementary subgroup schemes over extension fields. This explains their central role in this theory.

Definition 6.15   Let G be a finite group scheme over a field k. The π-*support*, denoted $\pi\text{-}\mathrm{supp}_G(M)$, of a kG-module M is the subset of $\mathrm{Proj}\, H^*(G,k)$ consisting of primes $\mathfrak{p}(\alpha)$, for $\alpha\colon K[t]/(t^p) \to KG_K$ a π-point such that $\alpha^*(M_K)$ is not projective.

The π-support is a "generator-invariant" generalisation of Carlson's rank variety for any finite group scheme. On the other hand, exactly as for finite groups, one has a canonical action of $H^*(G,k)$ on $\mathrm{StMod}(kG)$ and hence a notion of support for kG-modules; see Section 6.2, especially 6.5. The following result (see [46, Proposition 6.8], [23, Theorem 6.1]) reconciles these two notions. In doing so, it puts the results for elementary abelian p-groups [2], finite groups [11], restricted Lie algebras [44], and infinitesimal group schemes [56, 53] into a uniform statement.

Theorem 6.16   *One has* $\mathrm{supp}_G(M) = \pi\text{-}\mathrm{supp}_G(M)$ *for any kG-module* M.   □

6.6   Finite supergroup schemes

The definition of affine superschemes is parallel to that of affine group schemes. It is obtained replacing the underlying category of vector spaces over k by the category of super vector spaces. A *super vector space* over k is a $\mathbb{Z}/2$-graded vector space $V = V_0 \oplus V_1$. The tensor product of two such is given by

$$(V\otimes W)_0 = V_0\otimes W_0 \oplus V_1\otimes W_1 \quad \text{and} \quad (V\otimes W)_1 = V_0\otimes W_1 \oplus V_1\otimes W_0\,.$$

This tensor product has a symmetric braiding $V \otimes W \cong W \otimes V$ given by sending $v \otimes w$ to $(-1)^{|v||w|}w \otimes v$, where $|v|$ denotes 0 if $v \in V_0$ and 1 if $v \in V_1$. Thus the category $\mathsf{SVec}(k)$ of super vector spaces is a symmetric monoidal abelian category. A *commutative superalgebra* over k is a commutative algebra in this category. Thus it consists of an object A together with a multiplication $A \times A \to A$ which is commutative with respect to the symmetric braiding, associative, and unital. These form a category $\mathsf{CSAlg}(k)$.

An *affine superscheme* over k is a representable functor from commutative superalgebras over k to sets. The representing object of an affine superscheme S is the coordinate ring $k[S]$. An *affine supergroup scheme* is a functor $G\colon \mathsf{CSAlg}(k) \to \mathsf{Grp}$ whose composite with the forgetful functor $\mathsf{Grp} \to \mathsf{Set}$ is representable. The representing object, the coordinate ring $k[G]$ of G, is a commutative Hopf superalgebra. This way, we obtain a contravariant equivalence of categories from affine supergroup schemes to commutative Hopf superalgebras. A Hopf superalgebra need not be a Hopf algebra, for the diagonal map need not be a map of ungraded algebras.

A *finite supergroup scheme* is an affine supergroup scheme G with the property that the coordinate ring $k[G]$ is finite dimensional over k. In this case, we may dualise to get the *group algebra* $kG = \text{Hom}_k(k[G], k)$, which has the structure of cocommutative Hopf superalgebra. This way, we obtain a covariant equivalence of categories from finite supergroup schemes to finite dimensional cocommutative Hopf superalgebras.

Examples of finite supergroup schemes include finite groups, finite group schemes, exterior algebras, as well as Frobenius kernels of affine supergroup schemes such as the general linear ones $\text{GL}(a|b)$ and the orthosymplectic ones $\text{OSp}(a|2b)$. We write $\mathbb{G}_a^-$ for the finite supergroup scheme whose group algebra is an exterior algebra on one primitive element, in degree 1. This is the simplest example of a finite supergroup scheme which is not a finite group scheme.

A finite supergroup scheme G is *unipotent* if the kernel of the augmentation map $kG \to k$ is equal to the nil radical. This is equivalent to kG having exactly two simple modules, the trivial module in even degree and the trivial module in odd degree. For example, $\mathbb{G}_a^-$ is unipotent.

The module category for a finite supergroup scheme is an abelian category, with a parity change functor Π. The action of an element $a \in kG$ on $\Pi m \in \Pi M$ is given by $a.\Pi m = (-1)^{|a|}\Pi(a.m) \in \Pi M$. The stable module category $\text{StMod}(kG)$ is a $\mathbb{Z}/2$-graded triangulated category. So it comes with an internal shift Π whose square is naturally isomorphic to the identity, and a cohomological shift Ω^{-1}. The distinguished triangles

$$M_1 \to M_2 \to M_3 \to \Omega^{-1}M_1$$

are those isomorphic to triangles coming from short exact sequences of kG-modules via a pushout diagram

$$
\begin{array}{ccccccccc}
0 & \longrightarrow & M_1 & \longrightarrow & M_2 & \longrightarrow & M_3 & \longrightarrow & 0 \\
& & \| & & \downarrow & & \downarrow & & \\
0 & \longrightarrow & M_1 & \longrightarrow & I & \longrightarrow & \Omega^{-1}M_1 & \longrightarrow & 0
\end{array}
$$

where I is an injective module into which M_2 embeds.

For a $\mathbb{Z}/2$-graded triangulated category T, we need a slight modification to the definition of centre, to take account of the grading. We define $Z^{n,j}(\mathsf{T})$ ($n \in \mathbb{Z}$, $j \in \mathbb{Z}/2$) to be the natural transformations η from the identity functor to $\Sigma^n \Pi^j$ satisfying $\eta \Sigma = (-1)^n \Sigma \eta$ and $\eta \Pi = (-1)^j \Pi \eta$. These form a $\mathbb{Z} \times \mathbb{Z}/2$-graded ring. It is graded commutative, in the sense that if $x \in Z^{m,i}(\mathsf{T})$ and $y \in Z^{n,j}(\mathsf{T})$ then $yx = (-1)^{mn}(-1)^{ij}xy$. The cohomology ring $H^{*,*}(G, k)$ is also $\mathbb{Z} \times \mathbb{Z}/2$-graded commutative in this sense. For example, $H^{*,*}(\mathbb{G}_a^-, k)$ is a polynomial ring on a single generator in degree $(1, 1)$.

There is a canonical action of $H^{*,*}(G, k)$ on $\mathrm{StMod}(kG)$, so it makes sense to try to stratify $\mathrm{StMod}(kG)$ as a $\mathbb{Z}/2$-graded tensor triangulated category using this action. The definition of localising subcategory needs to be modified to take account of the extra grading; we only consider localising subcategories closed under the operation Π.

Definition 6.17   We define $\mathrm{Proj}\, H^{*,*}(G, k)$ to consist of the prime ideals which are homogeneous with respect to both gradings, with the Zariski topology, and as usual we exclude the maximal ideal of elements generated by homogeneous elements whose degree is not $(0,0)$. Since elements of degree (even,odd) or (odd,even) square to zero, these elements are contained in every prime ideal. Modulo these elements, the $\mathbb{Z}/2$-grading is just the mod two reduction of the $\mathbb{Z}$-grading, and so we can ignore it. So we write $H(G, k)$ for the singly graded ring whose degree i component is the degree $(i, 0)$ or $(i, 1)$ component of $H^{*,*}(G, k)$ according as i is even or odd. This is a graded ring which becomes strictly commutative if we reduce modulo nilpotents, and at the level of topological spaces, $\mathrm{Proj}\, H(G, k)$ may be identified with $\mathrm{Proj}\, H^{*,*}(G, k)$. Thus for example $H(\mathbb{G}_a^-, k)$ is a polynomial ring on a generator in degree one.

Definition 6.18   For a kG-module M, let $\mathrm{supp}_G(M)$ denote the support defined via the action of $H(G, k)$ on $\mathrm{StMod}(kG)$, as in Section 6.2.

Remark 6.19   By definition, $\mathrm{supp}_G(M)$ is a subset of $\mathrm{Proj}\, H(G, k)$. Recall that when M is finite-dimensional, $\mathrm{supp}_G(M)$ is a closed subset defined by the annihilator of $\mathrm{Ext}^{*,*}_{kG}(M, M)$ as a module over $H(G, k)$. Moreover, if G is unipotent, we can take the annihilator of $H^{*,*}(G, M)$.

6.7 Elementary supergroup schemes

The definition of elementary supergroup schemes is more complicated than for finite group schemes, and we have so far only addressed the unipotent case. Drupieski and Kujawa [38, 39, 40] suggest that similar definitions may suffice for arbitrary finite supergroup schemes.

In [24] we construct a family of finite connected unipotent supergroup schemes $E_{m,n}^-$ with $m, n \geq 1$ related to the Witt vectors and declare a supergroup scheme to be *elementary* if it is isomorphic to a quotient of some $E_{m,n}^- \times (\mathbb{Z}/p)^s$. These are classified; see [24, Remark 8.14] and compare with Definition 6.14.

Theorem 6.20 *Each elementary supergroup scheme is isomorphic to one of:*

I. $\mathbb{G}_{a(n)} \times (\mathbb{Z}/p)^s$ *with* $n, s \geq 0$,

II. $\mathbb{G}_{a(n)} \times \mathbb{G}_a^- \times (\mathbb{Z}/p)^s$ *with* $n, s \geq 0$,

III. *(i)* $E_{m,n}^- \times (\mathbb{Z}/p)^s$ *with* $m \geq 2$, $n \geq 1$, $s \geq 0$,

(ii) $E_{m,n,\mu}^- \times (\mathbb{Z}/p)^s$ *with* $m, n \geq 1$, $0 \neq \mu \in k^\times/(k^\times)^2$ *and* $s \geq 0$.

Here, $E_{m,n,\mu}^-$ *is a quotient of* $E_{m+1,n+1}^-$ *by a subgroup isomorphic to* $\mathbb{G}_{a(1)}$, *and only depends on the image of* μ *in* $k^\times/(k^\times)^2$. $\square$

Definition 6.21 The supergroup schemes of type III are said to be *Witt elementary*.

The role played by elementary supergroup schemes is explained by the following analogue from [24] of the theorems of Quillen and Chouinard for finite groups.

Theorem 6.22 *Let G be a unipotent finite supergroup scheme over a field k of characteristic $p \geq 3$. Then the following hold:*

(i) An element $x \in H^{*,*}(G, k)$ is nilpotent if and only if for every extension field K of k and every elementary sub-supergroup scheme E of G_K, the restriction of x_K, the image of x in $H^{*,*}(G_K, K)$, to $H^{*,*}(E, K)$ is nilpotent.

(ii) A kG-module M is projective if and only if for every extension field K of k and every elementary sub-supergroup scheme E of G_K, the restriction of M_K to E is projective. $\square$

Fortunately, if we ignore the comultiplicative structure of these elementary supergroup schemes, the $\mathbb{Z}/2$-graded algebra structure is easy to describe. The list corresponds to the one from Theorem 6.20.

Proposition 6.23 *If E is an elementary supergroup scheme over k then the algebra kE is isomorphic to one of the following:*

(i) a tensor product of copies of $k[s]/(s^p)$,

(ii) a tensor product of copies of $k[s]/(s^p)$ and one copy of $k[\sigma]/(\sigma^2)$,

(iii) a tensor product of copies of $k[s]/(s^p)$ and one copy of $k[s,\sigma]/(s^{p^m}, s^p - \sigma^2)$, where $m \geqslant 1$,

with $|s|$ even and $|\sigma|$ odd.

In particular, we have

$$kE^-_{m,n} \cong \frac{k[s_1, \ldots, s_n, \sigma]}{(s_1^p, \ldots, s_{n-1}^p, s_n^{p^m}, s^p - \sigma^2)} \tag{6.7.1}$$

with coproduct defined by

$$\Delta(s_i) = S_{i-1}(s_1 \otimes 1, \ldots, s_i \otimes 1, 1 \otimes s_1, \ldots, 1 \otimes s_i) \qquad (i \geqslant 1)$$
$$\Delta(\sigma) = \sigma \otimes 1 + 1 \otimes \sigma,$$

where the S_i are the maps coming from the comultiplication in $\mathbb{G}_a$.

Although an elementary supergroup scheme is not necessarily commutative as a $\mathbb{Z}/2$-graded algebra, because of the factors of type 6.23 (iii), it is nonetheless commutative in the ungraded sense.

Remark 6.24   For non-unipotent finite supergroup schemes, these elementary supergroup schemes are definitely not sufficient for detection. Drupieski and Kujawa [38] introduced a slightly more general set of supergroup schemes which they show suffice for $\mathrm{GL}(a|b)_{(r)}$. Following their notation, if f is a *p-polynomial*, meaning a polynomial of the form $f(t) = \sum_i a_i t^{p^i}$, we shall write $\mathbb{M}_{n;f}$ for the supergroup scheme defined by replacing $s_n^{p^m}$ by $f(s_n)$ in (6.7.1), and with the same comultiplication as $E^-_{m,n}$. Similarly, $\mathbb{M}_{n;f,\mu}$ is obtained in the same way from $E^-_{m,n-1,\mu}$. The only unipotent ones among these are our $E^-_{m,n}$ and $E^-_{m,n,\mu}$. These are all quotients of a supergroup scheme $\mathbb{M}_n$. Set

$$k\mathbb{M}_n \cong \frac{k[s_1, \ldots, s_{n-1}, \sigma][[s_n]]}{(s_1^p, \ldots, s_{n-1}^p, \sigma^2 - s_n^p)},$$

$$k\mathbb{M}_{n;f,\mu} \cong \frac{k[s_1, \ldots, s_n, \sigma]}{(s_1^p, \ldots, s_{n-1}^p, f(s_n) + \mu s_1, \sigma^2 - s_n^p)}.$$

Thus $k\mathbb{M}_{n;f,\mu}$ is the group ring over k of the corresponding supergroup scheme. We shall suppress the field from the notation, so that we also regard $\mathbb{M}_n$ as a profinite supergroup scheme over any extension field K of k. Similarly, if f is a p-polynomial with coefficients in K, and $\mu \in K$, we shall write $\mathbb{M}_{n;f,\mu}$. Finally, if $f = t^{p^m}$ and $\mu = 0$, we shall just write $\mathbb{M}_{n;m}$; this is the same as $E^-_{m,n}$.

6.8 π-points, π-supports and rank varieties

To generalise the theory of π-points from finite group schemes to finite supergroup schemes, instead of flat maps from $k[t]/(t^p)$, we consider maps of finite flat (or equivalently, projective) dimension from the superalgebra

$$A_k := \frac{k[t, \tau]}{(t^p - \tau^2)} \qquad (\tau \text{ odd}, t \text{ even})$$

We aim for theorems about unipotent finite supergroup schemes, but we stay more general for now in the interest of later developments, and also because we shall need to discuss $\mathrm{GL}(a|b)_{(n)}$ as part of the proof. So we shall include the elementary supergroup schemes $\mathbb{M}_{n;f,\mu}$ from Remark 6.24, but note that if G is unipotent then the only ones that occur are the ones listed at the beginning of Section 6.7.

We view A_k as a cocommutative Hopf superalgebra over k with τ and t primitive:

$$\Delta(\tau) = \tau \otimes 1 + 1 \otimes \tau \quad \text{and} \quad \Delta(t) = t \otimes 1 + 1 \otimes t.$$

This defines a homomorphism of algebras since

$$\begin{aligned}
\Delta(\tau^2) &= (\tau \otimes 1 + 1 \otimes \tau)^2 \\
&= \tau^2 \otimes 1 + \tau \otimes \tau - \tau \otimes \tau + 1 \otimes \tau^2 \\
&= t^p \otimes 1 + 1 \otimes t^p \\
&= \Delta(t^p).
\end{aligned}$$

The cohomology ring of the k-algebra A_k is easy to compute:

$$\mathrm{Ext}_{A_k}^{*,*}(k, k) = k[\eta] \otimes \Lambda(u)$$

where $|\eta| = (1, 1)$ and $|u| = (1, 0)$. In particular, modulo its radical it is a domain.

Definition 6.25 A π-*point* α of a finite supergroup scheme G is given as follows. We choose an extension field K of k, an elementary sub-supergroup scheme E of G_K, and a map of superalgebras, but not necessarily respecting the coproduct,

$$\alpha \colon A_K \to KE \subseteq KG_K$$

of finite flat dimension.

We put an equivalence relation on π-points, analogous to the one in Section 6.5.

Definition 6.26   We say that π-points $\alpha\colon A_K \to KG_K$ and $\beta\colon A_L \to LG_L$ are equivalent if, for all finite dimensional kG-modules M, the module $\alpha^*(M_K)$ has finite flat dimension if and only if $\beta^*(M_L)$ has finite flat dimension. We write $\Pi(G)$ for the set of equivalence classes of π-points of G.

Definition 6.27   We say a π-point α is *K-rational* if it is defined over the field K, so that it is a map $\alpha\colon A_K \to KG_K$.

If E is an elementary sub-supergroup scheme of G_K, and $\alpha\colon A_K \to KE$ is a π-point, then the radical of the kernel of restriction

$$H^{*,*}(G,k) \subseteq H^{*,*}(G_K,K) \to H^{*,*}(E,K) \to \mathrm{Ext}^{*,*}_{A_K}(K,K)$$

is a prime ideal $\mathfrak{p}(\alpha)$, for the target, modulo its radical, is a domain.

Lemma 6.28   *If α and β are equivalent π-points, then $\mathfrak{p}(\alpha) = \mathfrak{p}(\beta)$.*

Proof   A proof of this result is given in Proposition 6.8 of [26] for elementary supergroup schemes; it involves Carlson's L_ζ modules. The same argument applies without any change to the general case. $\square$

Definition 6.29   By the preceding lemma, one gets a well-defined map

$$\Phi_G\colon \Pi(G) \longrightarrow \mathrm{Proj}\, H^{*,*}(G,k)$$

where α is sent to $\mathfrak{p}(\alpha)$. By [26, Theorem 6.9], see also [26, §8], it is bijective when G is an elementary supergroup scheme. See Section 6.9 for the general case.

Definition 6.30   Let G be a finite *unipotent* supergroup scheme over a field k of characteristic $p \geqslant 3$. The *π-support* of a kG-module M, denoted $\pi\text{-}\mathrm{supp}_G(M)$, is the subset of $\Pi(G)$ consisting of equivalences classes of π-points $\alpha\colon A_K \to KG$ such that $\alpha^*(M_K)$ has infinite flat dimension. This is well-defined by design.

If E is an elementary supergroup scheme, then by Proposition 6.23 the group algebra kE is isomorphic to one of the following algebras:

I. $kE \cong k[s_1,\ldots,s_n]/(s_1^p,\ldots,s_n^p)$,

II. $kE \cong k[s_1,\ldots,s_n,\sigma]/(s_1^p,\ldots,s_n^p,\sigma^2)$,

III. $kE \cong k[s_1,\ldots,s_n,\sigma]/(s_1^p,\ldots,s_{n-1}^p,s_n^{p^m},s_n^p-\sigma^2)$ for some $m \geqslant 2, n \geqslant 1$.

In each case, a set of representatives for equivalence classes of π-points of E can be prescribed by choosing an extension field K of k and a point $0 \neq \lambda = (\lambda_1, \ldots, \lambda_{n+1})$ in $\mathbb{A}^{n+1}(K)$. The π-point

$$\alpha_\lambda \colon A_K \to KE_K$$

is as follows: In case I, which is covered by Carlson's theory of rank varieties, set

$$\alpha_\lambda(t) = \lambda_1 s_1 + \cdots + \lambda_n s_n \quad \text{and} \quad \alpha_\lambda(\tau) = 0.$$

In the second case, set

$$\alpha_\lambda(t) = \lambda_1 s_1 + \cdots + \lambda_n s_n \quad \text{and} \quad \alpha_\lambda(\tau) = \lambda_{n+1}\sigma.$$

See [26, §8]. In the last case, the π-point is defined by

$$\alpha_\lambda(t) = \lambda_1 s_1 + \cdots + \lambda_{n-1}s_{n-1} + \lambda_{n-1}s_{n-1} + \lambda_n s_n^{p^{m-1}} + \lambda_{n+1}^2 s_n,$$
$$\alpha_\lambda(\tau) = \lambda_{n+1}^p \sigma.$$

$$(6.8.1)$$

Note that

$$(\alpha_\lambda(t))^p = \lambda_{n+1}^{2p} s_n^p = \lambda_{n+1}^{2p}\sigma^2 = (\alpha_\lambda(\tau))^2$$

in KE_K, so this does indeed define a homomorphism of algebras.

The result below, which reproduces [26, Theorem 4.12], extends Theorem 6.11.

Theorem 6.31 *Let E be an elementary supergroup scheme defined over a field k. An E-module M is projective if and only if $\alpha_\lambda^*(M_K)$ has finite flat dimension for every extension field K of k and every $0 \neq \lambda \in \mathbb{A}^{n+1}(K)$.* □

Combining with Theorem 6.22 yields that π-points detect projectivity of modules.

Theorem 6.32 *Let G be a unipotent finite supergroup scheme over a field k of characteristic $p \geqslant 3$. A kG-module M is projective if and only if $\pi\text{-supp}_G(M) = \varnothing$.*

Proof If M is projective, K is an extension field of k, and E is a sub-supergroup scheme of G_K, then $\text{res}_{G_K,E}(M_K)$ is projective. So if $\alpha \colon A_K \to E \to G_K$ is a π-point then $\alpha^*(M_K)$ has finite flat dimension. Thus $\pi\text{-supp}_G(M)$ is empty.

Conversely, if $\pi\text{-supp}_G(M) = \varnothing$ and E is a sub-supergroup scheme of G_K for some extension field K, then $\text{res}_{G_K,E}(M_K)$ has empty π-support. By Theorem 6.31, $\text{res}_{G_K,E}(M_K)$ is projective for every such K and E. It follows from Theorem 6.22 that M is a projective kG-module. □

A crucial property of π-support is the tensor product formula.

Theorem 6.33   *Let G be a unipotent finite supergroup scheme over a field k of characteristic $p \geqslant 3$, and let M and N be kG-modules. As subsets of $\Pi(G)$*

$$\pi\text{-}\mathrm{supp}_G(M \otimes_k N) = \pi\text{-}\mathrm{supp}_G(M) \cap \pi\text{-}\mathrm{supp}_G(N).$$

Proof   For any elementary sub-supergroup scheme E of G_K, the restriction functor

$$\mathrm{res}_{G_K,E} \colon \mathrm{Mod}\, G_K \to \mathrm{Mod}\, E$$

commutes with tensor product. Hence, it suffices to prove the formula for elementary supergroup schemes. This is done in [26, Theorem 7.6].   □

For finite-dimensional modules over elementary supergroups π-support has a "rank variety" interpretation, analogous to Carlson's original construction recalled in Definition 6.12.

Let $k[Y_1, \ldots, Y_{n+1}]$ be the coordinate ring of $\mathbb{A}^{n+1}(k)$. Since the map

$$\Phi_E \colon \Pi(E) \to \mathrm{Proj}\, H^{*,*}(E, k) \cong \mathbb{P}^n(k)$$

is a homeomorphism, for any $\mathfrak{p} \in \mathrm{Proj}\, k[Y_1, \ldots, Y_{n+1}]$ there is a "generic" point $\lambda \in \mathbb{A}^{n+1}(K)$ such that $\Phi_E(a_\lambda) = \mathfrak{p}$ where a_λ is the π-point defined in (6.8.1).

Definition 6.34   Let k be a field of characteristic $p \geqslant 3$, and let E be a Witt elementary supergroup scheme over k. If M is an E-module, define $\mathcal{V}_E^r(M)$ to be the set of homogeneous primes $\mathfrak{p} \in \mathrm{Proj}\, k[Y_1, \ldots, Y_{n+1}]$ such that if $\lambda \in \mathbb{A}^{n+1}(K)$ is generic for $\mathfrak{p}$ then $\alpha_\lambda^*(M_K)$ has infinite flat dimension as a $K A_K$-module.

It follows from [26] that Φ_E takes the π-support of M bijectively to $\mathcal{V}_E^r(M)$.

Here is a free resolution of the trivial A_K-module K:

$$\cdots \to \Pi A_K \oplus A_K \xrightarrow{\left(\begin{smallmatrix} \tau & t \\ -t^{p-1} & -\tau \end{smallmatrix}\right)} A_K \oplus \Pi A_K \xrightarrow{\left(\begin{smallmatrix} \tau & t \\ -t^{p-1} & -\tau \end{smallmatrix}\right)} \Pi A_K \oplus A_K \xrightarrow{(\tau,t)} A_K \to 0.$$

This is periodic from degree one onwards, with period one. So for a π-point $\alpha \colon A_K \to K E_K$ and a E-module M, it follows that $\alpha^*(M_K)$ has finite projective dimension as a A_K-module if and only if we have $\mathrm{Ext}_{A_K}^1(K, \alpha^*(M_K)) = 0$. Taking homomorphisms from the above resolution to $\alpha^*(M_K)$, this is true if and only if

$$\Pi M_K \oplus M_K \xrightarrow{\begin{pmatrix} \alpha_\lambda(\tau) & \alpha_\lambda(t) \\ -\alpha_\lambda^{p-1}(t) & -\alpha_\lambda(\tau) \end{pmatrix}} M_K \oplus \Pi M_K \xrightarrow{\begin{pmatrix} \alpha_\lambda(\tau) & \alpha_\lambda(t) \\ -\alpha_\lambda^{p-1}(t) & -\alpha_\lambda(\tau) \end{pmatrix}} \Pi M_K \oplus M_K$$

is exact. In particular, if M is a finitely generated E-module, of dimension d, we can think of the action of

$$\begin{pmatrix} \alpha_\lambda(\tau) & \alpha_\lambda(t) \\ -\alpha_\lambda^{p-1}(t) & -\alpha_\lambda(\tau) \end{pmatrix} \tag{6.8.2}$$

on a direct sum of two copies of M_K as a $2d \times 2d$ matrix whose square is zero. Exactness is therefore just the condition that this matrix has rank equal to d, which is the largest possible for a square zero matrix of this size. This condition fails if and only if every $d \times d$ minor of the matrix vanishes. In particular, this is a set of homogeneous polynomial conditions, and therefore defines a closed homogeneous subvariety of $\mathbb{A}^{n+1}(K)$. So the following is an analogue of Carlson's rank variety.

Definition 6.35 If k is algebraically closed, and M is a finitely generated E-module of dimension d, we define the *rank variety* $V_E^r(M)$ to be $\{0\}$ together with the set of $\lambda \in \mathbb{A}^{n+1}(k)$ such that $\alpha_\lambda^*(M)$ has infinite flat dimension. Equivalently, $V_E^r(M)$ is the set of points λ where the $2d \times 2d$ matrix representing the action of (6.8.2) on M has rank strictly less than d.

If we extend the field from k to K, the polynomial equations defining the variety $V_{E_K}^r(M_K)$ are exactly the same as those defining $V_E^r(M)$. It follows that a prime $\mathfrak{p}$ is in $\mathcal{V}_E^r(M)$ if and only if a generic point for $\mathfrak{p}$ in a suitable $\mathbb{A}^{n+1}(K)$ is contained in the zero set of these polynomials. This happens if and only if $\mathfrak{p}$ is contained in the radical ideal defining $V_E^r(M)$.

Theorem 6.36 *For M a finite dimensional kG-module, $\mathcal{V}_E^r(M)$ is the Zariski closed subset of $\operatorname{Proj} k[Y_1, \ldots, Y_{n+1}]$ defined by the rank variety $V_E^r(M)$.*

On the other hand, it can be shown that for infinitely generated E-modules, every subset of $\operatorname{Proj} k[Y_1, \ldots, Y_{n+1}]$ occurs as $\mathcal{V}_E^r(M)$ for suitably constructed M.

6.9 Φ_G is a homeomorphism

In this section we show that the map $\Phi_G \colon \Pi(G) \to \operatorname{Proj} H^{*,*}(G, k)$, introduced in Definition 6.30, is bijective for a unipotent finite supergroup scheme G. Surjectivity will follow from Theorem 6.22 whereas for injectivity we have to recall some recent results of Drupieski and Kujawa. We shall use supergroup schemes $\mathbb{M}_n$ and $\mathbb{M}_{n;f,\mu}$ introduced in Remark 6.24.

Let $\mathrm{Hom}_{\mathrm{sgs}/k}(\mathbb{M}_n, G)$ be a functor from commutative k-superalgebras to sets defined as

$$\mathrm{Hom}_{\mathrm{sgs}/k}(\mathbb{M}_n, G)(R) = \mathrm{Hom}_{\mathrm{sgs}/R}(\mathbb{M}_{n,R}, G_R).$$

Definition 6.37 It is shown in [38, Theorem 3.3.6] that if G is any affine supergroup scheme of finite type then $\mathrm{Hom}_{\mathrm{sgs}/k}(\mathbb{M}_n, G)$ has the structure of a connected affine superscheme of finite type over k, which we denote $\mathbf{V}_n(G)$. Similarly, $\mathrm{Hom}_{\mathrm{sgs}/k}(\mathbb{M}_{n;f,\mu}, G)$ is an affine subsuperscheme of $\mathbf{V}_n(G)$, which we denote $\mathbf{V}_{n;f,\mu}(G)$. In the case $f = t^{p^m}$ and $\mu = 0$, we shall write $\mathbf{V}_{n;m}(G)$.

The K-points of $\mathbf{V}_n(G)$ are maps $\mathbb{M}_{n,K} \to G_K$ as above. In particular, we can identify $\mathbf{V}_n(G)$ with $\mathbf{V}_n(G_{(n)})$.

In §6.2 of [38], the authors construct a map of affine superschemes

$$\Psi_G \colon \mathbf{V}_n(G) \to \mathrm{Spec}\, H^{*,*}(G, k)$$

coming from a map of coordinate rings

$$\psi_G \colon H^{*,*}(G, k) \to k[\mathbf{V}_n(G)].$$

Let $\mathrm{GL}(a|b)$ be the general linear supergroup; see, for example, [37]. For a connected finite supergroup scheme $\mathrm{GL}(a|b)_{(n)}$ which is the Frobenius kernel of $\mathrm{GL}(a|b)$, we have the following. There is a map constructed in §6 of [38]

$$\overline{\phi} \colon k[\mathbf{V}_n(\mathrm{GL}(a|b))]_{\mathrm{ev}} = k[\mathbf{V}_n(\mathrm{GL}(a|b)_{(n)})]_{\mathrm{ev}} \to H^{*,0}(\mathrm{GL}(a|b)_{(n)})$$

with the property that the composite

$$k[\mathbf{V}_n(\mathrm{GL}(a|b))]_{\mathrm{ev}} \xrightarrow{\overline{\phi}} H(\mathrm{GL}(a|b)_{(n)}) \xrightarrow{\psi_{\mathrm{GL}(a|b)_{(n)}}} k[\mathbf{V}_{n;f,\mu}(\mathrm{GL}(a|b))]_{\mathrm{ev}}$$

is the Frobenius morphism F^n followed by the quotient map

$$k[\mathbf{V}_n(\mathrm{GL}(a|b))]_{\mathrm{ev}} \to k[\mathbf{V}_{n;f,\mu}(\mathrm{GL}(a|b))]_{\mathrm{ev}}.$$

This has the following consequence, as pointed out in Theorem 6.2.3 of [38].

Theorem 6.38 *The image of map*

$$\psi_{\mathrm{GL}(a|b)_{(n)}} \colon H^{*,*}(\mathrm{GL}(a|b)_{(n)}, k) \to k[\mathbf{V}_{n;f,\mu}(\mathrm{GL}(a|b)_{(n)})]$$

contains the p^nth power of every element of $k[\mathbf{V}_{n;f,\mu}(\mathrm{GL}(a|b)_{(n)})]$. The map

$$\Psi_{\mathrm{GL}(a|b)_{(n)}} : \mathbf{V}_n(\mathrm{GL}(a|b)_{(n)}) \to \mathrm{Spec}\, H^{*,*}(\mathrm{GL}(a|b)_{(n)}, k)$$

is injective on K-points for all extension fields K of k. □

The following is the analogue of [56, Theorem 5.2]; see also [45, Proposition 3.8].

Theorem 6.39 *Let G be connected finite supergroup scheme of height at most n. Then for m large enough, the image of the map*

$$\psi_G : H^{*,*}(G, k) \to k[\mathbf{V}_{n;m}(G)]$$

contains the p^nth power of every element of $k[\mathbf{V}_{n;m}(G)]$. Consequently, the map $\Psi_G : \mathbf{V}_{n;m}(G) \to \mathrm{Spec}\, H^{*,*}(G, k)$ is injective.

Proof Choose an embedding $G \to \mathrm{GL}(a|b)_{(r)}$ for suitable a, b and r. Then we have the following diagram

$$
\begin{array}{ccc}
H^{*,*}(\mathrm{GL}(a|b)_{(n)}, k) & \xrightarrow{\psi_{\mathrm{GL}(a|b)_{(n)}}} & k[\mathbf{V}_{n;m}(\mathrm{GL}(a|b)_{(n)})] \\
\downarrow & & \downarrow \\
H^{*,*}(G, k) & \xrightarrow{\psi_G} & k[\mathbf{V}_{n;m}(G)]
\end{array}
$$

The upper horizontal map surjects onto p^nth powers by Theorem 6.38, and the right hand vertical map is surjective since

$$\mathbf{V}_{n;m}(G) \to \mathbf{V}_{n;m}(\mathrm{GL}(a|b)_{(n)})$$

is a closed embedding by [38, Theorem 3.3.6]. It follows that the lower horizontal map surjects onto p^nth powers. □

For the algebra $A = k[t, \tau]/(t^p - \tau^2)$, we have a map

$$e : A \to k\mathbb{M}_n \to k\mathbb{M}_{n;m} \tag{6.9.1}$$

given by $e(\tau) = \sigma$, $e(t) = s_n$. Let $\mathbb{M}_n \to G_K$ be a homomorphism of supergroup schemes. Since G is unipotent, the group algebra KG_K is finite dimensional and local. Hence, any element in the augmentation ideal is nilpotent. Therefore, the homomorphism factors through some $\mathbb{M}_{n;m}$. We compose to get a π-point of G:

$$e_G : A_K \to K\mathbb{M}_n \to K\mathbb{M}_{n;m} \to KG_K.$$

Thus we have maps

$$\mathbb{P}\operatorname{Hom}_{\mathsf{sgs}/k}(\mathbb{M}_{n;m}, G) \hookrightarrow \mathbb{P}\operatorname{Hom}_{\mathsf{sgs}/k}(\mathbb{M}_n, G) \to \Pi(G) \xrightarrow{\Phi_G} \operatorname{Proj} H^{*,*}(G, k),$$

where we denote by $\mathbb{P}X$ the projectivisation of a conical affine super-scheme X. It follows from [38, §6.2] that the composition

$$\mathbb{P}\operatorname{Hom}_{\mathsf{sgs}}(\mathbb{M}_{n;m}, G) \to \operatorname{Proj} H^{*,*}(G, k)$$

is induced by the map Ψ_G.

Proposition 6.40   *The distinguished π-point $e\colon A \to k\mathbb{M}_{n:m}$ induces a bijection*

$$e^*\colon \operatorname{Proj} k[\mathbf{V}_{n;m}(\mathbb{M}_{n;m})] \to \Pi(\mathbb{M}_{n;m}).$$

Proof   This follows from Lemma 3.3.2 in [38]. $\square$

Remark 6.41   Proposition 6.40 is effectively saying that any equivalence class of π-points of $\mathbb{M}_{n;m}$ has a unique (up to a scalar) representative of the form $\phi \circ e$ where ϕ is an endomorphism of the supergroup scheme $\mathbb{M}_{n;m}$.

This observation almost immediately extends to any finite connected unipotent supergroup scheme G.

Corollary 6.42   *For any finite connected unipotent supergroup scheme G of height n, the π-point $e_G\colon A \to kG$ induces a surjection*

$$e_G^*\colon \operatorname{Proj} k[\mathbf{V}_n(G)] \to \Pi(G).$$

Proof   Let $\alpha\colon A_K \to KG_K$ be a π-point. Then α factors through an elementary supergroup $i\colon E \hookrightarrow G_K$. We consider the case $E = E_{m,n}^-$, $m \geq 1$, the other three cases are similar.

We factor $\alpha = \iota \circ \alpha'\colon A_K \to KE_{m,n}^- \to KG_K$. By Proposition 6.40 there exists an endomorphism $\phi\colon E_{m,n}^- \to E_{m,n}^-$ such that α' is equivalent to $\phi \circ e\colon A_K \to E_{m,n}^-$ as a π-point of $E_{m,n}^-$. Hence, α is equivalent to $i \circ \phi \circ e = e_G^*(i \circ \phi)$ as a π-point of G. Hence e_G^* is surjective. $\square$

Corollary 6.43   *For G a finite connected unipotent supergroup scheme, the map $\Phi_G\colon \Pi(G) \to \operatorname{Proj} H^{*,*}(G, k)$ is injective.*

Proof   Let G be of height n and choose large enough m such that any map $\mathbb{M}_n \to G$ factors through $\mathbb{M}_{n:m}$. Then the composition

$$\Psi_G\colon \operatorname{Proj} k[\mathbf{V}_{n:m}(G)] \xrightarrow{\ e_G^*\ } \Pi(G) \xrightarrow{\ \Phi_G\ } \operatorname{Proj} H^{*,*}(G, k)$$

is bijective by Theorem 6.39 and the first map is surjective by Corollary 6.42. Hence, Φ_G is injective. $\square$

To prove the main result, Theorem 6.48, of this section we wish to extend Corollary 6.43 to all—not necessarily connected—unipotent finite supergroup schemes. The strategy for the reduction from the general to the connected case closely follows [45]. We outline the key steps of the argument citing the proofs from [45] where they apply verbatim.

Let E be an elementary abelian p-group, which we can view as a supergroup scheme concentrated in even degree. Denote by $\sigma_{\mathsf{E}} \in H^*(\mathsf{E}, k)$ the cohomology class defined as

$$\sigma_{\mathsf{E}} = \prod_{0 \neq \xi \in H^1(\mathsf{E}, \mathbb{F}_p)} \beta(\xi),$$

where β denotes the Bockstein homomorphism. The key property of σ_{E} is that as a function on $\operatorname{Proj} H^*(\mathsf{E}, k)$ it vanishes on any subvariety $\operatorname{Proj} H^*(\mathsf{E}', k) \subset \operatorname{Proj} H^*(\mathsf{E}, k)$ for a proper subgroup E' of E.

For G a finite supergroup scheme with a group of connected components $\pi_0(G) = \mathsf{E}$ an elementary abelian p-group, we have a natural projection $\phi \colon G \to \mathsf{E}$. We set

$$\sigma_G = \phi^*(\sigma_{\mathsf{E}}) \in H^{*,0}(G, k).$$

The proof of the following result relies on the construction of the *Evens norm* [43] and goes exactly as in [45, 4.4].

Proposition 6.44 *Let $G = G^0 \rtimes \pi$ be a finite supergroup scheme with group of connected components π. Let E be an elementary abelian p-subgroup of π. Then we have an injective map*

$$\operatorname{Spec}(H^{*,*}(G^0 \rtimes \mathsf{E}, k)[\sigma^{-1}_{G^0 \rtimes \mathsf{E}}])/N_\pi(\mathsf{E}) \to \operatorname{Spec} H^{*,*}(G, k)$$

induced by the embedding of supergroups $G^0 \rtimes \mathsf{E} \hookrightarrow G$. □

Proposition 6.45 *[45, 4.5] Let $G = G^0 \rtimes \pi$ and let $(G^0)^\pi$ be the subsupergroup scheme of invariants of the connected component G^0 under the action of π. Then the image of the map in cohomology induced by the natural embedding of group schemes $(G^0)^\pi \times \pi \hookrightarrow G^0 \rtimes \pi = G$,*

$$H^{*,*}(G, k) \to H^{*,*}((G^0)^\pi \times \pi, k), \tag{6.9.2}$$

contains the $p^{|\pi|}$th power of every element. □

Corollary 6.46 *Let $G = G^0 \rtimes \mathsf{E}$ be a finite supergroup scheme with group of connected components E. The natural map*

$$\operatorname{Proj} H^{*,*}((G^0)^\pi \times \mathsf{E}, k) \to \operatorname{Proj} H^{*,*}(G, k)$$

is an embedding of topological spaces. □

The final preparatory result that we need underlines the connection between rank and support varieties. Let $\alpha\colon A \to kG$ be a k-rational π-point and $\mathfrak{p}(\alpha)$ be the corresponding homogeneous prime ideal in $H(G, k)$. Since α is defined over k, $\mathfrak{p}(\alpha)$ defines a closed point on $\operatorname{Proj} H(G, k)$ and we can localise $H^{*,*}(G, M)$ at $\mathfrak{p}(\alpha)$.

Lemma 6.47   *Let G be a unipotent finite supergroup scheme defined over an algebraically closed field k, and M a finite-dimensional kG-module. If α is a k-rational π-point, then $\alpha^*(M)$ has finite flat dimension if and only if $H^{*,*}(G, M)_{\mathfrak{p}(\alpha)} = 0$.*

Proof   This is similar to the second half of the proof of [45, Theorem 4.8].   □

Theorem 6.48   *If G is a unipotent finite supergroup scheme then the map*

$$\Phi_G\colon \Pi(G) \to \operatorname{Proj} H^{*,*}(G, k)$$

is bijective. If k is algebraically closed then for any finite-dimensional kG-module M, the map Φ_G takes equivalence classes of k-rational π-points in $\pi\text{-}\mathrm{supp}_G(M)$ bijectively onto the closed points of $\mathrm{supp}_G(M)$.

Proof   Surjectivity follows from Theorem 6.22, together with [26, Theorem 6.9] that shows that Φ_G is bijective for G elementary.

We have thus to show that if $\Phi_G(\alpha) = \Phi_G(\beta)$ for π-points α, β, then α is equivalent to β. We first consider the case when $\alpha\colon A \to kG$ and $\beta\colon A \to kG$ are both defined over the ground field k which we assume to be algebraically closed.

In this case the proof is essentially the same as the proof of Theorem 4.6 of [45] and proceeds in a series of reductions. We sketch the main steps here and refer the reader to [45] for details.

Since k is perfect, we can write $G = G^0 \rtimes \pi$ where G^0 is connected and π is a finite group (the group of connected components of G). By definition, α factors through some elementary supergroup $E_\alpha = E_\alpha^0 \times \pi_0(E_\alpha)$ and similarly β factors through $E_\beta = E_\beta^0 \times \pi_0(E_\beta)$. By choosing representatives α, β within their equivalence classes in such a way that $\pi_0(E_\alpha)$ and $\pi_0(E_\beta)$ are minimal, we can apply Proposition 6.44 to show that $\pi_0(E_\beta)$ is conjugate to $\pi_0(E_\alpha)$ by an element in $N_\pi(\pi_0(E_\alpha))$. Since conjugation preserves equivalence of π-points, we can now assume that both π-points factor through $G^0 \rtimes \pi_0(E_\alpha)$. Hence, we reduce to the case when $G = G^0 \rtimes \mathsf{E}$ with the group of connected components elementary abelian.

Since α, β factor through elementary sub-supergroup schemes of $G = G^0 \rtimes \mathsf{E}$, they both must factor through $(G^0)^{\mathsf{E}} \times \mathsf{E}$. Corollary 6.46 now implies that we can assume that $G = (G^0)^{\mathsf{E}} \times \mathsf{E}$ which completes the second reduction step. Finally, since the group algebra of $(G^0)^{\mathsf{E}} \times \mathsf{E}$ is isomorphic to a group algebra of a connected unipotent finite supergroup scheme, the statement follows from Corollary 6.43.

The statement about Φ_G identifying equivalences classes of π-points defined over k and closed points of $\mathrm{supp}_G(M)$ is easily seen to be equivalent to Lemma 6.47.

We finally prove injectivity of Φ_G for any π-points.

Let α, β be π-points such that $\mathfrak{p}(\alpha) = \mathfrak{p}(\beta)$. Since extending scalars does not affect $\mathfrak{p}(\alpha)$, we can assume that $\alpha, \beta \colon A_K \to KG_K$ are defined over the same field K which is algebraically closed. Suppose α and β are not equivalent. Then there exists a finite dimensional kG-module M such that $\alpha^*(M)$ has infinite flat dimension but $\beta^*(M)$ does not, or vice versa. Let $\mathfrak{p}(\alpha, K)$ be the radical of the kernel of the map $\alpha^* \colon H(G_K, K) \to H(A_K, K)$ which is a point in $\mathrm{Proj}\, H(G_K, K)$ lying over $\mathfrak{p}(\alpha)$, and similarly for $\mathfrak{p}(\beta, K)$. Since Φ_{G_K} takes π-points defined over K to the K-closed points of $\mathrm{supp}_{G_K}(M_K)$, we conclude that $\mathfrak{p}(\alpha, K) \in \mathrm{supp}_{G_K}(M_K)$. Hence, $\mathrm{Ann}_{H(G_K, K)}\, H(G_K, M_K) \subset \mathfrak{p}(\alpha, K)$ (see Remark 6.19); and similarly, $\mathrm{Ann}_{H(G_K, K)}\, H(G_K, M_K) \not\subset \mathfrak{p}(\beta, K)$. But for an ideal $\mathfrak{I}$ of $H(G, k)$ we have $\mathfrak{I} \subset \mathfrak{p}(\alpha) = \mathfrak{p}(\alpha, K) \cap H(G, k)$ if and only if $\mathfrak{I} \subset \mathfrak{p}_{\alpha, K}$. We conclude that $\mathfrak{p}(\alpha) \neq \mathfrak{p}(\beta)$ which is a contradiction. Hence, Φ_G is injective. $\square$

Remark 6.49 We can endow $\Pi(G)$ with the structure of the topological space by choosing π-$\mathrm{supp}_G(M)$ to be a closed set exactly when M is a finite dimensional kG-module. With this topology Φ_G becomes a homeomorphism.

Remark 6.50 A remarkable property of a connected finite group scheme G of height one (equivalently, a p-restricted Lie algebra) is that any equivalence class of π-points has a representative $K[t]/t^p \to KG$ which is actually a map of Hopf algebras. This goes back to the theorem of Suslin, Friedlander, and Bendel who identified $\mathrm{Spec}\, H^*(G, k)$ with the restricted nullcone of $\mathrm{Lie}\, G$ ([55, Lemma 1.2], [56, Theorem 5.2]). This implies, in particular, that the restriction functor defined by such a special "Hopf" π-point $\alpha \colon K[t]/t^p \to KG$

$$\alpha^* \colon \mathrm{StMod}\, KG \to \mathrm{StMod}\, K[t]/t^p$$

is a *tensor* functor.

We observe that the same phenomenon holds in the super case. Let G be a connected unipotent supergroup scheme of height 1. Then any elementary supergroup subscheme of G as listed in Theorem 6.20 must have trivial discrete part $(\mathbb{Z}/p)^s$ $(s = 0)$ and also have $n = 1$. Explicitly, we can only have $\mathbb{G}_{a(1)}$, $\mathbb{G}_{a(1)} \times \mathbb{G}_a^-$, and $E_{m,1}^-$, $E_{m,1,\mu}^-$. We consider the last two cases since the first two are easier.

In the case of $E = E_{m,1}^-$ (and also $E_{m,1,\mu}^-$) the special π-point representative $\alpha_\lambda \colon A_K \to kE$ as described in (6.8.1) will take the form

$$
\begin{aligned}
\alpha_\lambda(t) &= \lambda_1 s_1^{p^{m-1}} + \lambda_2^2 s_1, \\
\alpha_\lambda(\tau) &= \lambda_2^p \sigma.
\end{aligned}
\tag{6.9.3}
$$

Since s_1 and σ are primitive, α_λ defines a Hopf superalgebra map.

6.10 Stratification

As always k will be a field of characteristic $p \geqslant 3$ and G a unipotent finite supergroup scheme over k. Knowing that Φ_G is bijective we can apply the techniques developed for finite group schemes to identify π-supp and supp for *all* kG-modules M, thereby proving the analogue of the result of Avrunin and Scott (Theorem 6.10) for unipotent supergroup schemes. The arguments for the results stated in this section are similar to the ones for finite group schemes which appear in [23, 46], so we give only a brief outline of how they go, referring the reader to those papers for details.

We first compute π-supports of the local cohomology modules introduced in Section 6.2. From now on we identity $\operatorname{Proj} H^{*,*}(G, k)$ with $\Pi(G)$, using Theorem 6.48.

Proposition 6.51 *Let G be a unipotent finite supergroup scheme. If V is a specialisation closed subset of $\operatorname{Proj} H^{*,*}(G, k)$ then*

(i) $\pi\text{-}\mathrm{supp}_G(\Gamma_V(k)) = V$,

(ii) $\pi\text{-}\mathrm{supp}_G(L_V(k))$ *is the complement of V, and*

(iii) $\pi\text{-}\mathrm{supp}_G(\Gamma_{\mathfrak{p}}(k)) = \{\mathfrak{p}\}$.

Proof The proof of this is the same as in Proposition 6.6 of [46]. For the last statement, we choose specialisation closed subsets V and W as in Definition 6.3, and use the tensor product formula given in Theorem 6.33. $\qquad \square$

Given Theorems 6.22 and 6.33, and Proposition 6.51, one can argue as in the proof of [23, Theorem 6.1] to establish the result below.

Theorem 6.52 *Let G be a unipotent finite supergroup scheme. For any kG-module M one has $\operatorname{supp}_G(M) = \pi\text{-}\operatorname{supp}_G(M)$.* $\square$

At this point cosupport can no longer be ignored: We write $\operatorname{cosupp}_G(M)$ for the cosupport of a kG-module M, defined as in 6.4, using the action of $H^{*,*}(G, k)$ on $\operatorname{StMod}(kG)$. The π-*cosupport* of M consists of equivalence classes of π-points $\alpha\colon A_K \to KG_K$ such that $\alpha^*(\operatorname{Hom}_k(K, M))$ has infinite flat dimension; compare with Definition 6.30. It has to be checked that this is well defined; see [26, Theorem 4.12] for the case of elementary supergroup schemes. Then, using [24, Theorem 11.3] one can verify that π-cosupport, like π-support, detects projectivity.

Theorem 6.53 *Let G be a unipotent finite supergroup scheme. A kG-module M is projective if and only if $\pi\text{-}\operatorname{cosupp}_G(M) = \varnothing$.* $\square$

Here is the analogue of the tensor-product formula 6.33 for π-cosupport, and it is proved in the same way: reduce to the super elementary case and apply [26, Theorem 7.6].

Theorem 6.54 *Let G be a unipotent finite supergroup scheme over a field k of characteristic $p \geqslant 3$, and let M and N be kG-modules. As subsets of $\Pi(G)$ we have*

$$\pi\text{-}\operatorname{cosupp}_G \operatorname{Hom}_k(M, N) = \pi\text{-}\operatorname{supp}_G(M) \cap \pi\text{-}\operatorname{cosupp}_G(N). \qquad \square$$

It is now a simple matter to establish that $\operatorname{StMod}(kG)$ is stratified by the action of $H^{*,*}(G, k)$. As explained in Section 6.2 this yields the classification of localising subcategories of $\operatorname{StMod}(kG)$, namely, Theorem 6.1.

Theorem 6.55 *Let G be a unipotent finite supergroup scheme over k. For point $\mathfrak{p}$ in $\operatorname{Proj} H^{*,*}(G, k)$, the localising subcategory $\Gamma_{\mathfrak{p}} \operatorname{StMod}(kG)$ of $\operatorname{StMod}(kG)$ is minimal. In particular, there is a bijection, defined by $\pi\text{-}\operatorname{supp}_G(-)$, between localising subcategories of $\operatorname{StMod}(kG)$ and subsets of $\operatorname{Proj} H^{*,*}(G, k)$.*

Proof Since G is unipotent every localising subcategory of $\operatorname{StMod}(kG)$ is tensor ideal. Thus minimality of $\Gamma_{\mathfrak{p}} \operatorname{StMod}(kG)$ is tantamount to the statement that for any non-projective modules M, N in this subcategory, the kG-module $\operatorname{Hom}_k(M, N)$ is not projective; see [20, Lemma 3.9].

We can verify this in two ways: Using Proposition 6.51, and Theorems 6.53 and 6.54, one can mimic the proof of [22, Theorem 6.1] to prove that

$$\pi\text{-}\operatorname{cosupp}_G(M) = \operatorname{cosupp}_G(M) \quad \text{for any } kG\text{-module } M.$$

Fix kG-modules M and N with $\operatorname{supp}_G(M) = \{\mathfrak{p}\} = \operatorname{supp}_G(N)$. Then $\mathfrak{p}$ is also in $\pi\text{-}\operatorname{supp}_G(M)$, by Theorem 6.52. Moreover, $\mathfrak{p}$ is in the cosupport of N because $\operatorname{supp}_G(N)$ and $\operatorname{cosupp}_G(N)$ have the same maximal elements [21, Theorem 4.13], and hence also in $\pi\text{-}\operatorname{cosupp}_G(N)$. Then using Theorem 6.54 one gets

$$\pi\text{-}\operatorname{cosupp}_G \operatorname{Hom}_k(M, N) = \pi\text{-}\operatorname{supp}_G(M) \cap \pi\text{-}\operatorname{cosupp}_G(N) = \{\mathfrak{p}\}.$$

Thus $\operatorname{Hom}_k(M, N)$ is not projective, as desired.

Here is another way, which circumvents the cosupport detection theorem: Let $\mathfrak{m}$ be a closed point of $\operatorname{Proj} H^{*,*}(G, k)$. We argue as in [23, Proposition 6.3]: Choose a finite field extension K of k and a π-point $\alpha\colon A_K \to KG$ representing $\mathfrak{m}$. For any kG-module M there is an isomorphism of G_K-modules

$$\operatorname{Hom}_k(K, k) \otimes_k M \cong \operatorname{Hom}_k(K, M).$$

Since the module on the left is a direct sum of copies of M_K, we deduce that $\mathfrak{m}$ is in $\operatorname{supp}_G(M)$ if and only if it is in $\operatorname{cosupp}_G(M)$; this is the crucial observation. For then given non-projective M, N in $\Gamma_{\mathfrak{m}} \operatorname{StMod}(kG)$ it allows us to conclude that $\mathfrak{m}$ is also in $\operatorname{cosupp}_G(N)$, and hence Theorem 6.54 yields

$$\pi\text{-}\operatorname{cosupp}_G(\operatorname{Hom}_k(M, N)) = \{\mathfrak{m}\}.$$

Applying the observation once again, we conclude that $\mathfrak{m}$ is in the π-support of $\operatorname{Hom}_k(M, N)$, so it is not projective, by Theorem 6.32.

This settles the desired minimality at closed points in $\operatorname{Proj} H^{*,*}(G, k)$. A technique of reduction to closed points from [23, §8], see also [25, §3], allows us to treat general prime ideals. Note that the extra grading on $H^{*,*}(G, k)$ introduces no new complications because its projective spectrum is the same as that of the singly graded ring $H(G, k)$ discussed in Definition 6.17.

This finishes the proof of the theorem. □

As noted above, one can prove Theorem 6.55 without first establishing that $\pi\text{-}\operatorname{cosupp}_G(M) = \operatorname{cosupp}_G(M)$ for a kG-module M. In fact one can deduce this equality, and hence that π-cosupport detects projectivity, from the theorem above. This is what is done in [23, Part IV], where the reader can find further applications of the theorem. We mention one such application, the classification of thick subcategories in stmod kG. We introduce more terminology before stating the result.

A full triangulated subcategory $\mathfrak{C}$ in stmod kG is called *thick* if it closed under taking direct summands.

Theorem 6.56 *Let G be a unipotent finite supergroup scheme over k. There is a bijection, defined by $\pi\text{-}\mathrm{supp}_G(-)$, between thick subcategories of* stmod(kG) *and specialisation closed subsets of* $\mathrm{Proj}\, H^{*,*}(G,k)$.

The proof goes exactly as in [23, Theorem 10.3]. The result can be restated in the language of tensor triangular geometry by saying that the Balmer spectrum of stmod kG is homeomorphic to $\mathrm{Proj}\, H^{*,*}(G,k)$ (see [3, Theorem 5.2]).

Remark 6.57 Theorem 6.56 can be contrasted with calculations of Boe, Kujawa, and Nakano for a certain stable category of finite dimensional modules over the Lie superalgebra $\mathfrak{gl}(m|n)$ in characteristic 0 ([29]). There, the cohomology ring is more straightforward but does not carry enough information about the stable category. The authors achieve the classification of thick tensor ideal subcategories via a combination of reducing to certain elementary subalgebras and invariant theory.

6.11 Local duality

Let now G be any (not necessarily unipotent) finite supergroup scheme over a field k of positive characteristic $p \geqslant 3$. The group algebra kG is Frobenius; thus there is an isomorphism of kG-modules

$$\mathrm{Hom}_k(kG, k) \cong kG \otimes_k \delta_G$$

where δ_G is a one-dimensional kG-module and a super analogue of the modular function for finite group schemes; see [28, §4]. In particular, it has a parity, denoted ϵ_G, that records its internal degree. The isomorphism above leads to a duality on the stable category of G, namely for all M, N in stmod(kG), there is a natural isomorphism

$$\mathrm{Hom}_k(\underline{\mathrm{Hom}}_{kG}(M,N), k) \cong \underline{\mathrm{Hom}}_{kG}(N, \Omega(M \otimes_k \delta_G)). \qquad (6.11.1)$$

The isomorphism above can be deduced from Auslander's defect formula; see [25, §4] where this is done for finite group schemes. When G is a finite group, the isomorphism above is nothing but classical Tate duality. Building on the Tate duality theorem, one can mimic the arguments in

[25], or better yet [27] that treats general finite dimensional Gorenstein algebras, to get:

Theorem 6.58   *Set $H(G) := H^{*,*}(G,k)$ and fix $\mathfrak{p}$ in $\operatorname{Proj} H(G)$. The kG-module $\Gamma_\mathfrak{p}(\delta_G)$ is a dualising object in $\Gamma_\mathfrak{p} \operatorname{StMod}(kG)$, in that, for any kG-module M there is a natural isomorphism*

$$\widehat{\operatorname{Ext}}_{kG}^{i,*}(M, \Gamma_\mathfrak{p}(\delta_G)) \cong \operatorname{Hom}_{H(G)}(H^{*-d-i,*+\epsilon_G}(G,M), I(\mathfrak{p}))$$

where $d = \dim H(G)/\mathfrak{p}$ and $I(\mathfrak{p})$ is the injective hull of the residue field at $\mathfrak{p}$.   □

From this one gets, for example, a local cohomology spectral sequence:

$$H_\mathfrak{p}^{s,t,j} H^{*,*}(G,M)_\mathfrak{p} \Rightarrow H_{-s-t-d,j+\epsilon_G}(G, M \otimes \delta_G),$$

discovered by Greenlees and Lyubeznik in the context of finite groups [48, Theorem 2.1], [49, Lemma 7.1].

References

[1] M. Auslander. Functors and morphisms determined by objects. In *Representation theory of algebras (Proc. Conf., Temple Univ., Philadelphia, Pa., 1976)*, pages 1–244. Lecture Notes in Pure Appl. Math., Vol. 37, 1978.

[2] G. S. Avrunin and L. L. Scott. A Quillen stratification theorem for modules. *Bull. Amer. Math. Soc. (N.S.)*, 6(1):75–78, 1982.

[3] P. Balmer. The spectrum of prime ideals in tensor triangulated categories. *J. Reine Angew. Math.*, 588:149–168, 2005.

[4] P. Balmer. Picard groups in triangular geometry and applications to modular representation theory. *Trans. Amer. Math. Soc.*, 362(7):3677–3690, 2010.

[5] P. Balmer and G. Favi. Gluing techniques in triangular geometry. *Q. J. Math.*, 58(4):415–441, 2007.

[6] T. Barthel, D. Heard, and G. Valenzuela. Local duality for structured ring spectra. *J. Pure Appl. Algebra*, 222(2):433–463, 2018.

[7] C. P. Bendel. Cohomology and projectivity of modules for finite group schemes. *Math. Proc. Cambridge Philos. Soc.*, 131(3):405–425, 2001.

[8] D. J. Benson. *Representations and cohomology. II*, volume 31 of *Cambridge Studies in Advanced Mathematics*. Cambridge University Press, Cambridge, 1991. Cohomology of groups and modules.

[9] D. J. Benson. Modules with injective cohomology, and local duality for a finite group. *New York J. Math.*, 7:201–215, 2001.

[10] D. J. Benson. Idempotent kG-modules with injective cohomology. *J. Pure Appl. Algebra*, 212(7):1744–1746, 2008.

[11] D. J. Benson, J. F. Carlson, and J. Rickard. Complexity and varieties for infinitely generated modules. II. *Math. Proc. Cambridge Philos. Soc.*, 120(4):597–615, 1996.

[12] D. J. Benson and J. P. C. Greenlees. Commutative algebra for cohomology rings of classifying spaces of compact Lie groups. *J. Pure Appl. Algebra*, 122(1-2):41–53, 1997.

[13] D. J. Benson and J. P. C. Greenlees. Commutative algebra for cohomology rings of virtual duality groups. *J. Algebra*, 192(2):678–700, 1997.

[14] D. J. Benson and J. P. C. Greenlees. Localization and duality in topology and modular representation theory. *J. Pure Appl. Algebra*, 212(7):1716–1743, 2008.

[15] D. J. Benson and J. P. C. Greenlees. Stratifying the derived category of cochains on *BG* for *G* a compact Lie group. *J. Pure Appl. Algebra*, 218(4):642–650, 2014.

[16] D. J. Benson, S. Iyengar, and H. Krause. *Representations of finite groups: local cohomology and support*, volume 43 of *Oberwolfach Seminars*. Birkhäuser/Springer Basel AG, Basel, 2012.

[17] D. J. Benson, S. B. Iyengar, and H. Krause. Local cohomology and support for triangulated categories. *Ann. Sci. Éc. Norm. Supér. (4)*, 41(4):573–619, 2008.

[18] D. J. Benson, S. B. Iyengar, and H. Krause. Localising subcategories for cochains on the classifying space of a finite group. *C. R. Math. Acad. Sci. Paris*, 349(17-18):953–956, 2011.

[19] D. J. Benson, S. B. Iyengar, and H. Krause. Stratifying modular representations of finite groups. *Ann. of Math. (2)*, 174(3):1643–1684, 2011.

[20] D. J. Benson, S. B. Iyengar, and H. Krause. Stratifying triangulated categories. *J. Topol.*, 4(3):641–666, 2011.

[21] D. J. Benson, S. B. Iyengar, and H. Krause. Colocalizing subcategories and cosupport. *J. Reine Angew. Math.*, 673:161–207, 2012.

[22] D. J. Benson, S. B. Iyengar, H. Krause, and J. Pevtsova. Stratification and π-cosupport: finite groups. *Math. Z.*, 287(3-4):947–965, 2017.

[23] D. J. Benson, S. B. Iyengar, H. Krause, and J. Pevtsova. Stratification for module categories of finite group schemes. *J. Amer. Math. Soc.*, 31(1):265–302, 2018.

[24] D. J. Benson, S. B. Iyengar, H. Krause, and J. Pevtsova. Detecting nilpotence and projectivity over finite supergroup schemes. *Selecta Math. New Ser. 27*, 25, 2021.

[25] D. J. Benson, S. B. Iyengar, H. Krause, and J. Pevtsova. Local duality for representations of finite group schemes. *Compos. Math.*, 155(2):424–453, 2019.

[26] D. J. Benson, S. B. Iyengar, H. Krause, and J. Pevtsova. Rank varieties and π-points for elementary supergroup schemes. To appear in *Trans. Amer. Math. Soc.*

[27] D. J. Benson, S. B. Iyengar, H. Krause, and J. Pevtsova. Local duality for the singularity category of a finite dimensional Gorenstein algebra. *Nagoya Math. J.*, 244:1–24, 2021.

[28] D. J. Benson and J. Pevtsova. Representations and cohomology of a family of finite supergroup schemes. *J. Algebra*, 561:84–110, 2020.

[29] B. D. Boe, J. R. Kujawa, and D. K. Nakano. Tensor triangular geometry for classical Lie superalgebras. *Adv. Math.*, 314:228–277, 2017.

[30] R.-O. Buchweitz. Maximal Cohen–Macaulay modules and Tate cohomology. To appear in *Math. Surveys and Monographs, Amer. Math. Soc.* https://tspace.library.utoronto.ca/handle/1807/16682, 1986.

[31] J. F. Carlson. The varieties and the cohomology ring of a module. *J. Algebra*, 85(1):104–143, 1983.

[32] L. G. Chouinard. Projectivity and relative projectivity over group rings. *J. Pure Appl. Algebra*, 7(3):287–302, 1976.

[33] E. Dade. Endo-permutation modules over p-groups. II. *Ann. of Math. (2)*, 108(2):317–346, 1978.

[34] I. Dell'Ambrogio. Tensor triangular geometry and KK-theory. *J. Homotopy Relat. Struct.*, 5(1):319–358, 2010.

[35] I. Dell'Ambrogio and G. Stevenson. On the derived category of a graded commutative Noetherian ring. *J. Algebra*, 373:356–376, 2013.

[36] C. Drupieski and J. Kujawa. Support schemes for infinitesimal unipotent supergroups. Preprint, 2018.

[37] C. M. Drupieski. Cohomological finite-generation for finite supergroup schemes. *Adv. Math.*, 288:1360–1432, 2016.

[38] C. M. Drupieski and J. R. Kujawa. Graded analogues of one-parameter subgroups and applications to the cohomology of $GL_{m|n(r)}$. *Adv. Math.*, 348:277–352, 2019.

[39] C. M. Drupieski and J. R. Kujawa. On support varieties for Lie superalgebras and finite supergroup schemes. *J. Algebra*, 525:64–110, 2019.

[40] C. M. Drupieski and J. R. Kujawa. On the cohomological spectrum and support varieties for infinitesimal unipotent supergroup schemes. In *Advances in algebra*, volume 277 of *Springer Proc. Math. Stat.*, pages 121–167. Springer, 2019.

[41] W. G. Dwyer and J. P. C. Greenlees. Complete modules and torsion modules. *Amer. J. Math.*, 124(1):199–220, 2002.

[42] W. G. Dwyer, J. P. C. Greenlees, and S. Iyengar. Duality in algebra and topology. *Adv. Math.*, 200(2):357–402, 2006.

[43] L. Evens. The cohomology ring of a finite group. *Trans. Amer. Math. Soc.*, 101:224–239, 1961.

[44] E. M. Friedlander and B. J. Parshall. Support varieties for restricted Lie algebras. *Invent. Math.*, 86(3):553–562, 1986.

[45] E. M. Friedlander and J. Pevtsova. Representation-theoretic support spaces for finite group schemes. *Amer. J. Math.*, 127(2):379–420, 2005.

[46] E. M. Friedlander and J. Pevtsova. Π-supports for modules for finite group schemes. *Duke Math. J.*, 139(2):317–368, 2007.

[47] E. M. Friedlander and A. Suslin. Cohomology of finite group schemes over a field. *Invent. Math.*, 127(2):209–270, 1997.

[48] J. P. C. Greenlees. Commutative algebra in group cohomology. *J. Pure Appl. Algebra*, 98(2):151–162, 1995.

[49] J. P. C. Greenlees and G. Lyubeznik. Rings with a local cohomology theorem and applications to cohomology rings of groups. *J. Pure Appl. Algebra*, 149(3):267–285, 2000.

[50] J. P. C. Greenlees and J. P. May. Derived functors of I-adic completion and local homology. *J. Algebra*, 149(2):438–453, 1992.

[51] M. Hovey, J. H. Palmieri, and N. P. Strickland. Axiomatic stable homotopy theory. *Mem. Amer. Math. Soc.*, 128(610):x+114, 1997.

[52] J. C. Jantzen. *Representations of algebraic groups*, volume 107 of *Mathematical Surveys and Monographs*. American Mathematical Society, Providence, RI, second edition, 2003.

[53] J. Pevtsova. Infinite dimensional modules for Frobenius kernels. *J. Pure Appl. Algebra*, 173(1):59–86, 2002.

[54] D. Quillen. The spectrum of an equivariant cohomology ring. I, II. *Ann. of Math. (2)*, 94:549–572; ibid. (2) 94 (1971), 573–602, 1971.

[55] A. Suslin, E. M. Friedlander, and C. P. Bendel. Infinitesimal 1-parameter subgroups and cohomology. *J. Amer. Math. Soc.*, 10(3):693–728, 1997.

[56] A. Suslin, E. M. Friedlander, and C. P. Bendel. Support varieties for infinitesimal group schemes. *J. Amer. Math. Soc.*, 10(3):729–759, 1997.

7
Bi-incomplete Tambara Functors

Andrew J. Blumberg[a]

Michael A. Hill[b]

Abstract

For an equivariant commutative ring spectrum R, $\pi_0 R$ has algebraic structure reflecting the presence of both additive transfers and multiplicative norms. The additive structure gives rise to a Mackey functor and the multiplicative structure yields the additional structure of a Tambara functor. If R is an N_∞ ring spectrum in the category of genuine G-spectra, then all possible additive transfers are present and $\pi_0 R$ has the structure of an incomplete Tambara functor. However, if R is an N_∞ ring spectrum in a category of incomplete G-spectra, the situation is more subtle.

In this chapter, we study the algebraic theory of Tambara structures on incomplete Mackey functors, which we call bi-incomplete Tambara functors. Just as incomplete Tambara functors have compatibility conditions that control which systems of norms are possible, bi-incomplete Tambara functors have algebraic constraints arising from the possible interactions of transfers and norms. We give a complete description of the possible interactions between the additive and multiplicative structures.

7.1 Introduction

The complexity of the equivariant stable category for a finite group G is a consequence of the desideratum that the orbits G/H must be dualizable. In contrast to the non-equivariant setting, there are many possible variants of the equivariant stable category determined by which

[a] Department of Mathematics, Columbia University
[b] Department of Mathematics, University of California Los Angeles

orbits are dualizable. Classically, this structure is captured by a universe, an infinite-dimensional G-inner product space that contains infinitely many copies of a collection of finite-dimensional G-inner product spaces including $\mathbb{R}^n$ for each n. A result of Lewis tells us that G/H is dualizable in the equivariant stable category structured by U if G/H embeds in U [18]. Another way of saying this is that the universe controls which transfer maps exist. On π_0, a shadow of this is reflected in the Mackey functor structure.

Equivariant commutative ring spectra have traditionally also been controlled by a universe in the form of the action of $\mathcal{L}_G(U)$, the G-equivariant linear isometries operad for a universe U [17]. Just as the "additive" structure of the equivariant stable category is expressed by the presence of transfer maps, the multiplicative structure encoded in the operad can be described in terms of the multiplicative norms introduced in Hill–Hopkins–Ravenel [13]. Building on this, we defined the notion of an N_∞-operad and showed that these operads control the transfers and norms in the equivariant stable setting [5]. Moreover, we showed that the data of these operads is essentially algebraic, encoded in *indexing systems*. An important aspect of this perspective is that indexing systems capture a more general range of possible compatible systems of norms or transfers than universes. We review the definitions and combinatorics of indexing systems (and associated "indexing categories") in Section 7.2.1.

This algebra becomes most concrete when we pass to π_0. When we restrict to a model of the equivariant stable category that only has some transfer maps (e.g., the $\mathcal{O}$-stable categories of [7]), π_0 has the structure of an *incomplete Mackey functor*. That is, for any indexing system we have a notion of an incomplete Mackey functor associated to that indexing system.

When working with N_∞ ring spectra, it is standard in the subject to assume that the additive structure is complete and study variation in the multiplicative structure. Then on π_0 we obtain various kinds of incomplete Tambara functors, which are Mackey functors equipped with additional norm maps satisfying certain compatibilities [6]. Since incomplete Tambara functors are less familiar than incomplete Mackey functors, we give a concrete description.

Tambara functors can be expressed in terms of a particular equivariant Lawvere theory, being a diagram category indexed by a category of "polynomials" or "bispans" (see [25] and [24]).

Definition (Definition 7.11)   Let $\mathcal{P}^G$ be the category with objects finite G-sets and where the morphisms from S to T are isomorphism classes of "polynomials": diagrams of the form

$$T_h \circ N_g \circ R_f := S \xleftarrow{f} U_1 \xrightarrow{g} U_2 \xrightarrow{h} T$$

with f, g, and h maps in $\mathcal{S}et^G$.

The composition rules for these are somewhat involved; we review the theory of polynomials in detail in Section 7.2.2.

Incomplete Tambara functors are defined by restricting the collection of maps g, parameterizing the "norm" N_g in the polynomials, to lie in a subcategory of $\mathcal{S}et^G$. *A priori*, this just describes a subgraph of $\mathcal{P}^G$; unpacking the requirements for this subgraph to be a category led us to the definition of an indexing category. When these maps are in an indexing category $\mathcal{O}$, then this gives the category $\mathcal{P}_{\mathcal{O}}^G$ of polynomials with exponents in $\mathcal{O}$.

The natural follow-up question is to determine what happens when we vary the "additive" structure as well, i.e., restricting the map h parameterizing the "transfer" T_h to also lie in some subcategory of $\mathcal{S}et^G$. It is clear we at least need to restrict to considering indexing categories here as well, but additional compatibility will be required.

Definition (Definition 7.26)   Let $\mathcal{O}_a$ and $\mathcal{O}_m$ be indexing categories, and let $\mathcal{P}_{\mathcal{O}_a, \mathcal{O}_m}^G$ be the wide directed subgraph of $\mathcal{P}^G$ so that the arrows from S to T are the isomorphism classes of polynomials

$$T_h \circ N_g \circ R_f := S \xleftarrow{f} U_1 \xrightarrow{g} U_2 \xrightarrow{h} T$$

with $g \in \mathcal{O}_m$ and $h \in \mathcal{O}_a$.

In these terms, the following is the main question studied in this chapter.

Main Question   What compatibility must we have between the additive indexing category $\mathcal{O}_a$ and the multiplicative indexing category $\mathcal{O}_m$ so that the subgraph $\mathcal{P}_{\mathcal{O}_a, \mathcal{O}_m}^G$ of $\mathcal{P}^G$ is a subcategory?

The subtle point here arises from the "equivariant distributive property" which records how to take a norm of a sum or a transfer. Following Mazur, we call this kind of interchange "Tambara reciprocity" [20], and a key feature is that the formulae depend only on G and its subgroups. In general, this will involve transfers and norms connecting many intermediate subgroups.

Example 7.1 For $G = C_4$ with generator γ, the norm $N_e^{C_4}$ associated to the unique map $C_4 \to *$ satisfies

$$N_e^{C_4}(a + b) = N_e^{C_4}(a) + N_e^{C_4}(b) + tr_{C_2}^{C_4}\left(N_e^{C_2}(a) \cdot \gamma N_e^{C_2}(b)\right)$$
$$+ tr_e^{C_4}(a \cdot \gamma a \cdot \gamma^2 a \cdot \gamma^3 b + a \cdot \gamma a \cdot \gamma^2 b \cdot \gamma^3 b + a \cdot \gamma b \cdot \gamma^2 b \cdot \gamma^3 b).$$

Incomplete Tambara functors are specified by including only a subset of the possible norm maps. The required compatibility check ensures that if we have the norm N_K^H, then we must also have any of the norms that occur in any Tambara reciprocity formula. If we also include only some of the transfers, then we have much more stringent conditions: as the example shows, we run into several different transfers in the Tambara reciprocity formulae.

In general, distributivity of the twisted product, parameterized by some g, over twisted sums is recorded by the "dependent product" Π_g (we review this in Definition 7.16 below). This allows us to concisely state the required compatibility data.

Definition (Definition 7.29) Let $\mathcal{O}_a$ and $\mathcal{O}_m$ be indexing categories. The indexing category $\mathcal{O}_m$ *distributes over* $\mathcal{O}_a$ if for all maps $g \colon S \to T$ in $\mathcal{O}_m$, we have

$$\Pi_g\big((\mathcal{O}_a)_{/S}\big) \subseteq (\mathcal{O}_a)_{/T}.$$

In this situation, we will say that the pair $(\mathcal{O}_a, \mathcal{O}_m)$ is *compatible*.

The first main theorem, proved in Section 7.3, guarantees that this is the right notion.

Theorem (Theorem 7.30) If $(\mathcal{O}_a, \mathcal{O}_m)$ is compatible, then $\mathcal{P}_{\mathcal{O}_a, \mathcal{O}_m}^G$ is a subcategory of $\mathcal{P}^G$.

This gives rise to the following basic definition of a bi-incomplete Tambara functor.

Definition (Definition 7.34) Let $(\mathcal{O}_a, \mathcal{O}_m)$ be a compatible pair of indexing categories. An $(\mathcal{O}_a, \mathcal{O}_m)$-semi-Tambara functor is a product preserving functor

$$\underline{R} \colon \mathcal{P}_{\mathcal{O}_a, \mathcal{O}_m}^G \to \mathcal{S}et.$$

An $(\mathcal{O}_a, \mathcal{O}_m)$-Tambara functor is an $(\mathcal{O}_a, \mathcal{O}_m)$-semi-Tambara functor $\underline{R}$ such that for all finite G-sets T, $\underline{R}(T)$ is an abelian group.

We write $(\mathcal{O}_a, \mathcal{O}_m)$-$\mathcal{T}amb$ to denote the category of $(\mathcal{O}_a, \mathcal{O}_m)$-Tambara functors.

Bi-incomplete Tambara functors have good categorical properties analogous to the properties of (incomplete) Tambara functors. We review these in Section 7.4, including a discussion of the additively incomplete box product.

In Section 7.5, we express the notion of compatibility in more concrete terms, eventually reducing verification of compatibility to checking a tractable combinatorial condition.

Theorem (Theorem 7.65)   Let $\mathcal{O}_a$ and $\mathcal{O}_m$ be indexing categories. Then $(\mathcal{O}_a, \mathcal{O}_m)$ is compatible if and only if for every pair of subgroups $K \subseteq H$ such that H/K is an admissible H-set for $\mathcal{O}_m$ and for every admissible K-set T for $\mathcal{O}_a$, the coinduced H-set $\mathrm{Map}^K(H, T)$ is admissible for $\mathcal{O}_a$.

This formulation makes it clear that there are harsh necessary conditions on a pair $(\mathcal{O}_a, \mathcal{O}_m)$ for them to be compatible; we explore these in Section 7.6. These conditions alone rule out about half of all possible pairs!

Proposition (Proposition 7.69)   If $(\mathcal{O}_a, \mathcal{O}_m)$ is a compatible pair and H/K is an $\mathcal{O}_m$-admissible set for H, then for every $L \subseteq H$ such that $K \subseteq L$, the H-set H/L is $\mathcal{O}_a$-admissible.

To get a sense for how this plays out in practice, we analyze the classical examples coming from equivariant little disks and equivariant linear isometries operads. On the one hand, we find that the little disks do not necessarily interact well with each other:

Corollary (Corollary 7.81)   For any non-simple group G, there is a universe U such that the indexing category associated to the little disks in U is not compatible with itself.

On the other hand, as is implicit in the classical literature, the indexing category for the linear isometries operad on a universe U is always compatible with the indexing category corresponding to the little disks operad for U.

Proposition (Proposition 7.82)   Let U be a universe for G, let $\mathcal{O}_a$ be the indexing category associated to the little disks operad for U and let $\mathcal{O}_m$ be the indexing category associated to the linear isometries operad for U. Then $(\mathcal{O}_a, \mathcal{O}_m)$ is compatible.

We close in Section 7.7 by proving some basic change-of-group results and then putting forward a series of conjectures about an "external" form of bi-incomplete Tambara functors. The thesis work of Mazur and of

Hoyer [20, 15] showed that Tambara functors were Mackey functors with additional structure, i.e., external norm maps. Our conjectures outline how such a description should work for bi-incomplete Tambara functors, exhibiting them as ordinary incomplete Mackey functors together with additional structure.

A key step is producing additively incomplete versions of the norm functor.

Conjecture (Conjecture 7.90)   If $(\mathcal{O}_a, \mathcal{O}_m)$ is a compatible pair of indexing categories, and H/K is a $\mathcal{O}_m$-admissible H-set, then there is a norm functor

$$N_K^H : i_K^* \mathcal{O}_a\text{-}\mathcal{M}ackey \to i_H^* \mathcal{O}_a\text{-}\mathcal{M}ackey$$

that is symmetric monoidal with respect to the box product.

These would assemble into the incomplete Mackey functor version of the multiplicative symmetric monoidal Mackey functor structure on the G-Mackey functors. In particular, we would hope to have the analogue of the Hoyer–Mazur theorem that Tambara functors are G-commutative monoids in Mackey functors.

Conjecture (Conjecture 7.94)   For any compatible pair of indexing categories $(\mathcal{O}_a, \mathcal{O}_m)$, there is an equivalence of categories between $\mathcal{O}_m$-commutative monoids in $\mathcal{O}_a$-Mackey functors and $(\mathcal{O}_a, \mathcal{O}_m)$-Tambara functors.

Acknowledgements

We offer our sincere, heartfelt thanks to John Greenlees. John offered his generous support to both of us as young people in the field, and we have found his beautiful mathematics deeply influential. In addition, the myriad ways he has contributed to the homotopy theory community has been an inspiration.

We thank Magdalena Kędziorek for her support and understanding through this project, and we also thank the referee for an incredibly fast and detailed report.

Finally, we thank Mike Hopkins, Tyler Lawson, Mike Mandell, Peter May, and Jonathan Rubin for many interesting and helpful conversations about this and related mathematics.

This material is based upon work supported by the National Science Foundation under Grant No. DMS-1812064 and DMS-1811189.

7.2 Indexing systems, subcategories, and polynomials

The purpose of this section is to review the basic algebraic framework for incomplete Mackey and Tambara functors. Throughout the section, we will fix a finite group G and denote by $\mathcal{S}et^G$ the category of finite G-sets.

7.2.1 Indexing systems and categories

We begin with the definition of an indexing system; this is the basic categorical structure that organizes the possible relationships between transfers for different finite G-sets.

Definition 7.2   A *symmetric monoidal coefficient system* is a functor

$$\underline{C}\colon \mathcal{O}rb_G^{op} \to \mathcal{S}ym$$

from the opposite of the orbit category of G to the category of symmetric monoidal categories and strong symmetric monoidal functors.

Example 7.3   The fundamental example of a symmetric monoidal coefficient system is the functor which assigns to G/H the category of finite H-sets, with the symmetric monoidal structure induced by disjoint union; we denote this by $\underline{Set}^{\amalg}$.

Example 7.4   There is also a multiplicative version which assigns to G/H the category of finite H-sets with the Cartesian symmetric monoidal structure; we will denote this by $\underline{Set}^{\times}$

Definition 7.5   An *indexing system* is a full symmetric monoidal sub-coefficient system $\mathcal{O}$ of $\underline{Set}^{\amalg}$ that contains all trivial sets and is closed under

(1) levelwise finite limits and
(2) "self-induction": if $H/K \in \mathcal{O}(G/H)$ and $T \in \mathcal{O}(G/K)$, then

$$H \underset{K}{\times} T \in \mathcal{O}(G/H).$$

The collection of indexing systems for G forms a poset ordered by inclusion. This poset has a least and a greatest element.

Example 7.6   The poset of indexing systems has a least element: $\mathcal{O}^{tr}$, for which $\mathcal{O}^{tr}(G/H)$ is always the subcategory of $\mathcal{S}et^H$ of H-sets with a trivial H-action.

Example 7.7 The symmetric monoidal coefficient system $\underline{Set}^{\amalg}$ itself is an indexing system, $\mathcal{O}^{gen}$, and this is the maximal element in the poset of indexing systems.

We will often write $\mathcal{O}(H)$ for $\mathcal{O}(G/H)$.

Definition 7.8 We say that an H-set T is admissible if $T \in \mathcal{O}(H)$.

Indexing systems admit an intrinsic formulation, which we can interpret equivalently as gluing together all of $\mathcal{O}(H)$, viewed as subcategories of slice categories of Set^G via the equivalences

$$Set^H \xrightarrow[\simeq]{\underset{H}{G \times (\text{-})}} Set^G_{/(G/H)}.$$

Definition 7.9 An *indexing category* is a wide, pullback stable, finite coproduct complete subcategory $\mathcal{O}$ of the category of finite G-sets.

Indexing categories also form a poset under inclusion, and pullback stability and finite coproduct completeness guarantee that the assignment

$$G/H \mapsto \mathcal{O}_{/(G/H)}$$

gives us a map of posets from the poset of indexing categories to the poset of indexing systems.

Theorem 7.10 ([6, Theorem 1.4]) *The map from the poset of indexing categories to that of indexing systems is an isomorphism.*

Because of this isomorphism, we will engage in mild abuse of notation and use the same symbols to denote both indexing categories and the associated indexing systems.

7.2.2 Polynomials

The point of indexing categories, as opposed to indexing systems, is that this reformulation provides a convenient formalism for parameterizing norms for incomplete Tambara functors [6] and for incomplete Mackey functors [7]. Specifically, we employ a categorification of the notion of polynomials (also called "bispans" [24]). Although this definition works in the context of locally cartesian closed categories (by work of Gambino–Kock [10] and Weber [26]), we focus here for concreteness on Set^G.

Definition 7.11 The category $\mathcal{P}^G$ of *polynomials* has objects finite G-sets and morphisms between objects S and T the isomorphism classes of polynomials

$$T_h \circ N_g \circ R_f := S \xleftarrow{f} U_1 \xrightarrow{g} U_2 \xrightarrow{h} T$$

where an isomorphism of polynomials is specified by two isomorphisms $U_1 \to U_1'$ and $U_2 \to U_2'$ that make the evident diagrams commute.

The composition in the category is a little elaborate; we spend the rest of this subsection unpacking and explaining it.

Remark 7.12 The category $\mathcal{P}^G$ is obtained from a 2-category where the category of morphisms is the category with objects polynomials and morphisms maps of polynomials as in the definition, where the internally described square is a pullback square.

The category of polynomials can be presented in terms of generators and relations, which is often technically convenient. Specifically, as indicated, $T_h \circ N_g \circ R_f$ is a composite of basic maps:

Definition 7.13 Let $f\colon S \to T$ be a map of finite G-sets. Then we define the following morphisms

$$R_f = T \xleftarrow{f} S \xrightarrow{\mathrm{id}} S \xrightarrow{\mathrm{id}} S$$

$$N_f = S \xleftarrow{\mathrm{id}} S \xrightarrow{f} T \xrightarrow{\mathrm{id}} T$$

$$T_f = S \xleftarrow{\mathrm{id}} S \xrightarrow{\mathrm{id}} S \xrightarrow{f} T.$$

These are stand-ins for the three basic ways we might build the polynomials on a sequence of symbols:

R Repeat or drop some collection of the variables, then

N multiply collections of them (with "N" for the Galois theoretic norm), and

T sum up the result (with "T" for the transfer or trace).

Heuristically, the map N should be thought of as "multiply together the fibers over a point" and T as "sum together the fibers".

Composition in this category is most easily expressed by showing how to transform an arbitrary string of composable combinations of T, N, and Rs into one of the form in Definition 7.11.

First, we consider composing Ts with Ts, etc. In terms of our heuristic, we can duplicate, multiply, or add things either in stages or all at once.

Proposition 7.14 *For any composable maps $f\colon S \to T$ and $g\colon T \to U$, we have*

$$R_{g\circ f} = R_f \circ R_g$$
$$T_{g\circ f} = T_g \circ T_f$$
$$N_{g\circ f} = N_g \circ N_f.$$

In other words, T and N extend to covariant functors $\mathcal{S}et^G \to \mathcal{P}^G$, while R extends to a contravariant one.

Next, we encode the heuristic that adding or multiplying and then duplicating is the same as first duplicating and then adding or multiplying coordinatewise. This is expressed in terms of pullbacks, as follows.

Proposition 7.15 *Given a pullback diagram*

$$
\begin{array}{ccc}
T' & \xrightarrow{\;g'\;} & T \\
{\scriptstyle f'}\downarrow & & \downarrow{\scriptstyle f} \\
S' & \xrightarrow[\;g\;]{} & S,
\end{array}
$$

we have identities

$$R_f \circ T_g = T_{g'} \circ R_{f'} \text{ and } R_f \circ N_g = N_{g'} \circ R_{f'}.$$

The most complicated interchange is swapping T and N, which is a form of generalized distributivity. This uses the dependent product, which is heuristically the "product over the fibers".

Definition 7.16 If $g\colon S \to T$ is a map of finite G-sets, the *dependent product along g* is the functor

$$\Pi_g\colon \mathcal{S}et^G_{/S} \to \mathcal{S}et^G_{/T}$$

that is the right adjoint to the pullback along g.

Recall that an exponential diagram is a diagram isomorphic to one of the form

$$
\begin{array}{ccc}
T \xleftarrow{\;h\;} S \xleftarrow{\;f'\;} & T \underset{U}{\times} \Pi_g(S) \\
{\scriptstyle g}\downarrow & & \downarrow{\scriptstyle g'} \\
U \xleftarrow{\hspace{3.5em}} & \Pi_g(S), \\
 \qquad\; h'
\end{array}
$$

where

(1) h' is dependent product of h along g,

(2) g' is the pullback of g along h', and

(3) where f' is the counit of the pullback-dependent product adjunction.

Notice that only the maps h and g can vary freely; all of the other pieces are pseudofunctorially determined.

Proposition 7.17   *Given an exponential diagram*

$$
\begin{array}{ccc}
T \xleftarrow{\;h\;} S & \xleftarrow{\;f'\;} & T \underset{U}{\times} \Pi_g(S) \\
\Big\downarrow{g} & & \Big\downarrow{g'} \\
U \xleftarrow{\hspace{2.2cm}}_{h'} & & \Pi_g(S),
\end{array}
$$

then

$$
N_g \circ T_h = T_{h'} \circ N_{g'} \circ R_{f'}.
$$

Put another way, the exponential diagram records concisely how to express a general formula for the product of a sum.

Finally, we record an unexpected categorical result that arises from the asymmetry in the roles of R, S, and T.

Proposition 7.18   *For any objects S and T, the maps*

$$
\pi_S = [S \amalg T \leftarrow S \xrightarrow{=} S \xrightarrow{=} S] \text{ and } \pi_T = [S \amalg T \leftarrow T \xrightarrow{=} T \xrightarrow{=} T]
$$

are the projection maps witnessing $S \amalg T$ as the categorical product in $\mathcal{P}^G$.

In other words, regarding R as a functor

$$
R \colon \mathcal{S}et^{G,op} \to \mathcal{P}^G,
$$

it is product preserving.

7.2.3 Incomplete Tambara functors

A natural question to ask is when the subset of polynomials with maps in a restricted subcategory of finite G-sets itself forms a category. The key observation is that this occurs precisely when the subcategory in question is an indexing category [6, 2.10].

Theorem 7.19   *Let $\mathcal{O}$ be an indexing category. The category $\mathcal{P}^G_{\mathcal{O}}$ of polynomials with exponents in $\mathcal{O}$ has objects finite G-sets and morphisms between objects S and T the isomorphism classes of bispans*

$$
S \xleftarrow{\;f\;} U_1 \xrightarrow{\;g\;} U_2 \xrightarrow{\;h\;} T,
$$

where g is an arrow in $\mathcal{O}$. Disjoint union of finite G-sets is the categorical product.

With this definition in hand, we can define incomplete Tambara functors as follows.

Definition 7.20 ([6, Definition 4.1]) Let $\mathcal{O}$ be an indexing category. An $\mathcal{O}$-*semi-Tambara functor* is a product-preserving functor

$$\mathcal{P}_{\mathcal{O}}^G \to \mathcal{S}et.$$

An $\mathcal{O}$-*Tambara functor* is an $\mathcal{O}$-semi-Tambara functor which is valued in abelian groups.

7.2.4 Incomplete Mackey functors

Tambara's category of polynomials is a multiplicative generalization of Lindner's category of spans that records "linear functions" [19]. When we are interested in just the additive structure, we can choose the map g to be the identity and then we recover the Mackey version of the incomplete polynomials from [7].

Definition 7.21 ([7, Definition 2.23]) Let $\mathcal{O}$ be an indexing category. Then $\mathcal{A}_{\mathcal{O}}^G$ is the category with objects finite G-sets and morphisms isomorphism classes of spans

$$\mathcal{A}_{\mathcal{O}}^G(S,T) = \left\{ [S \leftarrow U \xrightarrow{h} T] \mid h \in \mathcal{O} \right\},$$

with composition given by pullback.

Remark 7.22 Once again, this is the quotient category associated to a 2-category in which we keep track of the coherence of iterated pullbacks.

Example 7.23 When $\mathcal{O} = \mathcal{S}et^G$, the category $\mathcal{A}_{\mathcal{O}}^G$ is the usual Lindner category of spans $\mathcal{A}^G$.

In the case of $\mathcal{O} = \mathcal{S}et^G$, this category is isomorphic to the subcategory of polynomials where the norm maps are only along isomorphisms, and this showed how to extract the underlying additive Mackey functor of an incomplete Tambara functor.

The disjoint union is again the product in $\mathcal{A}_{\mathcal{O}}^G$ (and in fact, here it is the biproduct).

Definition 7.24 ([7, Definition 2.24])   An $\mathcal{O}$-Mackey functor is a product preserving functor

$$\underline{M}\colon \mathcal{A}_{\mathcal{O}}^G \to \mathcal{S}et.$$

A semi-Mackey functor is a Mackey functor if it is group complete.

A map of $\mathcal{O}$-Mackey functors is a natural transformation. We will denote the category of $\mathcal{O}$-Mackey functors by $\mathcal{O}$-$\mathcal{M}ackey$.

Definition 7.25   The group completion of the representable $\mathcal{A}_{\mathcal{O}}(*,\text{-})$ is the $\mathcal{O}$-Burnside Mackey functor: $\underline{A}_{\mathcal{O}}$. For any subgroup $H \subseteq G$,

$$\underline{A}_{\mathcal{O}}(G/H) = \mathbb{Z}\big\{[H/K] \mid H/K \in \pi_0\mathcal{O}(H)\big\}$$

is the group completion of the commutative monoid $\pi_0\mathcal{O}(H)$ of isomorphism classes of admissible H-sets.

7.3 Additive Incompleteness

We now study the common generalization of incomplete Mackey and Tambara functors in terms of polynomials with additional restrictions.

Definition 7.26   If $\mathcal{O}_a$ and $\mathcal{O}_m$ are indexing categories, then let $\mathcal{P}^G_{\mathcal{O}_a,\mathcal{O}_m}$ be the wide subgraph of $\mathcal{P}^G$ with morphisms the isomorphism classes of polynomials

$$X \leftarrow S \xrightarrow{g} T \xrightarrow{h} Y,$$

with $g \in \mathcal{O}_m$ and $h \in \mathcal{O}_a$.

These are polynomials with norms parameterized by $\mathcal{O}_m$ and transfers parameterized by $\mathcal{O}_a$. We have no reason to believe, *a priori*, that this subgraph is actually a subcategory, and we can see what can go wrong with an explicit example.

Example 7.27   A $G = C_2$-Tambara functor is the following data:

(1) A commutative Green functor $\underline{R}$ and,
(2) a multiplicative map $n\colon \underline{R}(C_2/\{e\}) \to \underline{R}(C_2/C_2)$

that are required to satisfy the relations

(1) If $\overline{x}$ is the Weyl conjugate of x in $\underline{R}(C_2/\{e\})$, then

$$n(x) = n(\overline{x}) \text{ and } Res \circ n(x) = x \cdot \overline{x},$$

where Res is the restriction map $\underline{R}(C_2/C_2) \to \underline{R}(C_2/\{e\})$, and

(2) for any $a, b \in \underline{R}(C_2/\{e\})$, we have

$$n(a + b) = n(a) + n(b) + Tr(a \cdot \bar{b}),$$

where Tr is the transfer map $\underline{R}(C_2/\{e\}) \to \underline{R}(C_2/C_2)$.

In particular, the existence of the norm in this case necessitates the existence of the transfer map.

Remark 7.28 When dealing with specific classes of groups G, it might be reasonable to consider weaker compatibility requirements than the ones we study here. Specifically, it is not the case that for every group G we need to require transfers for all elements to consistently talk about norms. In fact, Georgakopoulos has shown that in that absence of torsion, the transfers themselves are forced by the existence of certain norms [11].

This puts some constraints on the additive indexing category. In fact, compatibility is a purely combinatorial condition.

Definition 7.29 Let $\mathcal{O}_a$ and $\mathcal{O}_m$ be indexing categories. The indexing system $\mathcal{O}_m$ *distributes over* $\mathcal{O}_a$ if for all maps $h \colon S \to T$ in $\mathcal{O}_m$, we have

$$\Pi_h\big((\mathcal{O}_a)_{/S}\big) \subseteq (\mathcal{O}_a)_{/T}.$$

In this situation, we will say that the pair $(\mathcal{O}_a, \mathcal{O}_m)$ is *compatible*.

The force of this definition is given by the following theorem.

Theorem 7.30 *If* $(\mathcal{O}_a, \mathcal{O}_m)$ *is compatible, then* $\mathcal{P}^G_{\mathcal{O}_a, \mathcal{O}_m}$ *is a subcategory of* $\mathcal{P}^G$.

Proof We need to verify the interchange formulae for T_h, N_g, and R_f, where $h \in \mathcal{O}_a$, $g \in \mathcal{O}_m$, and f is arbitrary.

For interchanges with R, assume we have pullback squares

$$
\begin{array}{ccc}
S' & \xrightarrow{f'} & S \\
{\scriptstyle g'}\downarrow & & \downarrow{\scriptstyle g} \\
T' & \xrightarrow{f} & T,
\end{array}
\qquad \text{and} \qquad
\begin{array}{ccc}
U' & \xrightarrow{f''} & U \\
{\scriptstyle h'}\downarrow & & \downarrow{\scriptstyle h} \\
T' & \xrightarrow{f} & T,
\end{array}
$$

where $g \in \mathcal{O}_m$ and $h \in \mathcal{O}_a$. By pullback stability, g' is also in $\mathcal{O}_m$ and h' in $\mathcal{O}_a$, and hence

$$R_f \circ N_g = N_{g'} \circ R_{f'} \text{ and } R_f \circ T_h = T_{h'} \circ R_{f''}$$

are again elements of $\mathcal{P}^G_{\mathcal{O}_a, \mathcal{O}_m}$.

Since both are subcategories, the compositions of Ts or Ns are also correct: for any composable pair $g_1, g_2 \in \mathcal{O}_m$ and $h_1, h_2 \in \mathcal{O}_a$,

$$N_{g_1} \circ N_{g_2} = N_{g_1 \circ g_2} \text{ and } T_{h_1} \circ T_{h_2} = T_{h_1 \circ h_2}$$

is of the desired form.

The key thing to check is therefore the interchange between T_h and N_g. For this, let

$$
\begin{array}{ccccc}
T & \xleftarrow{\ h\ } & S & \xleftarrow{\ f'\ } & T \underset{U}{\times} \Pi_g(S) \\
{\scriptstyle g} \big\downarrow & & & & \big\downarrow {\scriptstyle g'} \\
U & \xleftarrow{\hspace{3em} h' \hspace{3em}} & & & \Pi_g(S)
\end{array}
$$

be an exponential diagram expressing the interchange relation

$$N_g \circ T_h = T_{h'} \circ N_{g'} \circ R_{f'}.$$

The map g' is the pullback of g along h', and hence g being in $\mathcal{O}_m$ means that g' is in $\mathcal{O}_m$. The assumption that $(\mathcal{O}_a, \mathcal{O}_m)$ is compatible means exactly that h' is in $\mathcal{O}_a$, and hence the interchange is again in $\mathcal{P}^G_{\mathcal{O}_a, \mathcal{O}_m}$. $\qquad\square$

Example 7.31 For any $\mathcal{O}_m$, the pair $(\mathcal{O}^{gen}, \mathcal{O}_m)$ is compatible. The associated category of polynomials are precisely the "polynomials with exponents in $\mathcal{O}$" studied in [6].

Since the categories $\mathcal{O}_a$ and $\mathcal{O}_m$ are indexing categories, for any objects S and T, the maps

$$\pi_S = [S \amalg T \leftarrow S \xrightarrow{=} S \xrightarrow{=} S] \text{ and } \pi_T = [S \amalg T \leftarrow T \xrightarrow{=} T \xrightarrow{=} T]$$

are always in $\mathcal{P}^G_{\mathcal{O}_a, \mathcal{O}_m}$. These are the projection maps in the category $\mathcal{P}^G$, so we deduce the following.

Proposition 7.32 *If $(\mathcal{O}_a, \mathcal{O}_m)$ is compatible, then the disjoint union of finite G-sets is the categorical product in $\mathcal{P}^G_{\mathcal{O}_a, \mathcal{O}_m}$.*

In ordinary Mackey or incomplete Tambara functors, additive or multiplicative monoid structures on the mapping sets arise from transfering or norming along the fold maps. The proof of the usual case for Tambara functors (as in [25] or also in [24]) goes through without change in the bi-incomplete case, since the key features are that all of the terms are in fact again in the category in question. That is, in the bi-incomplete case, we simply skipped some transfers as well as norms.

Proposition 7.33   *Let $(\mathcal{O}_a, \mathcal{O}_m)$ be a compatible pair of indexing categories. Then the hom objects in $\mathcal{P}^G_{\mathcal{O}_a, \mathcal{O}_m}$ are naturally commutative semi-ring valued, with addition given by*

$$[S \leftarrow T_1 \rightarrow T_2 \rightarrow U] + [S \leftarrow V_1 \rightarrow V_2 \rightarrow U] = [S \leftarrow T_1 \amalg V_1 \rightarrow T_2 \amalg V_2 \rightarrow U],$$

and multiplication given by

$$[S \leftarrow T_1 \rightarrow T_2 \rightarrow U] \cdot [S \leftarrow V_1 \rightarrow V_2 \rightarrow U] =$$
$$\left[S \leftarrow \left((T_1 \underset{U}{\times} V_2) \amalg (T_2 \underset{U}{\times} V_1)\right) \rightarrow T_2 \underset{U}{\times} V_2 \rightarrow U\right].$$

The transfer maps are map of additive monoids and the norm maps are maps of multiplicative monoids.

This is formally exactly like what we see with Tambara and incomplete Tambara functor. There the hom objects are commutative semi-rings, and the restriction maps are ring homomorphisms.

Definition 7.34   Let $(\mathcal{O}_a, \mathcal{O}_m)$ be a compatible pair of indexing categories. An $(\mathcal{O}_a, \mathcal{O}_m)$-semi-Tambara functor is a product preserving functor

$$\underline{R} \colon \mathcal{P}^G_{\mathcal{O}_a, \mathcal{O}_m} \rightarrow \mathcal{S}et.$$

An $(\mathcal{O}_a, \mathcal{O}_m)$-Tambara functor is an $(\mathcal{O}_a, \mathcal{O}_m)$-semi-Tambara functor $\underline{R}$ such that for all finite G-sets T, $\underline{R}(T)$ is an abelian group.
  A map of $(\mathcal{O}_a, \mathcal{O}_m)$-Tambara functors is a natural transformation of product preserving functors.
  We will write $(\mathcal{O}_a, \mathcal{O}_m)$-$\mathcal{T}amb$ to denote the category of $(\mathcal{O}_a, \mathcal{O}_m)$-Tambara functors.

Implicit in the definition is that any product preserving functor

$$\mathcal{P}^G_{\mathcal{O}_a, \mathcal{O}_m} \rightarrow \mathcal{S}et$$

is naturally commutative semi-ring valued. This was a fundamental result of Tambara in the ordinary Tambara functor case [25]. Tambara also showed that the group completion of any semi-Tambara functor naturally has the structure of a Tambara functor. As in the incomplete Tambara functor case, the proof goes through without change.

Proposition 7.35 ([25], [24])   *Given any $(\mathcal{O}_a, \mathcal{O}_m)$-semi-Tambara functor $\underline{R}$, the object-wise group completion is an $(\mathcal{O}_a, \mathcal{O}_m)$-Tambara functor.*

More general compatibility

The proof of Theorem 7.30 only used that indexing categories are wide, pullback stable, together with a condition of closure under dependent product. That is, the same proof works to establish the analogous result in this more general case. It turns out that the wide subcategory of isomorphisms $\mathcal{S}et^G_{\cong}$, which is not an indexing category, will be necessary to define the "underlying" incomplete additive and multiplicative Mackey functors associated to a bi-incomplete Tambara functor.

Proposition 7.36 *Let $\mathcal{D}$ be any wide, pullback stable subcategory. Then*

(1) $\mathcal{S}et^G_{Iso}$ distributes over $\mathcal{D}$, and
(2) $\mathcal{D}$ distributes over $\mathcal{S}et^G_{Iso}$.

Proof For the first claim, note that the pullback along an isomorphism is an equivalence of categories, and hence the right adjoint, the dependent product, is naturally isomorphic to the pullback along the inverse to the isomorphism. Hence any pullback stable subcategory is closed under the dependent product along an isomorphism.

For the second claim, since dependent product is a functor, it takes isomorphisms to isomorphisms.   □

Note that $\mathcal{S}et^G_{Iso}$ is initial amongst all wide, pullback stable subcategories of $\mathcal{S}et^G$.

7.4 Categorical Properties

In this section, we describe the formal structure of the category of bi-incomplete Tambara functors.

7.4.1 The box product for incomplete Mackey functors

The box product of Mackey functors is essential in showing that Tambara functors have colimits, since this serves as a model for the Mackey functor underlying the coproduct. For coproducts in the bi-incomplete case, we need to build the box product of incomplete Mackey functors.

Lemma 7.37 *Indexing categories are closed under products: if $f_i \colon S_i \to T_i$ are in $\mathcal{O}$ with $i = 1, 2$, then*

$$f_1 \times f_2 \colon S_1 \times S_2 \to T_1 \times T_2$$

is as well.

Proof It suffices to show this when f_1 is the identity, since

$$f_1 \times f_2 = (1 \times f_2) \circ (f_1 \times 1),$$

and the twist map is an isomorphism (and hence in any indexing category). Since the Cartesian product distributes over the disjoint union and since indexing categories are pullback stable, we reduce to the case $S_1 = G/H$. This means we consider

$$1 \times f \colon G/H \times S \to G/H \times T.$$

We have a commutative square

$$
\begin{array}{ccc}
G/H \times S & \xleftarrow{\;\cong\;} & G \underset{H}{\times} i_H^* S \\[2pt]
\scriptstyle{1 \times f} \big\downarrow & & \big\downarrow \scriptstyle{1 \underset{H}{\times} i_H^* f} \\[2pt]
G/H \times T & \xrightarrow[\;\cong\;]{} & G \underset{H}{\times} i_H^* T,
\end{array}
$$

where the unlabeled maps are the natural "shearing" isomorphisms (the asymmetrical directions of the horizontal maps is to help make transparent that the right-hand map is in the image of induction). However, by [6, Proposition 3.13], induction preserves indexing categories, and by [6, Proposition 6.3], so does restriction. $\qquad\square$

The Cartesian product endows $\mathcal{A}^G$ with a symmetric monoidal structure (e.g., see the discussion in [24, Section 3]). Lemma 7.37 shows this is compatible with the inclusions

$$\mathcal{A}_{\mathcal{O}}^G \hookrightarrow \mathcal{A}^G$$

induced by $\mathcal{O} \hookrightarrow \mathcal{S}et^G$, where $\mathcal{A}^G$ is the Lindner category of Example 7.23.

Corollary 7.38 *The category $\mathcal{A}_{\mathcal{O}}^G$ is a symmetric monoidal subcategory of $\mathcal{A}^G$ under the Cartesian product.*

The box product on Mackey functors is the Day convolution product of the Cartesian product in $\mathcal{A}^G$ with the tensor product on $\mathcal{A}b$ [9]. We make an analogous definition for incomplete Mackey functors.

Definition 7.39 The box product on $\mathcal{O}$-Mackey functors is the left Kan extension of the tensor product along the Cartesian product: given $\underline{M}$ and $\underline{N}$, the box product is defined by

$$
\begin{array}{ccc}
\mathcal{A}_{\mathcal{O}}^{G} \times \mathcal{A}_{\mathcal{O}}^{G} & \xrightarrow{\underline{M} \otimes \underline{N}} & \mathcal{A}b \\
{\scriptstyle \times} \downarrow & \nearrow & \\
\mathcal{A}_{\mathcal{O}}^{G} & {\scriptstyle \underline{M} \square \underline{N}} &
\end{array}
$$

Remark 7.40 Strickland shows a very important generalization of this: we can actually take the left Kan extension not of the tensor product to $\mathcal{A}b$ but rather the Cartesian product to $\mathcal{S}et$.

This is essential for Tambara functors: Any product preserving functor from $\mathcal{P}^{G}$ to $\mathcal{A}b$ is necessarily zero, since a multiplication that is both linear and bilinear must be zero.

The definition as the left Kan extension gives a universal property of this incomplete box product completely analogous to the usual one for the box product: a map of $\mathcal{O}$-Mackey functors

$$\underline{M} \square \underline{N} \to \underline{P}$$

is the same data as a natural transformation of functors on $\mathcal{A}_{\mathcal{O}}^{\times 2}$

$$\underline{M}(\text{-}) \otimes \underline{N}(\text{-}) \to \underline{P}(\text{-} \times \text{-}).$$

We note there is a slight wrinkle with the usual Frobenius relation: the equations

$$a \otimes T_f(b) = T_f\big(R_f(a) \otimes b\big) \text{ and } T_f(b) \otimes a = T_f\big(b \otimes R_f(a)\big)$$

only make sense when f is a map in $\mathcal{O}$. When $\mathcal{O} = \mathcal{O}^{tr}$, then any map preserves isotropy and the Frobenius relation just expresses that the tensor is bilinear.

Example 7.41 If $\mathcal{O} = \mathcal{O}^{tr}$ is the initial indexing category of Example 7.6, then the box product is just the ordinary levelwise tensor product.

General properties of the Day convolution product show that this is a symmetric monoidal product on Mackey functors here too.

Proposition 7.42 *The box product is a symmetric monoidal product on $\mathcal{O}$-Mackey functors with unit the $\mathcal{O}$-Burnside Mackey functor.*

Definition 7.43 An $\mathcal{O}$-Green functor is a commutative monoid for the box product in $\mathcal{O}$-Mackey functors.

7.4.2 Colimits and limits

Strickland's careful treatment of the complete Tambara case actually goes through without change! At no point does Strickland use that we have transfers and norms for all maps in $\mathcal{S}et^G$, using instead the compatibility relations needed for particular given maps.

For coproducts, he uses that we can formally create the "norm of a transfer" in the box product by using the corresponding exponential diagram. We have to check that in both cases, we are working entirely in our restricted subcategory $\mathcal{P}^G_{\mathcal{O}_a,\mathcal{O}_m}$.

Proposition 7.44 ([24, Lemma 9.8]) *The coproduct of $(\mathcal{O}_a, \mathcal{O}_m)$-Tambara functors is the box product for $\mathcal{O}_a$-Mackey functors.*

For coequalizers, Strickland works very generally with relations in a wide collection of algebraic structures ([24, Definition 10.4]).

Proposition 7.45 ([24, Proposition 10.5]) *Coequalizers exist in $(\mathcal{O}_a, \mathcal{O}_m)$-Tambara functors.*

Since filtered colimits commute with finite products in $\mathcal{S}et$ and since $(\mathcal{O}_a, \mathcal{O}_m)$-Tambara functors are a full subcategory of a diagram category defined in terms of a product condition, we can deduce the existence of filtered colimits [24, Proposition 10.2]. Putting this all together, we conclude that all colimits exist.

Theorem 7.46 *The category of $(\mathcal{O}_a, \mathcal{O}_m)$-Tambara functors is cocomplete.*

Since the categories of bi-incomplete semi-Tambara functors are diagram categories in a complete category, they are automatically complete.

Theorem 7.47 *For any compatible pair $(\mathcal{O}_a, \mathcal{O}_m)$, the category of $(\mathcal{O}_a, \mathcal{O}_m)$-Tambara functors is complete, with limits formed objectwise.*

Proof The category of functors from a [skeletally] small category $\mathcal{J}$ to a complete category $\mathcal{C}$ is complete, with objects formed levelwise.

In our case, we need also ensure that the limit is again a product preserving functor, but this follows from limits and products both being categorical limits, and hence commuting. □

7.4.3 Forgetful Functors

Inclusions of indexing categories naturally give rise to inclusions of the corresponding wide subgraphs of polynomials. We only want to consider the case where the pair of indexing categories is compatible, so we focus on inclusions here.

Definition 7.48 Let $(\mathcal{O}_a, \mathcal{O}_m)$ and $(\mathcal{O}'_a, \mathcal{O}'_m)$ be two compatible pairs of indexing categories. We will write

$$(\mathcal{O}_a, \mathcal{O}_m) \subseteq (\mathcal{O}'_a, \mathcal{O}'_m)$$

and say that we have an inclusion of pairs if we have (not necessarily proper) inclusions

$$\mathcal{O}_a \subseteq \mathcal{O}'_a \text{ and } \mathcal{O}_m \subseteq \mathcal{O}'_m.$$

The following is immediate from the definitions.

Proposition 7.49 *If $(\mathcal{O}_a, \mathcal{O}_m) \subseteq (\mathcal{O}'_a, \mathcal{O}'_m)$ is an inclusion of compatible pairs of indexing categories, then the natural inclusion*

$$\mathcal{P}^G_{\mathcal{O}_a, \mathcal{O}_m} \hookrightarrow \mathcal{P}^G_{\mathcal{O}'_a, \mathcal{O}'_m}$$

is a product-preserving functor.

These inclusions give us "forgetful functors", since the composite of product-preserving functors is product-preserving:

Proposition 7.50 *If $(\mathcal{O}_a, \mathcal{O}_m) \subseteq (\mathcal{O}'_a, \mathcal{O}'_m)$ is an inclusion of compatible pairs of indexing categories, then precomposition with the inclusion of polynomials gives a forgetful functor*

$$(\mathcal{O}'_a, \mathcal{O}'_m)\text{-}\mathcal{T}amb \to (\mathcal{O}_a, \mathcal{O}_m)\text{-}\mathcal{T}amb.$$

Since limits are computed objectwise, the forgetful functors all commute with limits. Since the categories of $(\mathcal{O}_a, \mathcal{O}_m)$-Tambara functors are cocomplete diagram categories, we further deduce the existence of left-adjoints.

Proposition 7.51 *The forgetful functor*

$$(\mathcal{O}'_a, \mathcal{O}'_m)\text{-}\mathcal{T}amb \to (\mathcal{O}_a, \mathcal{O}_m)\text{-}\mathcal{T}amb$$

has a left-adjoint, the corresponding free functor.

Proof Since the forgetful functor commutes with limits, by the adjoint functor theorem, it suffices to show that we have a small set of projective generators. However, when

$$T = \coprod_{H \subseteq G} G/H,$$

then the Yoneda lemma shows that the (group completion) of the representable functor $\mathcal{P}^G_{\mathcal{O}_a, \mathcal{O}_m}(T, \text{-})$ is a projective generator. $\square$

Conceptually, these free functors freely adjoin any transfers and norms parameterized by $(\mathcal{O}'_a, \mathcal{O}'_m)$ but not in $(\mathcal{O}_a, \mathcal{O}_m)$.

Example 7.52 Let $\mathcal{O}_a = \mathcal{O}_m = \mathcal{O}^{tr}$. An $(\mathcal{O}_a, \mathcal{O}_m)$-Tambara functor here is just a coefficient system of commutative rings. Since $\mathcal{O}^{tr}$ is the initial indexing system, for any compatible pair $(\mathcal{O}_a, \mathcal{O}_m)$, we have an inclusion

$$(\mathcal{O}^{tr}, \mathcal{O}^{tr}) \subseteq (\mathcal{O}_a, \mathcal{O}_m).$$

The corresponding forgetful functor just records the underlying coefficient system of commutative rings.

For the corresponding left-adjoint, it is helpful to factor the inclusion into two steps. The pair $(\mathcal{O}_a, \mathcal{O}^{tr})$ is always compatible (see Theorem 7.62), and so we have inclusions of compatible pairs

$$(\mathcal{O}^{tr}, \mathcal{O}^{tr}) \subseteq (\mathcal{O}_a, \mathcal{O}^{tr}) \subseteq (\mathcal{O}_a, \mathcal{O}_m).$$

The left adjoint for first inclusion creates the free $\mathcal{O}_a$-Green functor, putting in all of the missing transfers in algebras and enforcing the Frobenius relation. The left adjoint for the second inclusion then freely puts in the norms.

It is worth noting here that this is the only order that works in general. If $\mathcal{O}_m$ is non-trivial, then a consequence of Corollary 7.70 below is that $(\mathcal{O}^{tr}, \mathcal{O}_m)$ is never compatible. In other words, we had to put in the missing transfers, and then we can put in the missing norms.

Underlying incomplete Mackey functors

The inclusions here make sense also for more general wide, pullback stable subcategories like $\mathcal{S}et^G_{\cong}$, as in Proposition 7.36. Here, we need to also note that Proposition 7.32 only used the wideness of the categories structuring the norms and transfers, since it used only the identity maps there. Using these more general forgetful functors, we can talk about "underlying" structures.

Example 7.53 Let $\mathcal{O}_a$ and $\mathcal{O}_m$ be compatible indexing categories. Then for any $(\mathcal{O}_a, \mathcal{O}_m)$-Tambara functor $\underline{R}$, we have

(1) an underlying additive $\mathcal{O}_a$-Mackey functor and
(2) an underlying multiplicative $\mathcal{O}_m$-semi-Mackey functor

which arise from the inclusions

$$(\mathcal{O}_a, \mathcal{S}et_{\cong}^G) \subseteq (\mathcal{O}_a, \mathcal{O}_m) \supseteq (\mathcal{S}et_{\cong}^G, \mathcal{O}_m).$$

The forgetful functor to the underlying additive $\mathcal{O}_a$-Mackey functor also has a left adjoint: this is the bi-incomplete version of the symmetric algebra. The forgetful functor to the underlying multiplicative $\mathcal{O}_m$-semi-Mackey functor is a little stranger, but it also has a left adjoint. This is a bi-incomplete version of Nakaoka's "Tambarization of a semi-Mackey functor" [21, Theorem 2.12].

7.5 Rewriting compatibility

The purpose of this section is to provide alternate conditions for compatibility that are easier to check in practice. The contravariant functoriality of the pullback gives covariant functoriality of the dependent product: if $f \colon S \to T$ and $g \colon T \to U$, then we have a natural isomorphism

$$\Pi_g \circ \Pi_f \cong \Pi_{g \circ f} \colon \mathcal{S}et_{/S}^G \to \mathcal{S}et_{/U}^G.$$

We use this to simplify the condition of compatibility: any map in $\mathcal{S}et^G$ can be written as a disjoint union of composites of fold maps and maps between orbits. Note here that there is also the possibility of some of the disjoint summands being empty. Putting these observations together, it suffices to consider the dependent products along

(1) the unique map $\varnothing \to T$,
(2) a disjoint union of maps $(S_1 \to T_1) \amalg (S_2 \to T_2)$,
(3) the fold map $S \amalg S \to S$, and
(4) maps of orbits $G/H \to G/K$.

Proposition 7.36 shows that the dependent product along an isomorphism preserves any pullback stable subcategory, so we will use this whenever it makes formulae easier.

7.5.1 Main reductions

We now begin a series of reductions of the condition from Definition 7.29.

Initial maps

Proposition 7.54 *The dependent product along $\iota\colon \varnothing \to T$ is always a terminal object $T \to T$ of the slice category over T.*

Proof The slice category over the empty sets is the full subcategory of initial objects in $\mathcal{S}et^G$. Every object in this category is uniquely isomorphic to every other object, and this means that every object is also a terminal object of this category. Being a right adjoint, the dependent product preserves terminal objects. $\qquad\square$

Corollary 7.55 *The dependent product along $\varnothing \to T$ preserves any indexing category.*

Disjoint Unions

Proposition 7.56 *Let $f\colon T \to T'$ and $g\colon U \to U'$. If $h\colon S \to T \amalg U$, let*

$$S_T = h^{-1}(T)$$

let $h_T : S_T \to T$ be the restriction of h, and similarly for S_U and h_U. Then we have a natural isomorphism

$$\Pi_{f\amalg g}(h) \cong (\Pi_f(h_T)) \coprod (\Pi_g(h_U)).$$

Proof The assignment

$$h \mapsto (h_T, h_U)$$

gives a functor

$$\mathcal{S}et^G_{/(T\amalg U)} \to \mathcal{S}et^G_{/T} \times \mathcal{S}et^G_{/U}.$$

This is an equivalence of categories, with inverse equivalence given by

$$\big((S_T \to T), (S_U \to U)\big) \mapsto (S_T \amalg S_U \to T \amalg U).$$

This product decomposition is also natural: we have a commutative diagram

$$
\begin{array}{ccc}
\mathcal{S}et^G_{/(T\amalg U)} & \xleftarrow{\ (f\amalg g)^*\ } & \mathcal{S}et^G_{/(T'\amalg U')} \\
\simeq\downarrow & & \uparrow\simeq \\
\mathcal{S}et^G_{/T} \times \mathcal{S}et^G_{/U} & \xleftarrow[(f^*,g^*)]{} & \mathcal{S}et^G_{/T'} \times \mathcal{S}et^G_{/U'}
\end{array}
$$

The result follows from noting that the left vertical map takes h to

(h_T, h_U), the right adjoint to the bottom map is (Π_f, Π_g), and right vertical map is the disjoint union. $\qquad\square$

Being closed under dependent products along disjoint unions follows from simply being closed under disjoint unions and the dependent products along the summands.

Corollary 7.57 *If an indexing system $\mathcal{O}$ is closed under dependent products along $f\colon T \to T'$ and $g\colon U \to U'$, then it is closed under the dependent product along $f \amalg g$.*

Fold maps

The dependent product along the fold map is closely connected to the categorical product in the slice categories.

Proposition 7.58 *Let $\nabla\colon T \amalg T \to T$ be the fold map, and $\iota_L, \iota_R\colon T \to T \amalg T$ be the left and right inclusions. We have a natural isomorphism*

$$\Pi_\nabla \cong \iota_L^* \underset{T}{\times} \iota_R^*.$$

Proof Given any $S \to T \amalg T$, let

$$S_L = \iota_L^*(S) \text{ and } S_R = \iota_R^*(S).$$

We have a natural isomorphism over $T \amalg T$:

$$S_L \amalg S_R \cong S$$

expressing the disjunctive property of maps to a disjoint union of sets. We now appeal to a direct construction of the dependent product: the fiber over a point $t \in T$ is the set of sections of S over $\nabla^{-1}(t)$. The set $\nabla^{-1}(t)$ is $\{t\} \amalg \{t\}$, and a section over this is by construction a pair (s_L, s_R), where $s_L \in S_L$ and $s_R \in S_R$ both map to t. This is the same data as a point in the fiber of $S_L \underset{T}{\times} S_R$ over t. $\qquad\square$

Lemma 7.59 *Slices of indexing categories are closed under fiber products.*

Proof This is a consequence of pullback stability. Let $f\colon S_1 \to T$ and $g\colon S_2 \to T$ both be in $\mathcal{O}$. Then the fiber product is defined by the pullback diagram

$$
\begin{array}{ccc}
S_1 \underset{T}{\times} S_2 & \xrightarrow{\ f'\ } & S_2 \\
{\scriptstyle g'}\big\downarrow & & \big\downarrow{\scriptstyle g} \\
S_1 & \xrightarrow{\ f\ } & T.
\end{array}
$$

Pullback stability of $\mathcal{O}$ guarantees that g' is in $\mathcal{O}$ as well, and hence the composite $f \circ g'$ is. □

Remark 7.60 Under the equivalences

$$
\mathcal{S}et^G_{/(G/H)} \simeq \mathcal{S}et^H,
$$

the fiber product in G-sets over G/H is sent to the ordinary product of H-sets. Lemma 7.59 is then the indexing category version of the statement "admissible H-sets are closed under products", which is [5, Lemma 4.11].

Corollary 7.61 *Any indexing category is closed under dependent products along fold maps.*

We pause here to note that these pieces alone are sufficient to show that the analogue of Green functors always works.

Theorem 7.62 *For any indexing category $\mathcal{O}_a$, the pair $(\mathcal{O}_a, \mathcal{O}^{tr})$ is compatible.*

Proof In $\mathcal{O}^{tr}$, the only allowed maps of orbits are isomorphisms, and hence the only conditions we needed to check to ensure compatibility are the first three. □

The $(\mathcal{O}_a, \mathcal{O}^{tr})$-Tambara functors are essentially $\mathcal{O}_a$-Green functors. In fact, Strickland's proof for the additively complete case $\mathcal{O}_a = \mathcal{O}^{gen}$ goes through without change in the incomplete case.

Proposition 7.63 ([24, Proposition 12.11]) *There is an equivalence of categories between $\mathcal{O}$-Green functors and $(\mathcal{O}, \mathcal{O}^{tr})$-Tambara functors.*

7.5.2 Admissibility

Our reductions show that the possible obstruction to compatibility of an additive and multiplicative indexing category is the dependent product along a map of orbits

$$
G/K \to G/H.
$$

It turns out that this is a surprisingly harsh condition. By Proposition 7.36, we may assume that $K \subseteq H$ and the map is the canonical quotient. We recall a proposition from [14].

Proposition 7.64 ([14, Proposition 2.3]) *Let $K \subseteq H$ be subgroups of G, and let $f: T \to G/K$ be a map of G-sets. Then the dependent product of f along the canonical quotient $G/K \to G/H$ is*

$$G \underset{H}{\times} \mathrm{Map}^K(H, T_e) \to G/H,$$

where $T_e = f^{-1}(eK)$ is the K-set corresponding to T under the equivalence of categories

$$\mathcal{S}et^K \simeq \mathcal{S}et^G_{/(G/K)}.$$

This gives us the last piece we need, so we collect all of our reductions into one statement.

Theorem 7.65 *Let $\mathcal{O}_a$ and $\mathcal{O}_m$ be indexing categories. Then $(\mathcal{O}_a, \mathcal{O}_m)$ is compatible if and only if for every pair of subgroups $K \subseteq H$ such that H/K is an admissible H-set for $\mathcal{O}_m$ and for every admissible K-set T for $\mathcal{O}_a$, the coinduced H-set $\mathrm{Map}^K(H, T)$ is admissible for $\mathcal{O}_a$.*

Although this may seem confusing, it records a very conceptual reformulation. Recall that we have symmetric monoidal Mackey functor extensions (in the sense of [8] or [12]):

(1) $\underline{Set}^{\amalg}$ has the coCartesian extension, where categorical transfers are induction, and

(2) $\underline{Set}^{\times}$ has the Cartesian extension, where the categorical transfers are coinduction.

An indexing system $\mathcal{O}_a$ is by definition a sub-symmetric monoidal coefficient system of $\underline{Set}^{\amalg}$. We have some closure under induction, but that is not relevant for compatibility. Additionally, since admissible sets are closed under products, we deduce that $\mathcal{O}_a$ is also a sub-symmetric monoidal coefficient system of the *Cartesian* symmetric monoidal Mackey functor $\underline{Set}^{\times}$. Theorem 7.65 is simply compatibility with the $\mathcal{O}_m$-Mackey structure.

Corollary 7.66 *A pair $(\mathcal{O}_a, \mathcal{O}_m)$ is compatible if and only if $\mathcal{O}_a$ is actually a sub-symmetric monoidal $\mathcal{O}_m$-Mackey functor of the Cartesian monoidal Mackey functor $\underline{Set}^{\times}$.*

7.6 Limits on Compatibility

For a general pair $(\mathcal{O}_a, \mathcal{O}_m)$ of indexing categories, it can be difficult to check that $\mathcal{O}_m$ distributes over $\mathcal{O}_a$. We make a few basic observations here.

7.6.1 The additive hull

It is straightforward to check that compatibility conditions are preserved by intersection.

Lemma 7.67 *If $(\mathcal{O}_a, \mathcal{O}_m)$ and $(\mathcal{O}_a', \mathcal{O}_m')$ are compatible pairs, then*

$$(\mathcal{O}_a \cap \mathcal{O}_a', \mathcal{O}_m \cap \mathcal{O}_m')$$

is a compatible pair.

From this, we deduce that there is always a kind of "compatible hull" of the additive indexing category making a compatible pair.

Proposition 7.68 *For any pair of indexing categories $\mathcal{O}_a$ and $\mathcal{O}_m$, there is a minimal $\overline{\mathcal{O}}_a$ containing $\mathcal{O}_a$ such that $(\overline{\mathcal{O}}_a, \mathcal{O}_m)$ is compatible.*

Proof Consider the set

$$\mathcal{E} = \left\{ (\mathcal{O}_a', \mathcal{O}_m') \mid (\mathcal{O}_a', \mathcal{O}_m') \text{ compatible } \& \ (\mathcal{O}_a, \mathcal{O}_m) \leq (\mathcal{O}_a', \mathcal{O}_m') \right\},$$

and let

$$\overline{\mathcal{O}}_a = \bigcap_{(\mathcal{O}_a', \mathcal{O}_m') \in \mathcal{E}} \mathcal{O}_a' \text{ and } \overline{\mathcal{O}}_m = \bigcap_{(\mathcal{O}_a', \mathcal{O}_m') \in \mathcal{E}} \mathcal{O}_m'.$$

The set $\mathcal{E}$ non-empty because $(\mathcal{O}^{gen}, \mathcal{O}_m)$ is always in this, and this also shows that $\overline{\mathcal{O}}_m = \mathcal{O}_m$. By Lemma 7.67, the pair $(\overline{\mathcal{O}}_a, \mathcal{O}_m)$ is compatible. $\qquad\square$

Proposition 7.69 *If $(\mathcal{O}_a, \mathcal{O}_m)$ is a compatible pair and H/K is an $\mathcal{O}_m$-admissible set for H, then for every $L \subseteq H$ such that K is sub-conjugate to L, the H-set H/L is $\mathcal{O}_a$-admissible.*

Proof For any subgroup K, the K-set

$$\{a, b\} := * \amalg *$$

is always admissible for any indexing system. By Theorem 7.65, if H/K is an $\mathcal{O}_m$-admissible H-set, then we must have that

$$\mathrm{Map}^K \left(H, \{a, b\} \right) \cong \mathrm{Map}\left(H/K, \{a, b\} \right) \in \mathcal{O}_a(H).$$

Since admissible sets are closed under conjugation, it suffices to consider the case that L contains K. Consider the function

$$f \colon H/K \to \{a, b\}$$

defined by

$$f(hK) = \begin{cases} a & h \in L \\ b & h \notin L. \end{cases}$$

Then the stabilizer of f is L, and hence we have a summand

$$H/L \cong H \cdot f \subseteq \mathrm{Map}^K \left(H, \{a, b\} \right),$$

which means H/L is admissible for $\mathcal{O}_a$, as desired. $\qquad\square$

Corollary 7.70 *If $(\mathcal{O}_a, \mathcal{O}_m)$ is a compatible pair, then $\mathcal{O}_a \geq \mathcal{O}_m$ in the partial order on indexing categories given by inclusion.*

Corollary 7.71 *If G is an admissible G-set for $\mathcal{O}_m$, then $\mathcal{O}_a = \mathcal{S}et^G$ is the terminal indexing category.*

Put another way, having the norm from the trivial subgroup to G can only happen for ordinary incomplete Tambara functors.

Corollary 7.70 also bounds sharply the number of compatible pairs.

Corollary 7.72 *Let $n(G)$ be the cardinality of the poset of indexing categories for G and let $c(G)$ be the number of pairs of indexing categories $(\mathcal{O}, \mathcal{O}')$ such that $\mathcal{O} \geq \mathcal{O}'$. Then we have*

$$\frac{\#\ compatible\ pairs}{\#\ all\ pairs} \leq \frac{c(G)}{n(G)^2} \leq \frac{1}{2} + \frac{1}{2n(G)}.$$

Proof The first bound follows immediately from Corollary 7.70 and the observation that there are $n(G)^2$ total possible pairs. The second bound follows from a combinatorial observation. If we consider all poset structures on the set of n elements, then the maximal number of pairs (a, b) with $a \geq b$ occurs when the poset is a total order. In this case, we have $\frac{1}{2}(n^2 + n)$, from which the second bound follows. $\qquad\square$

Example 7.73 For $G = C_p$, the poset of indexing categories is the total order on two elements, and we achieve the bound: 3/4 of the pairs are compatible.

For larger groups, we expect the bounds to be less than $1/2$, as the poset of indexing categories seem generically to contain many incomparable elements.

Example 7.74 For $G = C_{p^2}$, Balchin–Barnes–Roitzheim showed there are 5 indexing categories in the poset, and looking at them shows that there are 13 pairs $(\mathcal{O}, \mathcal{O}')$ with $\mathcal{O} \geq \mathcal{O}'$ [1, Theorem 2]. Of these, 12 are compatible: the indexing category for the little disks operad on $\infty(1 + \lambda)$, where λ is a 2-dimensional faithful representation of C_{p^2}, is not compatible with itself, due to Corollary 7.71.

Using the classifications of Balchin–Barnes–Roitzheim, of Balchin–Bearup–Pech–Roitzheim, and of Rubin, we can at least provide upper bounds on the number of compatible pairs. We stress that in these cases, we have used only Corollary 7.70 and Proposition 7.69, and we do not know if all of the possible pairs that conform to these also satisfy the conditions of Theorem 7.65.

Example 7.75 For $G = C_{p^3}$, Balchin–Barnes–Roitzheim showed there are 14 indexing categories [1, Theorem 2], and looking at the poset structure (as depicted in [22, Section 3.2]), we see 67 are comparable. Of these, 55 conform to the conditions of Proposition 7.69.

Example 7.76 For $G = C_{pq}$ with p and q distinct primes, Balchin–Bearup–Pech–Roitzheim and Rubin showed there are 10 indexing categories [2, Section 3], [22, Section 3.2]. Of the 100 possible pairs, 44 are comparable, and 39 conform to the conditions of Proposition 7.69.

To try to get sharper estimates on the number of compatible pairs, we first look at the edge-cases: when an indexing system is compatible with itself. Recall that an indexing category is "linear isometry-like" if whenever H/K is an admissible H-set, the sets H/L for any $K \subseteq L \subseteq H$ are also all admissible [22].

Corollary 7.77 *If $(\mathcal{O}, \mathcal{O})$ is compatible, then $\mathcal{O}$ is linear isometry-like.*

These conditions alone already put stringent constraints, and we have only used the "trivial" part of $\mathcal{O}_a$. Any additional stabilizer types will give more conditions.

Remark 7.78 It is not necessarily the case that if $(\mathcal{O}_a, \mathcal{O}_m)$ is compatible and $\mathcal{O}'_a \geq \mathcal{O}_a$, then $(\mathcal{O}'_a, \mathcal{O}_m)$ is compatible: additional stabilizers can show up.

7.6.2 Examples: little disks and linear isometries

The initial motivation for our study of N_∞-operads was the problem of understanding to what extent linear isometries and little disks operads failed to be equivalent, equivariantly. We recall the explicit conditions for admissibility for linear isometries and little disks operads.

Theorem 7.79 ([5, Theorems 4.18 & 4.19])   *Let U be a universe.*

A finite H-set T is admissible for the linear isometries operad for U if and only if there is an H-equivariant isometry

$$\mathbb{R} \cdot T \otimes U \hookrightarrow U,$$

where $\mathbb{R} \cdot T$ is the permutation representation generated by T.

A finite H-set T is admissible for the little disks operad for U if and only if there is an H-equivariant embedding

$$T \hookrightarrow U.$$

Proposition 7.80   *Let V be a faithful representation of G. If there exists a subgroup H such that G/H does not embed in $\infty(1+V)$, then the indexing category $\mathcal{O}$ associated to the little disks operad is not compatible with itself.*

Proof   Since V is a faithful representation of G, for any non-trivial subgroup $H \subseteq G$, the H-fixed points of V are a proper subspace of V. Since there are finitely many non-trivial subgroups, we find that the collection of all vectors in V with a non-trivial stabilizer is a finite union of hyperplanes and hence a proper subset. We therefore have vectors with a trivial stabilizer. Any of these gives an equivariant embedding $G \hookrightarrow V$, and Theorem 7.79 then says that G is an admissible G-set for $\mathcal{O}$.

Corollary 7.71 shows that if G is an admissible $\mathcal{O}_m$ set, then all H-sets are $\mathcal{O}_a$ admissible for all H. However, Theorem 7.79 also shows that the assumption that G/H does not embed is equivalent to the assertion that G/H is not admissible. This means $\mathcal{O}$ cannot be compatible with itself. $\square$

One key result in this direction was [5, Theorem 4.24], which showed the existence of little disks operads inequivalent to any linear isometries operad, provided the group is not simple. The key step was producing a representation of G such that G embeds but G/N does not for some normal subgroup: inducing up the reduced regular representation for N works.

Corollary 7.81   *For any non-simple group G, there is a representation V such that the indexing category for the little disks on V is not compatible with itself.*

On the other hand, as topology informs us, the indexing category for the linear isometries operad on a universe U is always compatible with its corresponding little disks. We can see this algebraically.

Proposition 7.82 *Let U be a universe for G, let $\mathcal{O}_a$ be the indexing category associated to the little disks operad for U and let $\mathcal{O}_m$ be the indexing category associated to the linear isometries operad for U. Then $(\mathcal{O}_a, \mathcal{O}_m)$ is compatible.*

Proof By Theorem 7.65, it suffices to show that the $\mathcal{O}_a$-admissible sets are closed under coinduction along $\mathcal{O}_m$-admissible sets.

When $T = H/K$, the isometric embedding condition of Theorem 7.79 can be rewritten as the existence of an isometric embedding

$$\mathbb{R}[H] \underset{\mathbb{R}[K]}{\otimes} U \hookrightarrow U.$$

Using the isomorphism of induction with coinduction in representations, we deduce our desired result. If T is a finite K-set that equivariantly embeds into $i_K^* U$, then a choice of such an embedding gives an embedding

$$\mathrm{Map}^K(H, T) \hookrightarrow \mathrm{Map}^K(H, i_K^* U) \cong \mathbb{R}[H] \underset{\mathbb{R}[K]}{\otimes} U \hookrightarrow U,$$

as desired. $\square$

7.6.3 The multiplicative hull

In Proposition 7.68, we saw that there is a *smallest* additive $\mathcal{O}_a$ compatible with any fixed multiplicative $\mathcal{O}_m$ using the intersection. The poset of indexing categories is actually a lattice, and intersection realizes the "meet". The join operation is more difficult to describe, but it has been identified by Rubin [23]. Rubin works with "transfer systems" or "norm systems", another equivalent form of the data of an indexing system defined independently by Rubin [22] and Balchin–Barnes–Roitzheim [1]. For convenience, we state the result in indexing categories.

Proposition 7.83 ([23, Proposition 3.1]) *The join of two indexing categories $\mathcal{O}$ and $\mathcal{O}'$ is just the finite coproduct complete subcategory generated by them.*

In other words, pullback stability is automatic. What this means for us, however, is that we can join together multiplicatively compatible indexing categories.

Proposition 7.84   *Let $\mathcal{O}_a$ be an indexing category. If $(\mathcal{O}_a, \mathcal{O}_m)$ and $(\mathcal{O}_a, \mathcal{O}'_m)$ are both compatible, then $\mathcal{O}_a$ is complatible with the join $\mathcal{O}_m \vee \mathcal{O}'_m$.*

Proof   One way to reinterpret Proposition 7.83 is that a map of orbits $G/H \to G/K$ is in the join if and only if it can be written as a composition of maps

$$G/H \xrightarrow{f_0} G/H_1 \xrightarrow{f_1} \ldots \xrightarrow{f_n} G/K,$$

with the maps f_i for $0 \le i \le n$ in at least one of $\mathcal{O}_m$ or $\mathcal{O}'_m$. We assumed that $\mathcal{O}_a$ was compatible with both $\mathcal{O}_m$ and $\mathcal{O}'_m$, and hence it is closed under dependent products along any of the maps in them. The proof of Theorem 7.65 shows it suffices to check closure on maps between orbits, so we are done.   □

This means we can find a *largest* multiplicative $\mathcal{O}_m$ compatible with a given $\mathcal{O}_a$.

Corollary 7.85   *For any $\mathcal{O}_a$, there is a largest $\mathcal{O}_m$ such that $(\mathcal{O}_a, \mathcal{O}_m)$ is compatible.*

Proof   We simply join together all $\mathcal{O}_m$ such that $(\mathcal{O}_a, \mathcal{O}_m)$ is compatible. The set of these is non-empty, because $\mathcal{O}^{tr}$ is compatible with all $\mathcal{O}_a$.   □

Remark 7.86   In contrast to Remark 7.78, if $(\mathcal{O}_a, \mathcal{O}_m)$ is compatible and $\mathcal{O}'_m \le \mathcal{O}_m$, then $(\mathcal{O}_a, \mathcal{O}'_m)$ is compatible. This reinforces an underlying theme that $\mathcal{O}_m$ puts constraints on $\mathcal{O}_a$, but not vice versa.

7.7 Change of group

A basic property of Mackey functors that makes computation easier is the fact that induction is naturally isomorphic to coinduction. This is a reflection of the Wirthmüller isomorphism of "genuine" equivariant stable homotopy theory, and it makes it relatively easy to form various kinds of resolutions we might want for homological algebra. We can view this as the G-equivariant version of "additive", since it means that finite sums indexed by the group are finite products indexed by the group. Work of Berman cleanly explains this point [3].

In incomplete Tambara functors, we have none of this. The coproduct and product do not agree, and while the forgetful functor to incomplete Tambara functors for a subgroup has both adjoints, they never agree. The coproduct, product, and right adjoints are all expressible as corresponding functors on the underlying Mackey functors (the latter ones because the forgetful functor commutes with limits). In certain cases, the left adjoint to the forgetful functor is also expressible in terms of a functor on Mackey functors: the norm. We close by giving a few easy results of this form for bi-incomplete Tambara functors, and then stating some conjectures for structure that would tie everything together.

7.7.1 Restriction functors & coinduction

One of the key technical lemmas used in studying incomplete Tambara functors was knowing when a pair of adjoint functors on a category extends to a pair of adjoint functors on polynomials in that category with some restricted class of norms ([6, Theorem 2.17]). The main ingredients were

(1) the image of induction forms an "essential sieve": if $f \colon S \to G \underset{H}{\times} T$ is a map, then this is isomorphic over $G \underset{H}{\times} T$ to a map of the form

$$G \underset{H}{\times} f' \colon G \underset{H}{\times} S' \to G \underset{H}{\times} T,$$

and

(2) an H-equivariant map f is in $i_H^* \mathcal{O}$ if and only if $G \underset{H}{\times} f$ is in $\mathcal{O}$.

At no point in the proof did we use the fact that all transfers exist, so we deduce that the result goes through in this setting without change.

Proposition 7.87 *Precomposition with induction and restriction, respectively give an adjoint pair of functors*

$$i_H^* \colon (\mathcal{O}_a, \mathcal{O}_m)\text{-}\mathcal{T}amb^G \rightleftarrows (i_H^* \mathcal{O}_a, i_H^* \mathcal{O}_m)\text{-}\mathcal{T}amb^H \colon \mathrm{CoInd}_H^G.$$

The key feature here is that coinduction and restriction are "the same" functor no matter which indexing categories we use: simply precompose with restriction or induction respectively. In particular, the values are determined when we forget all the way down to coefficient systems.

7.7.2 Induction

Since the restriction functors commute with limits, for formal reasons we know they have left adjoints. As in Mackey functors and incomplete Tambara functors, these left adjoints are concisely described as left Kan extensions.

Definition 7.88   Let

$$n_H^G : (i_H^* \mathcal{O}_a, i_H^* \mathcal{O}_m)\text{-}\mathcal{T}amb^H \to (\mathcal{O}_a, \mathcal{O}_m)\text{-}\mathcal{T}amb^G$$

be the left Kan extension along the induction functor

$$G \underset{H}{\times} (\text{-}) : \mathcal{P}_{i_H^* \mathcal{O}_a, i_H^* \mathcal{O}_m}^H \to \mathcal{P}_{\mathcal{O}_a, \mathcal{O}_m}^G.$$

Since induction is product preserving, the left Kan extension along it preserves product preserving functors [16]. It is formal that this gives the left adjoint.

Proposition 7.89   *The functor n_H^G is left-adjoint to the restriction functor.*

In Mackey functors, this left adjoint is actually isomorphic to the right adjoint. In $\mathcal{O}_a$-Mackey functors, this is not always the case. Instead, we have such an isomorphism when G/H is admissible (this is $\underline{\pi}_0$ of [7, Theorem 3.25]).

7.7.3 The norm & externalized forms

The key application of the theses of Mazur and of Hoyer was that the norm functor on Mackey functors describes the left adjoint on Tambara functors [20, 15], and hence we have a reinterpretation of Tambara functors as the G-commutative monoids in the category of Mackey functors [12]. A similar statement here would provide a clean, algebraic interpretation of Corollary 7.66. We present several conjectures here.

Conjecture 7.90   *If $(\mathcal{O}_a, \mathcal{O}_m)$ is a compatible pair, then for every admissible H/K for $\mathcal{O}_m$, coinduction restricts to a functor*

$$\mathrm{Map}^K(H, \text{-}) : i_K^* \mathcal{O}_a \to i_H^* \mathcal{O}_a.$$

In this case, we will also say "coinduction preserves $\mathcal{O}_a$".

Since these are wide subcategories, this conjecture is really a statement about the maps in the category being closed under coinduction. This is a twisted version of Lemma 7.37, and it gives the twisted version of Corollary 7.38.

Proposition 7.91   *If coinduction from K to H preserves $\mathcal{O}$, then it induces a functor*

$$\mathcal{A}_{i_K^* \mathcal{O}}^K \to \mathcal{A}_{i_H^* \mathcal{O}}^H.$$

This is the heart of generalizing the Hoyer–Mazur norm, and it gives us the following definition which we conjecture is correct.

Definition 7.92 If coinduction from K to H preserves $\mathcal{O}$, then left Kan extension gives a norm functor

$$_{\mathcal{O}}N_K^H : \mathcal{O}\text{-}\mathcal{M}ackey^K \to \mathcal{O}\text{-}\mathcal{M}ackey^H$$

that commutes with the box product.

Definition 7.92 says that the category of $\mathcal{O}_a$-Mackey functors is naturally a symmetric monoidal $\mathcal{O}_m$-Mackey functor under the box product and norms: the restriction maps are the ordinary restrictions (which are strong symmetric monoidal) and the transfer maps are the norms.

Warning 7.93 Just as the box product depends heavily on $\mathcal{O}$, ranging from the levelwise tensor product (for $\mathcal{O}^{tr}$) to the usual box product on Mackey functors, so too will any norms. Different $\mathcal{O}$ will give different definitions of the norm.

The symmetric monoidal $\mathcal{O}_m$-Mackey structure gives a collection of endofunctors of $i_H^* \mathcal{O}_a$-Mackey functors for all H: for any admissible H/K for $\mathcal{O}_m$, let

$$_{\mathcal{O}_a}N^{H/K} := _{\mathcal{O}_a}N_K^H i_K^*.$$

This functor is isomorphic to the left Kan extension along $\mathrm{Map}(H/K, \text{-})$ [4]. We should think of this as the categorical version of the formula expressing the action of the Burnside ring of finite H-sets on any Mackey functor $\underline{M}$ evaluated at G/H:

$$[H/K] \cdot m = tr_K^H res_K^H(m).$$

Multiplying together the norm functors gives us an endofunctor for any $\mathcal{O}_m$-admissible H-set T: writing T as $T = H/K_1 \amalg \cdots \amalg H/K_n$, let

$$_{\mathcal{O}_a}N^T(\underline{M}) := _{\mathcal{O}_a}N^{H/K_1}\underline{M}\square \ldots \square_{\mathcal{O}_a}N^{H/K_n}\underline{M}.$$

With this, we conjecture the general external form of a $(\mathcal{O}_a, \mathcal{O}_m)$-Tambara functor.

Conjecture 7.94 A $(\mathcal{O}_a, \mathcal{O}_m)$-Tambara functor is a $\mathcal{O}_m$-commutative monoid in the symmetric monoidal $\mathcal{O}_m$-Mackey functor of $\mathcal{O}_a$-Mackey functors.

Unpacking this, this would mean that a $(\mathcal{O}_a, \mathcal{O}_m)$-Tambara functor is the following data:

(1) A $\mathcal{O}_a$-Mackey functor $\underline{R}$,
(2) a unital commutative monoid structure on $\underline{R}$: $\underline{R} \square \underline{R} \to \underline{R}$, and
(3) for every map of $\mathcal{O}_m$-admissible H sets $S \to T$, an "external norm":
    a map of commutative monoids

$$\mathcal{O}_a N^S i_H^* \underline{R} \to \mathcal{O}_a N^T i_H^* \underline{R}.$$

These are required to be compatible in the sense that the norm map for $S \to T$ restricts to the norm map for the restriction of $S \to T$ and the norm maps compose to give the norm map for the composite.

References

[1] S. Balchin, D. Barnes, and C. Roitzheim. N_∞-operads and associahedra. arXiv.org:1905.03797, 2019.

[2] S. Balchin, D. Bearup, C. Pech, and C. Roitzheim. Equivariant homotopy commutativity for $G = C_{pqr}$. *Tbilisi Math. Journal*, Special Issue on Homotopy Theory, Spectra, and Structured Ring Spectra, 17–31, 2020.

[3] J. D. Berman. On the commutative algebra of categories. *Algebr. Geom. Topol.*, 18(5):2963–3012, 2018.

[4] A. J. Blumberg, T. Gerhardt, M. A. Hill, and T. Lawson. The Witt vectors for Green functors. *J. Algebra*, 537:197–244, 2019.

[5] A. J. Blumberg and M. A. Hill. Operadic multiplications in equivariant spectra, norms, and transfers. *Adv. Math.*, 285:658–708, 2015.

[6] A. J. Blumberg and M. A. Hill. Incomplete Tambara functors. *Algebr. Geom. Topol.*, 18(2):723–766, 2018.

[7] A. J. Blumberg and M. A. Hill. Equivariant stable categories for incomplete systems of transfers. *J. London Math. Soc*, 0:1–38, 2021.

[8] A. M. Bohmann and A. Osorno. Constructing equivariant spectra via categorical Mackey functors. *Algebr. Geom. Topol.*, 15(1):537–563, 2015.

[9] B. Day. On closed categories of functors. In *Reports of the Midwest Category Seminar, IV*, Lecture Notes in Mathematics, Vol. 137, pages 1–38. Springer, Berlin, 1970.

[10] N. Gambino and J. Kock. Polynomial functors and polynomial monads. *Math. Proc. Cambridge Philos. Soc.*, 154(1):153–192, 2013.

[11] N. Georgakopoulos. Norms determine transfers in the absence of torsion. In preparation, 2019.

[12] M. A. Hill and M. J. Hopkins. Equivariant symmetric monoidal structures. arxiv.org: 1610.03114, 2016.

[13] M. A. Hill, M. J. Hopkins, and D. C. Ravenel. On the nonexistence of elements of Kervaire invariant one. *Ann. of Math. (2)*, 184(1):1–262, 2016.

[14] M. A. Hill and K. Mazur. An equivariant tensor product on Mackey functors. *Journal of Pure and Applied Algebra*, 2019.

[15] R. Hoyer. *Two topics in stable homotopy theory*. PhD thesis, University of Chicago, 6 2014.

[16] G. M. Kelly and S. Lack. Finite-product-preserving functors, Kan extensions and strongly-finitary 2-monads. *Appl. Categ. Structures*, 1(1):85–94, 1993.

[17] L. G. Lewis, Jr., J. P. May, M. Steinberger, and J. E. McClure. *Equivariant stable homotopy theory*, volume 1213 of *Lecture Notes in Mathematics*. Springer-Verlag, Berlin, 1986. With contributions by J. E. McClure.

[18] L. G. Lewis, Jr. Change of universe functors in equivariant stable homotopy theory. *Fund. Math.*, 148(2):117–158, 1995.

[19] G. Lindner. A remark on Mackey-functors. *Manuscripta Math.*, 18(3):273–278, 1976.

[20] K. L. Mazur. *On the Structure of Mackey Functors and Tambara Functors*. PhD thesis, University of Virginia, 5 2013.

[21] H. Nakaoka. Tambarization of a Mackey functor and its application to the Witt-Burnside construction. *Adv. Math.*, 227(5):2107–2143, 2011.

[22] J. Rubin. Detecting Steiner and linear isometries operads. *Glasgow Mathematical Journal*, 63(2):307–342, 2021.

[23] J. Rubin. Operadic lifts of the algebra of indexing systems. *Journal of Pure and Applied Algebra*, 2021. Accepted.

[24] N. Strickland. Tambara functors. arXiv:1205.2516, 2012.

[25] D. Tambara. On multiplicative transfer. *Comm. Algebra*, 21(4):1393–1420, 1993.

[26] M. Weber. Polynomials in categories with pullbacks. *Theory Appl. Categ.*, 30:No. 16, 533–598, 2015.

8

Homotopy Limits of Model Categories, Revisited

Julia E. Bergner[a]

Abstract

The definition of the homotopy limit of a diagram of left Quillen functors of model categories has been useful in a number of applications. In this chapter we review its definition and summarize some of these applications. We conclude with a discussion of why we could work with right Quillen functors instead, but cannot work with a combination of the two.

To John Greenlees on the occasion of his 60th birthday.

8.1 Introduction

In the papers [5] and [6], we develop constructions for the homotopy pullback, and then more general homotopy limit, of a suitable diagram of left Quillen functors between model categories. In this chapter, we seek to revisit some of the ideas of those papers, with three main goals. First, we would like to try to make some of the exposition of those papers more accessible to the reader. Second, we would like to introduce some of the interesting examples that have been developed by other authors since those papers were written. Third, we discuss some possible approaches to working in a more flexible context, namely when we have diagrams with both left and right Quillen functors, and why they do not behave as we might wish.

A first question one might ask is why we would want a notion of homotopy limits of model categories in the first place. Just as in any

[a] Department of Mathematics, University of Virginia

other context in which we have various diagrams of mathematical objects, when we have Quillen pairs between model categories it is reasonable to ask whether more complicated diagrams of such have a corresponding model category with appropriate universal properties. Since the purpose of model categories is arguably to work in a good homotopy-theoretic setting, it is only natural that we would want such a construction to be homotopy invariant in some suitable way.

Often when we want to consider homotopy limits of diagrams, we work in an ambient model category, or more general homotopy theory, in which we have built-in methods for constructing them. However, there is no known "model category of model categories", so we must content ourselves with ad hoc constructions. Nonetheless, we would like some way to reassure ourselves that the homotopy limit construction that we define has some right to have that name. We can do so by migrating to a suitable model of $(\infty, 1)$-categories in which homotopy limits can be defined unambiguously. We do not revisit this issue in this chapter, but note that we work with the complete Segal space model in [5] and [6]; Harpaz has proved a similar result in the quasi-category model [15].

After an introduction to some necessary model category background, we begin the main part of the chapter with the construction of homotopy limits of model categories, beginning with the more tangible special case of homotopy pullbacks. We then present a sampling of examples for which this construction has proved useful. Since this chapter is being written for the proceedings of John Greenlees' birthday conference, we give particular attention to the example of adelic models for tensor-triangulated categories that he developed with Balchin [2]. We refer the reader to [18] and [19] for related constructions and examples thereof.

At the end of the chapter, we discuss the rigidity of this construction. Specifically, while it can be modified to the context of model categories and right Quillen functors, we do not get homotopy limits with the correct properties if we try to modify our construction to diagrams consisting of both kinds of functors.

In the paper [7], we consider a similar definition for homotopy colimits of model categories. However, for a number of reasons that construction is not as satisfactory as this one, so we do not include it here. We refer the reader to work of Harpaz and Prasma [16] for another approach to homotopy colimits.

Acknowledgements

We would like to thank Scott Balchin for discussions about this chapter, as well as the referee for many helpful comments.

The author was partially supported by NSF grant DMS-1906281.

8.2 Model categories and localizations

In this section we give a brief review of some model category notions and techniques that we need for our construction and some of our examples.

Recall that a *model category* $\mathcal{M}$ is a category with three distinguished classes of morphisms: weak equivalences, fibrations, and cofibrations, satisfying five axioms [12, 3.3], [20]. An object x in $\mathcal{M}$ is *fibrant* if the unique map $x \to *$ to the terminal object is a fibration. Dually, an object x in $\mathcal{M}$ is *cofibrant* if the unique map $\varnothing \to x$ from the initial object is a cofibration.

We also want to consider appropriate functors between model categories. Because of the dual nature of the fibrations and cofibrations, it is sensible to consider adjoint pairs of functors in this situation.

Definition 8.1

1 Let $\mathcal{M}$ and $\mathcal{N}$ be model categories. An adjoint pair of functors

$$F \colon \mathcal{M} \rightleftarrows \mathcal{N} \colon G$$

is a *Quillen pair* if the left adjoint F preserves cofibrations and the right adjoint G preserves fibrations. We refer to F as a *left Quillen functor* and similarly refer to G as a *right Quillen functor*.

2 A Quillen pair is a *Quillen equivalence* if, for any cofibrant object x of $\mathcal{M}$ and any fibrant object y of $\mathcal{N}$, a map $Fx \to y$ is a weak equivalence in $\mathcal{N}$ if and only if its adjoint map $x \to Gy$ is a weak equivalence in $\mathcal{M}$.

We note that an equivalent definition of Quillen pair is that the left adjoint F preserves cofibrations and acyclic cofibrations. Using this approach, we could consider left Quillen functors independently of having a right adjoint functor, and analogously consider right Quillen functors as those that preserve fibrations and acyclic fibrations.

Example 8.2   Recall that the simplicial indexing category Δ^{op} is defined to be the category whose objects are finite ordered sets

$[n] = \{0 \to 1 \to \cdots \to n\}$ and morphisms the opposites of the order-preserving maps between them. A *simplicial set* is a functor

$$K \colon \Delta^{\mathrm{op}} \to \mathcal{S}ets.$$

We denote by $\mathcal{S}Sets$ the category of simplicial sets, and this category has a natural model structure Quillen equivalent to the standard model structure on topological spaces originally due to Quillen [13, I.10]. A map of simplicial sets is a weak equivalence precisely if its geometric realization is a weak homotopy equivalence of topological spaces. The cofibrations are the monomorphisms, and the fibrations are the *Kan fibrations*, or maps with the right lifting property with respect to the horn inclusions $\Lambda^k[n] \to \Delta[n]$ for $k \geq 1$ and $0 \leq k \leq n$.

Although there are many other possible model structures on the category of simplicial sets, throughout this chapter we assume this model structure unless otherwise stated.

Example 8.3 Given a model category $\mathcal{M}$, there is also a model structure on the category $\mathcal{M}^{[1]}$ whose objects of $\mathcal{M}^{[1]}$ are morphisms of $\mathcal{M}$, and whose morphisms are given by pairs of morphisms in $\mathcal{M}$ making the appropriate square diagram commute. A morphism in $\mathcal{M}^{[1]}$ is a weak equivalence if its component maps are weak equivalences in $\mathcal{M}$, and cofibrations are defined analogously. More generally, $\mathcal{M}^{[n]}$ is the category with objects strings of n composable morphisms in $\mathcal{M}$; the model structure can be defined analogously.

We now consider some additional features that a model category might have, beginning with the notion of properness.

Definition 8.4 A model category $\mathcal{M}$ is *right proper* if every pullback of a weak equivalence along a fibration is a weak equivalence. It is *left proper* if every pushout of a weak equivalence along a cofibration is a weak equivalence. It is *proper* if it is both left and right proper.

Example 8.5 The model structure for simplicial sets from Example 8.2 is proper. The fact that it is left proper follows from the fact that all objects are cofibrant, while the fact that it is right proper requires a bit more verification [17, 13.1.13].

In practice, most of the model structures that we consider have the additional structure of being cofibrantly generated.

Definition 8.6 A model category $\mathcal{M}$ is *cofibrantly generated* if there exist sets I and J, called the *generating cofibrations* and *generating acyclic* *cofibrations*, respectively, each satisfying the small object argument [17, 10.5.15], such that a map in $\mathcal{M}$ is a fibration if and only if it has the right lifting property with respect to the maps in J, and a map is an acyclic fibration if and only if it has the right lifting property with respect to the maps in I.

Example 8.7   The model structure on simplicial sets is cofibrantly generated. The set I of generating cofibrations can be taken to be the set of boundary inclusions

$$\{\partial\Delta[n] \to \Delta[n] \mid n \geq 0\},$$

and the set J of generating acyclic cofibrations can be taken to be the set of horn inclusions

$$\{\Lambda^k[n] \to \Delta[n] \mid n \geq 1, 0 \leq k \leq n\}.$$

We now want to look at model categories which are combinatorial, which, loosely speaking, can be considered to be model categories that behave sufficiently like the model structure on simplicial sets. The formal definition requires some technical points to be developed, however. Since, in particular, they require the use of filtered colimits, we begin by recalling the definition of filtered diagrams, over which such colimits are taken.

Definition 8.8   Let λ be a regular cardinal.

1 A category $\mathcal{I}$ is λ-*filtered* if any subcategory with fewer than λ morphisms has a compatible cocone in $\mathcal{I}$.

2 An object c of a category $\mathcal{C}$ is λ-*small* if for any λ-filtered category $\mathcal{I}$ and diagram $X \colon \mathcal{I} \to \mathcal{C}$,

$$\mathrm{Hom}(c, \mathrm{colim}\, X_i) \cong \mathrm{colim}\, \mathrm{Hom}(c, X_i).$$

Definition 8.9   [8, 2.2] A category $\mathcal{C}$ is *locally presentable* if it is closed under colimits and if there is a regular cardinal λ and a set of objects A in $\mathcal{C}$ such that:

1 every object in A is small with respect to λ-filtered colimits, and

2 every object in $\mathcal{C}$ can be expressed as a λ-filtered colimit of elements of A.

In what follows, we use the terminology "filtered colimits" for λ-filtered colimits for some sufficiently large regular cardinal λ.

We can now state the definition of a combinatorial model category.

Definition 8.10 [8, 2.1] A model category is *combinatorial* if it is cofibrantly generated and locally presentable.

For the purposes of this chapter, we want to consider combinatorial model categories because they can often be localized with respect to sets of maps to obtain a new model structure on the same underlying category but with more weak equivalences. To formalize this idea, we need to consider mapping spaces.

It is often helpful, when working with a model category, to consider not only sets of morphisms between two given objects, but simplicial sets of morphisms.

Definition 8.11 A *simplicial category* or *category enriched in simplicial sets* is a category $\mathcal{C}$ such that for any objects x and y of $\mathcal{C}$, there is a simplicial set $\mathrm{Map}_{\mathcal{C}}(x, y)$ of morphisms from x to y, together with a composition law satisfying associativity and unitality. These simplicial sets are called *mapping spaces*.

Example 8.12 The category of simplicial sets has the structure of a simplicial category, where, given simplicial sets K and L, the mapping space $\mathrm{Map}(K, L)$ is defined to have n-simplices

$$\mathrm{Map}(K, L)_n = \mathrm{Hom}(K \times \Delta[n], L),$$

where $\Delta[n]$ is the standard n-simplex.

Any model category can be equipped with such mapping spaces, chosen in a homotopy invariant way, even if the underlying category does not have a canonical simplicial category structure. There are a number of ways to define such *homotopy mapping spaces*, including the simplicial localization constructions of Dwyer and Kan [10], [11], or the approach via framings of Hirschhorn [17, §19.1]. We denote the homotopy mapping space from an object x to any object y by $\mathrm{Map}^h(x, y)$.

We can now set up the necessary definitions to discuss localizations of model categories.

Definition 8.13 Let $\mathcal{M}$ be a model category and S a class of maps in $\mathcal{M}$.

1 An object x of $\mathcal{M}$ is *S-colocal* if it is cofibrant and if for every map $f \colon a \to b$ in $\mathcal{S}$, the induced map of homotopy mapping spaces

$$\mathrm{Map}^h_{\mathcal{M}}(x, a) \to \mathrm{Map}^h_{\mathcal{M}}(x, b)$$

is a weak equivalence of simplicial sets.

2 A map $f: d \to d$ is an *S-colocal equivalence* if, for every S-colocal object x, the induced map of homotopy mapping spaces

$$\mathrm{Map}^h_{\mathcal{M}}(x, c) \to \mathrm{Map}^h_{\mathcal{M}}(x, d)$$

is a weak equivalence of simplicial sets.

We often consider the following notion of colocal objects with respect to a set of objects.

Definition 8.14 Let $\mathcal{M}$ be a model category and K a set of objects in $\mathcal{M}$. A map $x \to y$ is a *K-local equivalence* if, for every object a in K, the induced map of homotopy mapping spaces

$$\mathrm{Map}^h(a, x) \to \mathrm{Map}^h(a, y)$$

is a weak equivalence of simplicial sets.

Definition 8.15 Let $\mathcal{M}$ be a model category, K a set of objects in $\mathcal{M}$, and S the class of K-colocal equivalences. The *right Bousfield localization* of $\mathcal{M}$ with respect to S, if it exists, is a model structure $\mathcal{R}_S\mathcal{M}$ on the underlying category of $\mathcal{M}$ such that:

1 the weak equivalences are the S-colocal equivalences, and
2 the fibrations are precisely the fibrations of $\mathcal{M}$.

Observe that we could consider such a localization as being with respect to the set K of objects, rather than with respect to the class S of maps. We have chosen to use the notation S to be parallel with the definition of left Bousfield localization below.

Finally, we bring together the different flavors of model categories we have considered to state the following existence theorem for right Bousfield localizations of model categories.

Theorem 8.16 *[17, 5.1.1], [4, §5] Let $\mathcal{M}$ be a right proper combinatorial model category, K a set of objects in $\mathcal{M}$, and S the class of K-colocal equivalences.*

1 *The right Bousfield localization $\mathcal{R}_S\mathcal{M}$ of $\mathcal{M}$ exists.*
2 *The cofibrant objects of $\mathcal{R}_S\mathcal{M}$ are precisely the K-colocal objects of $\mathcal{M}$.*
3 *The model category $\mathcal{R}_S\mathcal{M}$ is right proper.*

Much more common in the literature is the notion of left Bousfield localization, which we now review. It is not needed for our main construction, but does appear in some of the applications we discuss.

Definition 8.17 Let $\mathcal{M}$ be a model category and S a set of maps in $\mathcal{M}$.

1 A fibrant object z of $\mathcal{M}$ is *S-local* if for every map $a \to b$ in S, the induced map of simplicial sets

$$\mathrm{Map}^h(b, z) \to \mathrm{Map}^h(a, z)$$

is a weak equivalence.

2 A morphism $x \to y$ in $\mathcal{M}$ is an *S-local equivalence* if for any S-local object Z, the induced map

$$\mathrm{Map}^h(y, z) \to \mathrm{Map}^h(x, z)$$

is a weak equivalence of simplicial sets.

Definition 8.18 Let $\mathcal{M}$ be a model category and S a set of maps in $\mathcal{M}$. The *left Bousfield localization* of $\mathcal{M}$ with respect to S, if it exists, is a model structure $\mathcal{L}_S\mathcal{M}$ on the same underlying category as $\mathcal{M}$, such that:

1 the weak equivalences are the S-local equivalences, and
2 the cofibrations are exactly the cofibrations of $\mathcal{M}$.

Theorem 8.19 *[17, 4.1.1], [4, §4] Let $\mathcal{M}$ be a left proper combinatorial model category and S a set of maps in $\mathcal{M}$.*

1 *The left Bousfield localization $\mathcal{L}_S\mathcal{M}$ of $\mathcal{M}$ exists.*
2 *The model category $\mathcal{L}_S\mathcal{M}$ is left proper.*

8.3 Homotopy pullbacks of model categories

In this section, we recall the construction of the homotopy pullback of an appropriate diagram of model categories, and we discuss when it can be given the structure of a model category itself.

The following definition first appeared in the literature in [22]. There, and in our original paper [5], it was called a *homotopy fiber product* instead. Here, we distinguish more explicitly between the homotopy limit and a more general preliminary category of diagrams.

Definition 8.20 Suppose that

$$\mathcal{M}_1 \xrightarrow{F_1} \mathcal{M}_3 \xleftarrow{F_2} \mathcal{M}_2$$

is a diagram of left Quillen functors of model categories. Define their

lax homotopy pullback to be the category $\mathcal{M}$ whose objects are given by
5-tuples $(x_1, x_2, x_3; u, v)$ such that each x_i is an object of $\mathcal{M}_i$ fitting into
a diagram

$$F_1(x_1) \xrightarrow{\ u\ } x_3 \xleftarrow{\ v\ } F_2(x_2).$$

A morphism of $\mathcal{M}$, say $f \colon (x_1, x_2, x_3; u, v) \to (y_1, y_2, y_3; z, w)$, is given
by a triple of maps $f_i \colon x_i \to y_i$ for $i = 1, 2, 3$, such that the diagram

$$
\begin{array}{ccccc}
F_1(x_1) & \xrightarrow{\ u\ } & x_3 & \xleftarrow{\ v\ } & F_2(x_2) \\
\Big\downarrow{\scriptstyle F_1(f_1)} & & \Big\downarrow{\scriptstyle f_3} & & \Big\downarrow{\scriptstyle F_2(f_2)} \\
F_1(y_1) & \xrightarrow{\ z\ } & y_3 & \xleftarrow{\ w\ } & F_2(y_2)
\end{array}
$$

commutes.

Remark 8.21   For ease of notation, we have simply denoted this category by $\mathcal{M}$, rather than something more suggestive of a diagram category or even of a homotopy pullback. In the later section on homotopy limits, we give more explicit notation that one could also employ here if more clarity is needed.

The lax homotopy pullback $\mathcal{M}$ can be given the structure of a model category, where the weak equivalences and cofibrations are given levelwise. In other words, f is a weak equivalence if each map f_i is a weak equivalence in $\mathcal{M}_i$, and cofibrations are defined analogously.

However, the model category $\mathcal{M}$ is not yet what we want. We would like to require the maps u and v to be weak equivalences rather than arbitrary maps.

Definition 8.22   Suppose that

$$\mathcal{M}_1 \xrightarrow{\ F_1\ } \mathcal{M}_3 \xleftarrow{\ F_2\ } \mathcal{M}_2$$

is a diagram of left Quillen functors of model categories. Its *homotopy pullback* is the full subcategory of the lax homotopy pullback for which the maps u and v are weak equivalences.

However, this restriction comes at a price. On its own, the homotopy pullback cannot be given the structure of a model category because it is not generally closed under limits and colimits. The typical solution in this kind of situation is rather to find a localization of the model structure on the lax homotopy pullback $\mathcal{M}$ so that the fibrant and cofibrant objects have the maps u and v weak equivalences. In that direction, we have the following theorem.

Theorem 8.23 *[5, 3.1] Let $\mathcal{M}$ be the lax homotopy pullback of a diagram of left Quillen functors*

$$\mathcal{M}_1 \xrightarrow{\ F_1\ } \mathcal{M}_3 \xleftarrow{\ F_2\ } \mathcal{M}_2$$

where each of the categories $\mathcal{M}_i$ is combinatorial. Further assume that $\mathcal{M}$ is right proper. Then there exists a right Bousfield localization of $\mathcal{M}$ whose cofibrant objects $(x_1, x_2, y_2; u, v)$ have both u and v weak equivalences in $\mathcal{M}_3$.

Proof Since we have assumed that the categories $\mathcal{M}_1$, $\mathcal{M}_2$, and $\mathcal{M}_3$ are combinatorial, and in particular locally presentable, we can find, for each $i = 1, 2, 3$, a set $\mathcal{A}_i$ of objects of $\mathcal{M}_i$ generating all of $\mathcal{M}_i$ by filtered colimits. We further assume that the objects of each $\mathcal{A}_i$ are all cofibrant in the corresponding model category $\mathcal{M}_i$. For an explicit construction for such sets, we refer the reader to presentations of combinatorial model categories as given by Dugger [8].

Given $a_1 \in \mathcal{A}_1$ and $a_2 \in \mathcal{A}_2$, consider the class of all objects x_3 in $\mathcal{M}_3$ such that there are pairs of weak equivalences

$$F_1(a_1) \xrightarrow{\ \simeq\ } x_3 \xleftarrow{\ \simeq\ } F_2(a_2).$$

For every choice of a_1 and a_2 such that $F_1(a_1)$ is weakly equivalent to $F_2(a_2)$, choose a single cofibrant object x_3 in this weak equivalence class of objects; such a choice is possible since $\mathcal{A}_1$ and $\mathcal{A}_2$ have been chosen to be sets. Define the set $\mathcal{B}_3$ to be the union of this set of all such objects x_3 together with the generating set $\mathcal{A}_3$ for $\mathcal{M}_3$. To retain uniformity of notation going forward, for $i = 1, 2$, let $\mathcal{B}_i = \mathcal{A}_i$.

Consider the set

$$\{(x_1, x_2, x_3; u, v) \mid x_i \in \mathcal{B}_i, u, v \text{ weak equivalences in } \mathcal{M}_3\}$$

of objects of the lax homotopy limit $\mathcal{M}$. Taking filtered colimits of such objects, we can obtain the set $\mathcal{B}$ of all objects $(x_1, x_2, x_3; u, v)$ of $\mathcal{M}$ for which the maps u and v are weak equivalences. Although, as we discussed earlier, arbitrary colimits do not necessarily preserve weak equivalences, an important feature of filtered colimits is that they do [8, 7.3].

Now we take a right Bousfield localization of $\mathcal{M}$ with respect to the set $\mathcal{B}$. Since $\mathcal{M}$ is given by a diagram of combinatorial model categories, it is itself combinatorial, so if $\mathcal{M}$ is right proper, then this localization has a model structure by Theorem 8.16. The class of cofibrant objects of this model category is precisely the smallest class of cofibrant objects of $\mathcal{M}$

containing this set and closed under homotopy colimits and weak equivalences [17, 5.1.5, 5.1.6]. Thus, our only remaining step is to show that homotopy colimits of objects in $\mathcal{B}$ still have u and v weak equivalences.

Suppose that $\mathcal{C}$ is a small category and $X \colon \mathcal{C} \to \mathcal{M}$ a functor such that the objects in the image of $\mathcal{C}$ are in the set $\mathcal{B}$. In other words, for any object α of $\mathcal{C}$, we have

$$X(\alpha) = (x_1^\alpha, x_2^\alpha, x_3^\alpha; u^\alpha, v^\alpha)$$

with $x_i^\alpha \in \mathcal{B}_i$ and the maps u^α and v^α weak equivalences. Notice that for each $i = 1, 2, 3$, we have diagrams $X_i \colon \mathcal{C} \to \mathcal{M}_i$ such that for each α in $\mathcal{C}$, $X_i(\alpha) = x_i^\alpha$. Thus, the homotopy colimit of the original diagram X is the object

$$(\mathrm{hocolim}\, X_1, \mathrm{hocolim}\, X_2, \mathrm{hocolim}\, X_3; \mathrm{hocolim}(u), \mathrm{hocolim}(v))$$

in $\mathcal{M}$. Since we have assumed that each object x_i^α is in $\mathcal{B}_i$ and hence cofibrant, we get that the maps $\mathrm{hocolim}(u)$ and $\mathrm{hocolim}(v)$ are weak equivalences in $\mathcal{M}_3$ by [17, 19.4.2]. $\qquad\square$

Unfortunately, the right properness assumption in this theorem is quite restrictive. In particular, we have not been able to identify conditions on the model categories $\mathcal{M}_1$, $\mathcal{M}_2$, and $\mathcal{M}_3$ guaranteeing that the model category $\mathcal{M}$ is right proper. An option for weakening the structure of a model category to accommodate localizations of non-right proper model categories is given by Barwick [4]. Alternatively, when the conditions of this theorem are not satisfied, we can still use the levelwise model structure on the lax homotopy pullback $\mathcal{M}$ and simply restrict to the homotopy pullback as a subcategory when necessary.

Remark 8.24   One might ask why we have chosen to use left Quillen functors. We could make a dual definition in which we instead use right Quillen functors; the arrows u and v in the definition of the homotopy limit would then more naturally be taken to point in the opposite direction, and underlying model structure should then be taken to be the projective model structure, in which the weak equivalences and fibrations are taken to be levelwise. Being careful about such details, it is not hard to verify that this dual construction works perfectly well.

One might then wonder if one could find a sensible definition of a homotopy pullback of a left Quillen functor along a right Quillen functor. Because this discussion requires other homotopy limits of model categories, we defer it to Section 8.8, after we have discussed the definition of general homotopy limits.

8.4 Examples of homotopy pullbacks of model categories

Let us now consider some examples of this construction. We start with some basic general examples that we also described in [5].

Example 8.25 Let us consider the following special case of a homotopy pullback, the homotopy fiber of a map.

Let $F: \mathcal{M} \to \mathcal{N}$ be a left Quillen functor of model categories. Then the *homotopy fiber* of F is the homotopy fiber product of the diagram

where the map $* \to \mathcal{N}$ is necessarily the map from the trivial model category on a single object $*$ to the initial object $\varnothing$ of $\mathcal{N}$.

Using the definition of the homotopy pullback, the objects of the homotopy fiber are triples $(*, m, n; u, v)$, m is an object of $\mathcal{M}$, n is an object of $\mathcal{N}$, $u: \varnothing \to n$ is the unique such map, and $v: F(m) \to n$. Imposing our condition that u and v be weak equivalences, n must be weakly equivalent to the initial object of $\mathcal{N}$, and m is any object of $\mathcal{M}$ whose image under F is weakly equivalent to the initial object of $\mathcal{N}$.

Here, we see that the requirement that functors be left Quillen is prohibitively restrictive for many examples one might want to consider! Although we would like to look at the homotopy fiber over arbitrary objects, it is not possible under this hypothesis.

A further specialization illustrates its particularly odd nature still more. If we take the analogue of a loop space and define the "loop model category" as the homotopy pullback of the diagram

for any model category $\mathcal{M}$, we simply get the subcategory of $\mathcal{M}$ whose objects are weakly equivalent to the initial object.

Example 8.26 Toën's motivation for defining the homotopy pullback of model categories was to prove associativity of his derived Hall algebras [22]. In this context, we work with a stable model category $\mathcal{N}$; this extra assumption that the homotopy category is triangulated implies that $\mathcal{N}$ has a zero object 0 that is both initial and terminal.

Recall the model structure on the morphism category $\mathcal{N}^{[1]}$ from Example 8.3. Given an object of $\mathcal{N}^{[1]}$, namely a map $f\colon x \to y$ in $\mathcal{N}$, let $F\colon \mathcal{N}^{[1]} \to \mathcal{N}$ be the target map, so that $F(f\colon x \to y) = y$. Let $C\colon \mathcal{N}^{[1]} \to \mathcal{N}$ be the cone map, so that $C(f\colon x \to y) = y \amalg_x 0$. Using these functors, we get a diagram

$$
\begin{array}{c}
\mathcal{N}^{[1]} \\
\downarrow C \\
\mathcal{N}^{[1]} \xrightarrow{\ F\ } \mathcal{N}.
\end{array}
$$

The homotopy pullback of this diagram is Quillen equivalent to the model category $\mathcal{N}^{[2]}$ whose objects are pairs of composable morphisms in $\mathcal{N}$ [22, §4]. The idea is that an object of the homotopy pullback consists of

$$
(w \to z, x \to y, z', u, v),
$$

where

$$
F(w \to z) \xrightarrow[\simeq]{u} z' \xleftarrow[\simeq]{v} C(x \to y).
$$

Applying the definitions of F and C, we get

$$
z \xrightarrow{\simeq} z' \xleftarrow{\simeq} y \amalg_x 0.
$$

This data is essentially that of a distinguished triangle $x \to y \to z$ in the homotopy category, which is precisely an object of $\mathcal{N}^{[2]}$.

Example 8.27 The next example lifts a result in stable homotopy theory to the level of model categories, as first proved by Gutiérrez and Roitzheim [14].

Given a prime p, let $\mathbb{Z}_p$ denote the p-adic integers. Then the integers are given by the pullback

$$
\begin{array}{ccc}
\mathbb{Z} & \longrightarrow & \prod_p \mathbb{Z}_p \\
\downarrow & & \downarrow \\
\mathbb{Q} & \longrightarrow & \mathbb{Q} \otimes_{\mathbb{Z}} \left(\prod_p \mathbb{Z}_p \right).
\end{array}
$$

This "arithmetic fracture square" has further generalizations in number theory, but in homotopy theory has the following upgrade to the context of spectra.

Let E be a spectrum, and consider the localization functor L_E that inverts maps of spectra that are isomorphisms after applying E_*-homology. Furthermore, given any abelian group G, let MG denote its corresponding Moore spectrum, namely, the connective spectrum whose 0th homology group is G and whose other homotopy groups are trivial [1, III.6]. Suppose that J and K form a partition of the set of all prime numbers, and let $\mathbb{Z}_J$ and $\mathbb{Z}_K$ denote the set of integers localized at the sets J and K, respectively. Then a classical result is that any spectrum X is the homotopy pullback of the diagram of spectra

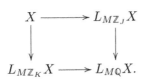

We now use the homotopy pullback of model categories to give a further upgrade to the level of model categories. Let Sp denote the model structure for symmetric spectra. As above, let E be any spectrum, and let $L_E Sp$ denote the left Bousfield localization of Sp with respect to the E_*-equivalences. As before, we consider the Moore spectra $M\mathbb{Z}_J$ and $M\mathbb{Z}_K$ associated to some partition J, K of the set of primes. Gutiérrez and Roitzheim prove that Sp is Quillen equivalent to the homotopy limit of the diagram of left Quillen functors

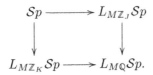

Indeed, they prove that this result holds for any left proper, combinatorial, stable model category $\mathcal{C}$ in place of the model structure Sp [14, 3.4]. See also [9] for details about how why such a localization is possible.

8.5 Homotopy limits of model categories

Let us now discuss how to generalize the homotopy pullback construction to more general homotopy limits of model categories. While the notation gets more complicated, the essential idea is the same.

If $\mathcal{D}$ is a small category, then a $\mathcal{D}$-shaped diagram $\mathcal{M}$ is given by a collection $\mathcal{M}_\alpha$, one for each object α of $\mathcal{D}$, together with left Quillen functors $F_{\alpha,\beta}^{\theta} \colon \mathcal{M}_\alpha \to \mathcal{M}_\beta$, compatible with one another in the sense that, if $\theta \colon \alpha \to \beta$ and $\delta \colon \beta \to \gamma$ are composable maps in $\mathcal{D}$, then

$$F_{\alpha,\gamma}^{\delta\theta} = F_{\beta,\gamma}^{\delta} \circ F_{\alpha,\beta}^{\theta}.$$

The superscript θ indexes different left Quillen functors between the same model categories, corresponding to distinct maps $\theta \colon \alpha \to \beta$. More precisely, if we consider the (large) category $\mathcal{MC}$ of model categories with left Quillen functors between them, or some small subcategory of it, then we can describe such a diagram as by a functor $\mathcal{M} \colon \mathcal{D} \to \mathcal{MC}$.

Definition 8.28   Let $\mathcal{M}$ be a $\mathcal{D}$-shaped diagram of model categories. Then the *lax homotopy limit* of $\mathcal{M}$, denoted by $\mathcal{L}_{\mathcal{D}}\mathcal{M}$, has objects families $(x_\alpha, u_{\alpha,\beta}^{\theta})$ where x_α is an object of $\mathcal{M}_\alpha$ and $u_{\alpha,\beta}^{\theta} \colon F_{\alpha,\beta}^{\theta}(x_\alpha) \to x_\beta$ is a morphism in $\mathcal{M}_\beta$, satisfying the compatibility condition

$$u_{\alpha,\gamma}^{\delta\theta} = u_{\beta,\gamma}^{\delta} \circ F_{\beta,\gamma}^{\delta}(u_{\alpha,\beta}^{\theta}).$$

Morphisms are component-wise maps of such families, making the appropriate diagrams commute.

The *homotopy limit* for $\mathcal{M}$, denoted by $\mathcal{L}im_{\mathcal{D}}\mathcal{M}$, is the full subcategory of $\mathcal{L}_{\mathcal{D}}\mathcal{M}$ whose objects satisfy the additional condition that all maps $u_{\alpha,\beta}^{\theta}$ are weak equivalences in their respective $\mathcal{M}_\beta$.

As for the lax homotopy pullback, the lax homotopy limit $\mathcal{L}_{\mathcal{D}}\mathcal{M}$ can be given the injective model structure, assuming that all model categories in the diagram are sufficiently nice. On the other hand, just as in the case of the homotopy pullback, $\mathcal{L}im_{\mathcal{D}}\mathcal{M}$ does not have the structure of a model category, since the weak equivalence requirement is not preserved by general limits and colimits. Again, we can ask whether we can find a localization of the more general model structure so that the fibrant-cofibrant objects do have the maps weak equivalences, and obtain the following result.

Theorem 8.29   *[6, 3.2] Let $\mathcal{L}_{\mathcal{D}}\mathcal{M}$ be the lax homotopy limit of a $\mathcal{D}$-diagram combinatorial model categories $\mathcal{M}_\alpha$, and assume that $\mathcal{L}_{\mathcal{D}}\mathcal{M}$ has the structure of a right proper model category. Then there exists a right Bousfield localization of $\mathcal{L}_{\mathcal{D}}\mathcal{M}$ whose cofibrant objects $(x_\alpha, u_{\alpha,\beta}^{\theta})$ have all x_α cofibrant and all maps $u_{\alpha,\beta}^{\theta}$ weak equivalences in $\mathcal{M}_\beta$.*

The proof is very similar to the one for homotopy pullbacks, so we omit it here, referring the reader to [6] for the details. The concern about whether a given diagram actually satisfies these conditions is also analogous to the situation for the homotopy pullback, as are the potential resolutions.

8.6 Examples of homotopy limits of model categories

We now consider some examples of homotopy limits of model categories, the first few of which were given in [6].

Example 8.30 Let $\mathcal{D}$ be the category $\bullet \to \bullet$. Then, a corresponding diagram of model categories has the form $F\colon \mathcal{M}_1 \to \mathcal{M}_2$. The homotopy limit has objects $(x_1, x_2; u)$ where x_i is an object of $\mathcal{M}_i$ for $i = 1, 2$, and $u\colon F(x_1) \to x_2$ is a weak equivalence in $\mathcal{M}_2$. Thus, the homotopy limit is given by the weak essential image of $\mathcal{M}_1$ in $\mathcal{M}_2$, i.e., the subcategory of $\mathcal{M}_2$ whose objects are weakly equivalent to objects in the image of F.

Example 8.31 Let us now consider an equalizer diagram $\bullet \rightrightarrows \bullet$. Such a diagram of model categories looks like

$$\mathcal{M}_1 \underset{F_2}{\overset{F_1}{\rightrightarrows}} \mathcal{M}_2.$$

The homotopy limit has objects $(x_1, x_2; u_1, u_2)$ with x_i an object of $\mathcal{M}_i$ for $i = 1, 2$ and u_1, u_2 weak equivalences

$$F_1(x_1) \overset{u_1}{\longrightarrow} x_2 \overset{u_2}{\longleftarrow} F_2(x_1).$$

Thus, the equalizer of model categories is equivalent to the subcategory of $\mathcal{M}_2$ which is in the weak essential image of both F_1 and F_2.

Example 8.32 As a more interesting example, we show that the category of simplicial sets is equivalent to the homotopy limit of model categories of *n-types*, or topological spaces with nontrivial homotopy groups concentrated in degrees n and below. We apply the homotopy limit construction in the context of simplicial sets, so that we have a combinatorial model structure, but can invoke the Quillen equivalence with topological spaces throughout to recover statements about topological spaces.

Consider the model category $\mathcal{SS}ets$ of topological spaces. For each natural number n, take a left Bousfield localization of $\mathcal{SS}ets$, denoted by $\mathcal{SS}ets_{\leq n}$, with respect to the set of maps

$$\{\partial\Delta[k] \to \Delta[k] \mid k > n + 1\}.$$

J. E. Bergner

The weak equivalences in $\mathcal{SSets}_{\leq n}$ are the *n-equivalences*, that is, maps $X \to Y$ such that the induced maps

$$\pi_i(|X|) \to \pi_i(|Y|)$$

are weak homotopy equivalences for all $i \leq n$. These model structures form a diagram of left Quillen functors

$$\cdots \to \mathcal{SSets}_{\leq 3} \to \mathcal{SSets}_{\leq 2} \to \mathcal{SSets}_{\leq 1} \to \mathcal{SSets}_{\leq 0}.$$

The homotopy limit of this diagram has objects

$$\cdots \to X_3 \to X_2 \to X_1 \to X_0$$

for which each map $X_{n+1} \to X_n$ is an *n*-equivalence, and morphisms given by *n*-equivalences $X_n \to Y_n$ for all $n \geq 0$ such that the resulting diagram commutes.

There is a functor $\mathcal{SSets} \to \mathcal{Lim}_n\mathcal{SSets}_{\leq n}$ taking a space X to the constant sequence on X. In the other direction, there is a functor sending a diagram of spaces

$$\cdots \to X_3 \to X_2 \to X_1 \to X_0$$

to its homotopy limit $\mathrm{holim}_n X_n$ in $\mathcal{SSets}$.

Taking the constant diagram on a simplicial set X, followed by its homotopy limit, recovers X. Composing in the other direction, we need to prove that a diagram of simplicial sets

$$\cdots \to X_3 \to X_2 \to X_1 \to X_0$$

is equivalent to the constant diagram given by $\mathrm{holim}_n X_n$. In other words, we want to show that the homotopy limit of this diagram is *n*-equivalent to X_n for all $n \geq 0$, a fact which follows from [17, 19.6.13].

The following example is very similar in nature and gives a variant of the Chromatic Convergence Theorem [21, 7.5.7] at the model category level.

Example 8.33 For a fixed prime p and finite spectrum X, let $L_n X$ denote the *n*-th chromatic localization of the spectrum X, namely, the localization with respect to the spectrum $E(n)$. The Chromatic Convergence Theorem states that X is equivalent to the homotopy limit of these localized spectra.

Now, let $\mathcal{Sp}$ denote the model category of symmetric spectra, and $L_n\mathcal{Sp}$ the left Bousfield localization of $\mathcal{Sp}$ with respect to the $E(n)$-equivalences. After restricting to sufficiently finitary diagrams of spectra (namely, those whose homotopy limits are finite spectra), there is a Quillen pair between Sp and the homotopy limit of the model categories $L_n Sp$. Unfortunately, although Gutiérrez and Roitzheim are able to prove that the unit of this adjunction is a weak equivalence, they are not able to verify what we would hope to be true, that this adjunction is in fact a Quillen equivalence [14, §2.2].

8.7 Adelic models of tensor-triangulated categories

In this section we summarize the role of the homotopy limit of model categories used by Balchin and Greenlees in [2]. We have made this example its own section due to the background content necessary to explain it.

The setting for this work is that of tensor-triangulated categories, and more generally, model categories whose homotopy categories have such a structure. In brief, a tensor-triangulated category is a triangulated category that also has the structure of a monoidal category. We refer to Balmer's paper for more detail [3].

Let us recall some notions from the general theory. Recall that a subcategory of a tensor-triangulated category is *thick* if it is closed under retracts and completion of distinguished triangles. It is an *ideal* if it is closed under completing triangles and tensoring with arbitrary elements of the ambient category.

We can now define the Balmer spectrum, which is one of the key structures of interest in this section.

Definition 8.34 Let $\mathcal{T}$ be a tensor-triangulated category.

1 A *prime ideal* in $\mathcal{T}$ is a proper thick ideal $\mathfrak{p}$ such that if $a \otimes b$ is an object of $\mathfrak{p}$, then either a or b is also an object of $\mathfrak{p}$.

2 The *Balmer spectrum* of $\mathcal{T}$ is the set of prime ideals of the triangulated subcategory of small objects in $\mathcal{T}$, denoted by $\mathrm{Spc}^\omega(\mathcal{T})$.

Here, we follow the naming convention of Balchin and Greenlees, in calling "small" objects that are often referred to as "compact" in the context of triangulated categories.

In fact, the Balmer spectrum has the structure of a topological space, using the Zariski topology, and as such can be assigned a dimension. It is *Noetherian* if its open sets satisfy the ascending chain condition.

We can now define our primary tensor-triangulated categories of interest.

Definition 8.35 A tensor-triangulated category is *finite-dimensional Noetherian* if it is rigidly small-generated and its Balmer spectrum is finite-dimensional and Noetherian.

We have not defined what it means to be rigidly small-generated here, since it is rather technical, but instead refer the reader to [2, §2.A] for more information. We shortly discuss in more depth the model category analogue that is of more interest to us here.

In that direction, as we stated at the beginning of this section, we are interested in model categories whose homotopy categories are the kinds of tensor-triangulated categories that we have been discussing and hence have well-behaved Balmer spectra. We begin with some additional structures on a model category. The first uses some possibly unfamiliar additional terms, for which we refer the reader to [2, §4.B].

Definition 8.36 A model category C is *rigidly compactly generated* if it is a stable, proper, compactly generated symmetric monoidal model category such that:

- the monoidal unit $\mathbb{1}$ is ω-cell compact and cofibrant, where ω denotes the smallest infinite ordinal, and
- there is a set of generating cofibrations of the form $S^n \otimes \mathcal{G} \to D^{n+1} \otimes \mathcal{G}$ where $\mathcal{G}$ is a set of ω-cell compact and ω-small cofibrant objects whose images in the homotopy category of C are rigid.

Definition 8.37 A model category C is *finite dimensional Noetherian* if it is a rigidly compactly generated symmetric monoidal model category such that the Balmer spectrum of its homotopy category is Noetherian and finite-dimensional. In particular, the homotopy category is a finite dimensional Noetherian tensor-triangulated category.

For the remainder of this section assume that C satisfies the conditions of this definition.

Consider the Balmer spectrum of C. Given an object M of C and Balmer prime $\mathfrak{p}$, we denote by $L_\mathfrak{p} M$ the localization of the object M with respect to $\mathfrak{p}$. We denote the completion of M at $\mathfrak{p}$ by $\Lambda_\mathfrak{p} M$. We refer the reader to [2] for a precise description of how to generalize these standard notions from commutative algebra to the particular model category setting considered here.

Let n be the dimension of the Balmer spectrum of the homotopy category of C where C is as above. Let $[n] = \{0 \leq 1 \leq \cdots \leq n\}$, and consider the cubical diagram $\mathcal{P}[n]$ given by the subsets of $[n]$. Further consider the diagram of nonempty subsets $\mathcal{P}_0[n]$.

Definition 8.38 Given any object M of $\mathcal{C}$, the *adelic diagram*

$$M_{\mathrm{ad}} : \mathcal{P}_0([n]) \to \mathcal{C}$$

is defined by, for any subset $(d_0 > \cdots > d_s)$ of $[n]$,

$$M_{\mathrm{ad}}(d_0 > \cdots > d_s) :=$$

$$\prod_{\dim(\mathfrak{p}_0)=d_0} L_{\mathfrak{p}_0} \prod_{\dim(\mathfrak{p}_1)=d_1, \mathfrak{p}_1 \subsetneq \mathfrak{p}_0} L_{\mathfrak{p}_1} \cdots \prod_{\dim(\mathfrak{p}_s)=d_s, \mathfrak{p}_s \subsetneq \mathfrak{p}_{s-1}} L_{\mathfrak{p}_s} \Lambda_{\mathfrak{p}_s} M.$$

The morphisms are as described in [2].

For example, when $n = 1$, we get the homotopy pullback diagram

$$\prod_{\dim(\mathfrak{p}_0)=0} L_{\mathfrak{p}_0} \Lambda_{\mathfrak{p}_0} M$$

$$\downarrow$$

$$\prod_{\dim(\mathfrak{p}_1)=1} L_{\mathfrak{p}_1} \Lambda_{\mathfrak{p}_1} M \longrightarrow \prod_{\dim(\mathfrak{p}_0)=0} L_{\mathfrak{p}_0} \prod_{\dim(\mathfrak{p}_1)=1} L_{\mathfrak{p}_1} \Lambda_{\mathfrak{p}_1} M$$

The following Adelic Approximation Theorem gives a description of the unit object in $\mathcal{C}$ as a homotopy limit of a cubical diagram. For simplicity, following the example just given, we state it in the 1-dimensional case which is given by a homotopy pullback; the general case requires more general homotopy limits.

Theorem 8.39 *[2, 8.1] Let $\mathcal{C}$ be an finite-dimensional Noetherian model category with 1-dimensional Balmer spectrum. Then the unit object $\mathbb{1}$ of $\mathcal{C}$ is the homotopy pullback of the diagram $\mathbb{1}_{\mathrm{ad}} : \mathcal{P}_0[1] \to \mathcal{C}$ given by*

$$\prod_{\dim(\mathfrak{p}_0)=0} L_{\mathfrak{p}_0} \Lambda_{\mathfrak{p}_0} \mathbb{1}$$

$$\downarrow$$

$$\prod_{\dim(\mathfrak{p}_1)=1} L_{\mathfrak{p}_1} \Lambda_{\mathfrak{p}_1} \mathbb{1} \longrightarrow \prod_{\dim(\mathfrak{p}_0)=0} L_{\mathfrak{p}_0} \prod_{\dim(\mathfrak{p}_1)=1} L_{\mathfrak{p}_1} \Lambda_{\mathfrak{p}_1} \mathbb{1}$$

We similarly state the following theorem in this simplified case.

Definition 8.40 [2, 9.1] Let $\mathcal{C}$ be a finite-dimensional Noetherian model category with 1-dimensional Balmer spectrum. We define the adelic module category $\mathbb{1}_{\mathrm{ad}} - \mathrm{mod}_{\mathcal{C}}$ to be the diagram of module categories of the appropriate adelic rings

$$\mathbb{1}_{ad}(0) - \mathrm{mod}_{\mathcal{C}}$$

$$\mathbb{1}_{ad}(1) - \mathrm{mod}_{\mathcal{C}} \longrightarrow \mathbb{1}_{ad}(0 < 1) - \mathrm{mod}_{\mathcal{C}}.$$

The morphisms are given by extension of scalars functors corresponding to the maps of rings, which are in particular left Quillen, and we can give the category of such diagrams the injective model structure.

Now, given an object of $\mathcal{C}$, we can tensor with the diagram $\mathbb{1}_{ad}$ to obtain an object of $\mathbb{1}_{ad} - \mathrm{mod}_{\mathcal{C}}$; given such a diagram, we can take its homotopy limit.

Proposition 8.41   *[2, 9.2] The tensor and limit functors define a Quillen pair*

$$\mathbb{1}_{ad} \otimes - : \mathcal{C} \leftrightarrows \mathbb{1}_{ad} - \mathrm{mod}_{\mathcal{C}} : \lim.$$

Finally, we come to the application of the homotopy limit construction to this situation.

Theorem 8.42   *[2, 9.3] Let $\mathcal{C}$ be an finite-dimensional Noetherian model category. Then there is a Quillen equivalence between $\mathcal{C}$ and the homotopy limit of its associated diagram of adelic module categories $\mathbb{1}_{ad} - \mathrm{mod}_{\mathcal{C}}$.*

Observe that $\mathcal{C}$ is not assumed to be combinatorial here; see [2, 4.20, 9.3] for a discussion of how to work around this obstacle.

8.8 On not mixing left and right Quillen functors

We now return to the question raised in Remark 8.24, namely, whether we can define a homotopy pullback of a diagram of model categories

$$\mathcal{M}_1 \xrightarrow{F_1} \mathcal{M}_3 \xleftarrow{G_2} \mathcal{M}_2,$$

where F_1 is a left Quillen functor and G_2 is a right Quillen functor.

In this section, we talk though possible solutions to this question, but ultimately why they are unsatisfactory. These ideas were developed in several discussions with Yuri Berest.

A first idea for how to mix the kinds of functors is to make the following definition.

Definition 8.43 Consider a diagram of model categories

$$\mathcal{M}_1 \overset{F_1}{\to} \mathcal{M}_3 \overset{G_2}{\leftarrow} \mathcal{M}_2$$

where F_1 is a left Quillen functor and G_2 is a right Quillen functor with left adjoint F_2. Define the *lax homotopy pullback* $\mathcal{L}$ to be the category whose objects are 5-tuples $(x_1, x_2, x_3; u, v)$ where each x_i is an object of $\mathcal{M}_i$, and

$$u \colon F_1(x_1) \to x_3, v \colon F_2(x_3) \to x_2.$$

A morphism

$$f \colon (x_1, x_2, x_3; u, v) \to (y_1, y_2, y_3; z, w)$$

is given by morphisms $f_i \colon x_i \to y_i$ in $\mathcal{M}_i$ such that the diagrams

$$
\begin{array}{ccc}
F_1(x_1) & \overset{u}{\longrightarrow} & x_3 \\
{\scriptstyle F_1(f_1)}\downarrow & & \downarrow{\scriptstyle f_3} \\
F_1(y_1) & \underset{z}{\longrightarrow} & y_3
\end{array}
\qquad
\begin{array}{ccc}
x_2 & \overset{v}{\longleftarrow} & F_2(x_3) \\
{\scriptstyle f_2}\downarrow & & \downarrow{\scriptstyle F_2(f_3)} \\
y_2 & \underset{w}{\longleftarrow} & F_2(y_3)
\end{array}
$$

commute. Give $\mathcal{L}$ the injective model structure, in which the weak equivalences and cofibrations are defined levelwise.

The *homotopy pullback* is the full subcategory of $\mathcal{L}$ consisting of objects $(x_1, x_2, x_3; u, v)$ such that u is a weak equivalence in $\mathcal{M}_3$ and v is a weak equivalence in $\mathcal{M}_2$.

The problem here is that we do not get the homotopy limit that we want. This definition is identical to that of the homotopy limit of the diagram

$$\mathcal{M}_1 \overset{F_1}{\to} \mathcal{M}_3 \overset{F_2}{\to} \mathcal{M}_2.$$

But this homotopy limit is equivalent to the model category $\mathcal{M}_1$, as the initial object in the diagram, so we do not get something capturing the structure we expect in a homotopy pullback.

However, the above definition is only one possibility. Perhaps the problem is that we asked for the map $F_2(x_3) \to x_2$ to be a weak equivalence; namely, asked for a weak equivalence after passing to the adjoint map. What if we instead ask for the map $x_3 \to G_2(x_2)$ to be a weak equivalence, and then use its adjoint to define the morphisms in the model category (so that cofibrations behave well)? So, here is an alternative definition.

Definition 8.44 Consider a diagram of model categories

$$\mathcal{M}_1 \overset{F_1}{\to} \mathcal{M}_3 \overset{G_2}{\leftarrow} \mathcal{M}_2$$

where F_1 is a left Quillen functor and G_2 is a right Quillen functor with left adjoint F_2. Define the *lax homotopy pullback* $\mathcal{L}$ to be the category whose objects are 5-tuples $(x_1, x_2, x_3; u, v)$ where each x_i is an object of $\mathcal{M}_i$, and

$$u\colon F_1(x_1) \to x_3, v\colon x_3 \to G_2(x_2).$$

A morphism

$$f\colon (x_1, x_2, x_3; u, v) \to (y_1, y_2, y_3; z, w)$$

is given by morphisms $f_i\colon x_i \to y_i$ in $\mathcal{M}_i$ such that the diagrams

$$
\begin{array}{ccc}
F_1(x_1) & \xrightarrow{\ u\ } & x_3 \\
{\scriptstyle F_1(f_1)}\downarrow & & \downarrow{\scriptstyle f_3} \\
F_1(y_1) & \xrightarrow[\ z\ ]{} & y_3
\end{array}
\qquad
\begin{array}{ccc}
x_2 & \xleftarrow{\ v'\ } & F_2(x_3) \\
{\scriptstyle f_2}\downarrow & & \downarrow{\scriptstyle F_2(f_3)} \\
y_2 & \xleftarrow[\ w'\ ]{} & F_2(y_3)
\end{array}
$$

commute, where v' denotes the adjoint to v and similarly for w'. Give $\mathcal{M}$ the injective model structure, in which the weak equivalences and cofibrations are defined levelwise.

The *homotopy pullback* is the full subcategory of $\mathcal{L}$ consisting of objects $(x_1, x_2, x_3; u, v)$ such that u and v are weak equivalences in $\mathcal{M}_3$.

However, we claim that this definition still does not work. The first step is to show that there is a model structure on $\mathcal{L}$ so that the cofibrant objects have u and v weak equivalences, obtained by right Bousfield localization on the injective model structure, via the same kind of proof as used above for Theorem 8.23.

As in that proof, we can use the combinatorial structure on each of the model categories $\mathcal{M}_i$ to produce an appropriate generating set of objects $(x_1, x_2, x_3; u, v)$ which respect to which we localize, and for which u and v are weak equivalences.

A key point in that proof, however, is that taking homotopy colimits of objects in this generating set preserves the property that the maps u and v are weak equivalences. In our new modified setting, when we look at morphisms, they are defined in terms of v', not in terms of v. If the maps v' were weak equivalences, then that property would be preserved. It does not seem to be the case, however, that homotopy colimits should preserve that v is a weak equivalence. We cannot even make any assumption about v' being a weak equivalence when v is, so it seems we lose all control over what we know about v.

In summary, we do not seem to have a good way to take homotopy pullbacks of diagrams consisting of both left and right Quillen functors; we expect that the situation for homotopy limits of similarly mixed diagrams is analogously problematic.

References

[1] J. F. Adams. *Stable homotopy and generalised homology.* University of Chicago Press, Chicago, Ill.-London, 1974. Chicago Lectures in Mathematics.

[2] S. Balchin and J. P. C. Greenlees. Adelic models of tensor-triangulated categories. *Adv. Math.*, 375:107339, 45, 2020.

[3] P. Balmer. The spectrum of prime ideals in tensor triangulated categories. *J. Reine Angew. Math.*, 588:149–168, 2005.

[4] C. Barwick. On left and right model categories and left and right Bousfield localizations. *Homology Homotopy Appl.*, 12(2):245–320, 2010.

[5] J. E. Bergner. Homotopy fiber products of homotopy theories. *Israel J. Math.*, 185:389–411, 2011.

[6] J. E. Bergner. Homotopy limits of model categories and more general homotopy theories. *Bull. Lond. Math. Soc.*, 44(2):311–322, 2012.

[7] J. E. Bergner. Homotopy colimits of model categories. In *An alpine expedition through algebraic topology*, volume 617 of *Contemp. Math.*, pages 31–37. Amer. Math. Soc., Providence, RI, 2014.

[8] D. Dugger. Combinatorial model categories have presentations. *Adv. Math.*, 164(1):177–201, 2001.

[9] D. Dugger. Spectral enrichments of model categories. *Homology Homotopy Appl.*, 8(1):1–30, 2006.

[10] W. G. Dwyer and D. M. Kan. Function complexes in homotopical algebra. *Topology*, 19(4):427–440, 1980.

[11] W. G. Dwyer and D. M. Kan. Simplicial localizations of categories. *J. Pure Appl. Algebra*, 17(3):267–284, 1980.

[12] W. G. Dwyer and J. Spaliński. Homotopy theories and model categories. In *Handbook of algebraic topology*, pages 73–126. North-Holland, Amsterdam, 1995.

[13] P. G. Goerss and J. F. Jardine. *Simplicial homotopy theory*, volume 174 of *Progress in Mathematics*. Birkhäuser Verlag, Basel, 1999.

[14] J. J. Gutiérrez and C. Roitzheim. Towers and fibered products of model structures. *Mediterr. J. Math.*, 13(6):3863–3886, 2016.

[15] Y. Harpaz. Lax limits of model categories. *Theory Appl. Categ.*, 35:Paper No. 25, 959–978, 2020.

[16] Y. Harpaz and M. Prasma. The Grothendieck construction for model categories. *Adv. Math.*, 281:1306–1363, 2015.

[17] P. S. Hirschhorn. *Model categories and their localizations*, volume 99 of *Mathematical Surveys and Monographs*. American Mathematical Society, Providence, RI, 2003.

[18] T. Hütteman and O. Röndigs. Twisted diagrams and homotopy sheaves. preprint available at math.AT/0805.4076, 2008.

[19] T. Hüttemann, J. R. Klein, W. Vogell, W. Waldhausen, and B. Williams. The "fundamental theorem" for the algebraic K-theory of spaces. I. *J. Pure Appl. Algebra*, 160(1):21–52, 2001.

[20] D. G. Quillen. *Homotopical algebra*. Lecture Notes in Mathematics, No. 43. Springer-Verlag, Berlin-New York, 1967.

[21] D. C. Ravenel. *Nilpotence and periodicity in stable homotopy theory*, volume 128 of *Annals of Mathematics Studies*. Princeton University Press, Princeton, NJ, 1992. Appendix C by Jeff Smith.

[22] B. Toën. Derived Hall algebras. *Duke Math. J.*, 135(3):587–615, 2006.